Advances in Intelligent Systems and Computing

Volume 1315

The series "Advances in Intelligent Systems and Computing" contains publications on theory, applications, and design methods of Intelligent Systems and Intelligent Computing. Virtually all disciplines such as engineering, natural sciences, computer and information science, ICT, economics, business, e-commerce, environment, healthcare, life science are covered. The list of topics spans all the areas of modern intelligent systems and computing such as: computational intelligence, soft computing including neural networks, fuzzy systems, evolutionary computing and the fusion of these paradigms, social intelligence, ambient intelligence, computational neuroscience, artificial life, virtual worlds and society, cognitive science and systems, Perception and Vision, DNA and immune based systems, self-organizing and adaptive systems, e-Learning and teaching, human-centered and human-centric computing, recommender systems, intelligent control, robotics and mechatronics including human-machine teaming, knowledge-based paradigms, learning paradigms, machine ethics, intelligent data analysis, knowledge management, intelligent agents, intelligent decision making and support, intelligent network security, trust management, interactive entertainment, Web intelligence and multimedia.

The publications within "Advances in Intelligent Systems and Computing" are primarily proceedings of important conferences, symposia and congresses. They cover significant recent developments in the field, both of a foundational and applicable character. An important characteristic feature of the series is the short publication time and world-wide distribution. This permits a rapid and broad dissemination of research results.

Indexed by SCOPUS, DBLP, EI Compendex, INSPEC, WTI Frankfurt eG, zbMATH, Japanese Science and Technology Agency (JST), SCImago.

All books published in the series are submitted for consideration in Web of Science.

More information about this series at http://www.springer.com/series/11156

Zhengbing Hu · Sergey Petoukhov ·
Matthew He
Editors

Advances in Artificial Systems for Medicine and Education IV

 Springer

Editors
Zhengbing Hu
International Center of Informatics
and Computer Science,
Faculty of Applied Mathematics
National Technical University of Ukraine
"Kyiv Polytechnic Institute"
Kyiv, Ukraine

Sergey Petoukhov
Laboratory of Biomechanical Systems
Mechanical Engineering Research Institute
of the Russian Academy of Sciences
Moscow, Russia

Matthew He
Halmos College of Arts and Sciences
Nova Southeastern University
Ft. Lauderdale, FL, USA

ISSN 2194-5357   ISSN 2194-5365   (electronic)
Advances in Intelligent Systems and Computing
ISBN 978-3-030-67132-7   ISBN 978-3-030-67133-4   (eBook)
https://doi.org/10.1007/978-3-030-67133-4

This Springer imprint is published by the registered company Springer Nature Switzerland AG
The registered company address is: Gewerbestrasse 11, 6330 Cham, Switzerland

Contents

Advances in Mathematics
and Bio-mathematics

Functional Systems Integrated with a Universal Agent of Artificial Intelligence and Higher Neurocategories

Georgy K. Tolokonnikov$^{(\boxtimes)}$

Federal Scientific Agro-Engineering Center VIM, Russian Academy of Sciences, 1st Institute Passage, 5, Moscow, Russia
admcit@mail.ru

Abstract. The paper proposes a hybrid functional system equipped with a universal AI agent that implements a strong AI variant. The system is a versatile prototype for medical prosthetics. The task is to find formalisms for modeling strong AI within neural networks based on higher-dimensional categories and convolutional polycategories. n-categories are reduced according to the theorem below to convolutional polycategories with convolutions of a special form. The necessity of using functional systems hierarchies based on the highest G-neurocategories of hierarchies is shown, examples of such hierarchies are given. The listed results are new and original.

Keywords: Functional systems · Systems theory · Categories · Higher-dimensional categories · Convolutional polycategories · Hierarchy of systems

1 Introduction

Categorical systems were introduced in [1] as polyarrows of a convolutional polycategory or neurocategory, convolutional polycategories were introduced in [2], the example of visual representation of convolutions $S = (s_1, s_2, \ldots)$ and polycategories is given in Fig. 1.

The system-forming factor of the theory of functional systems by P.K. Anokhin-K. V. Sudakov is modeled by a triple (F, Δ, S_r), respectively, with an endofunctor, convolution and categorical system corresponding to the result. A convolution unifies (applies to a set, see Fig. 1) a set of systems $S_i, i = 1, 2, \ldots$ called subsystems. For arbitrary categorical systems (F, Δ, S_r), it is natural to call the triple a *categorical functional system*. Anokhin-Sudakov's categorical functional systems will be called these triples, provided that the subsystems are selected from the set of systems defined by P.K. Anokhin's scheme and its degenerate version called the metabolic level functional system.

Individual organs or their aggregates are not functional systems, it is necessary to fix the result and build a set of organs for it participating in achieving the result. For example, the functional subsystem for maintaining the required concentration Ca^{2+} in

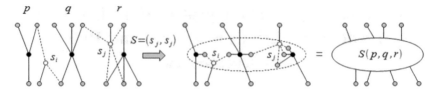

Fig. 1. Convolution S of polyarrows p, q, r (direction from bottom to top, not specified)

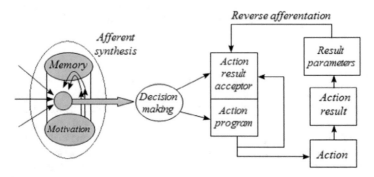

Fig. 2. Diagram of the functional system of the behavioral act

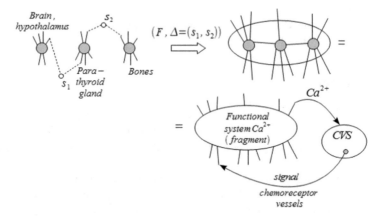

Fig. 3. Fragment of a functional system Ca^{2+}

the blood of the cardiovascular system (CVS) is as follows (a fragment corresponding to the restoration of a reduced concentration Ca^{2+}) (Fig. 3).

With a decrease in the concentration Ca^{2+}, the chemoreceptors of the CVS vessels transmit to the hypothalamus information about the decrease in concentration, the brain acts on the parathyroid gland, which secretes parathyroid hormone. This hormone "flushes" Ca^{2+} into the blood from the depot in the bones. Concentration Ca^{2+} returns to normal.

An organism consists of a set of organs (modeled by categorical systems), which are combined under the influence of system-forming factors into functional systems to achieve results, the results replace each other in a hierarchy, which we will discuss below. Natural intelligence, the central phenomenon that is designed to simulate by strong artificial intelligence (AI), is also organized as functional systems.

In the next paragraphs 2 and 3, this hybrid functional system is constructed with application to medical prosthetics; in paragraph 4, formalisms for strong AI based on neurocategories are examined, in paragraph 5, hierarchies of functional systems are discussed. In conclusion, a brief summary of the work and conclusions are given.

2   Functional System Coupled with a Universal Agent AI

A hybrid functional system with a built-in universal AI agent is proposed, with the most important attribute of a strong AI, namely, the ability to generate new algorithms that were not originally embedded in the machine. This system is borderline, the next steps of modeling intelligence will require the introduction into the blocks of afferent synthesis and the development of a solution, in fact, the introduction into the "I" of the personality.

J. McCarthy [3, 4] used the predicate calculus of classical logic (IP), the description of the environment and the system is carried out by selecting additional axioms, in the resulting formal system, the derivation of the calculus allows finding the path of the system transition from the initial state to target (achievement of the result). The Macarty model operates as follows. A person determines the parameters of the desired result in the form of the target state of the environment and the system (enters them into the PDR model), gives a description of the environment and the system in the language of the IP. The solver develops an algorithm for achieving the result, which is launched in the AD block. Effectors act, receptors measure and transmit parameters to the ARD model.

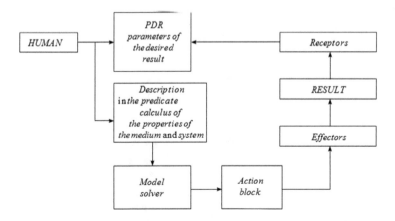

Fig. 4. Macarty model schematic

The diagrams in Fig. 2 and Fig. 4 are close. Work on the development of the parameters of the desired result (afferent synthesis and decision-making) is carried out by a person. If the result is not achieved by the algorithms developed by the solver, then the properties of the environment, systems, parameters of the required result are changed and loaded into the solver again. It turns out that the choice of IP for all tasks (results) is not optimal, often other calculi, selected taking into account the task, are more effective. Taking into account the choice of calculus, it is possible to generalize the Macarty scheme in an obvious way, using another calculus instead of IP.

Let's move on to a functional system with a universal AI agent Fig. 5. A universal agent [5] (see also [6])) is able to take on everything except the development of an image of the required result, uses theorems on universal calculus (E. Post 1943, S. Maslov 1964, Yu. Matiyasevich 1967) and their generalizations.

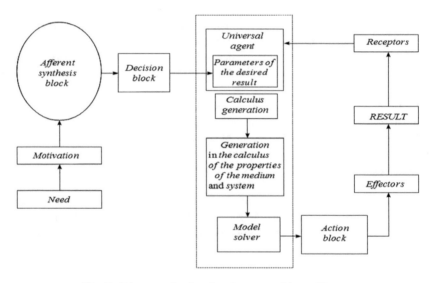

Fig. 5. Diagram of a functional system with an AI agent

These theorems are basically formulated in the same way. Let a certain class of calculi be given (Post calculus, etc.). The theorem states that there is a universal calculus that generates all possible calculi of a given class and all conclusions in them. The universal agent is, as far as we know, the first computer complex with strong artificial intelligence, the attributes of which include the ability to exhaustively develop new algorithms of behavior that are not included in the agent during its manufacture. Here is one of the theorems on the universal calculus, suitable for constructing a universal agent. Let $\{A, P, \alpha, \pi\}$ be Post's canonical calculus, A - alphabet of calculus, P - alphabet of calculus variables, α - set of axioms, π - inference rules, calculus is intended for derivation of all A_0-words. It is proved that there is a normal calculus over A_0 (in which there is only one axiom k and special one-premise inference rules $G_i p| = pG'_i$, p is a variable, G, G' are words in the alphabet of the calculus) equivalent over A_0 the original one, while the alphabet A' can be different, but always $A_0 \subseteq A'$.

In equivalent calculi over A, exactly the same collection of -words (theorems) is deducible. Let's call a word $k \triangle G_1 \to G'_1 \triangle \ldots \triangle G_n \to G'_n \triangle| = pG'$ in the alphabet $A \cup \{\triangle, \to, | =\}$ by the notation of the normal calculus with the axiom k and the rules $G_i p| = pG'_i$, i = 1,2,....

Post's Theorem on Universal Calculus. It is possible to construct a canonical calculus U in an alphabet $A \cup \{\triangle, \to, | =\}, A = \{a_1, \ldots, a_m\}$ in which a word is derivable if and only if it has the form NP, where N is the notation of some normal calculus in the alphabet A, and P is a word derivable in this calculus. The axiom of the calculus U is taken to be the word $\triangle | =$, the rules of inference are taken as follows

$$\frac{p\triangle| = p}{pa_i\triangle| = pa_i}, \quad \frac{p_1 p_2 \triangle| = p_1 p_2, q| = r}{qp_1 \to p_2\triangle| = r}, \quad \frac{q_1 \triangle p_1 \to p_2 \triangle q_2| = p_1 r}{q_1 \triangle p_1 \to p_2 \triangle q_2| = rp_2},$$

$p, p_1, p_2, q, q_1, q_2, r$ - variable calculus.

It is very simple to apply the theorem, defining with its help, both calculi and conclusions in them, and it can be done manually.

The agent's work is as follows. The person (or the ASB and the DB) writes the parameters of the desired result into the agent's memory, and also trains the agent of the subject area using the neural network or the HTM technology of J. Hawkins in order to discard the generated calculus and conclusions that are not related to this subject area. In fact, the agent himself can perform this function, as it is not difficult to understand. Now, the agent starts the calculus generator, receives the first based on it, finds an output to the result, starts the output program for the effectors, the receptors (can be artificial) supply information about the parameters of the result obtained. If the parameters coincide with the parameters of the acceptor, the result is achieved, the system can be slowed down (the need is satisfied, the motivation decreases). If they do not match, then the agent starts the calculus generator again (you can take into account the experience gained, or you don't have to worry - if there is an algorithm for solving the problem, then, according to the theorem, sooner or later the generator will work out this desired algorithm and the problem will be solved).

Both logic and calculus (including environment parameters, etc.), and conclusions in the universal agent are generated, nothing is required in advance, except for describing the parameters of the desired result. The agent will always achieve the result, without reasoning and emotions, so he is called a zombie agent. The problem is solved theoretically, but the problem of real programming the agent, taking into account the constraints for the generated calculi, is very difficult. However, this is already largely an engineering challenge.

So, the existing need generates motivation, which achieves the development of the ASB and the DB of the result image, the parameters of which are transferred to the universal agent. If the result, determined by its way, is achievable, then it will inevitably be achieved (for the functioning of the universal agent there are all the necessary resources).

3   Universal Prototype for Medical Prosthetic

During prosthetics, a lost or damaged organ is replaced with an artificial one (artificial kidney, etc.). This artificial organ becomes a kind of prosthesis for all functional systems in which it participates. It is not difficult in each specific case to determine which part of the functional system replaced by the prosthesis. So, you should focus on the general diagram of a functional system with a universal AI agent. Receptors, effectors, and everything related to a universal agent can be artificially executed. In other words, the universal agent itself, with this implementation, turns into a universal prosthesis. It is important to note that the energy, time and other resources required for the agent to function are available. Universality means that the functional system itself is not concretized, the reasoning is carried out for all functional systems at once. Everything possible is prosthetic, up to the border of this possible: the next steps expanding the model of a universal agent already refer to "prosthetics" in a sense of the "I" of a person, with needs, emotions, desires (image of the desired result) and consciousness.

4   Formalisms for Strong AI Based on Neurocategories

It is a well-known hypothesis that a network of living neurons in the brain with traffic from spikes is not limited to reflexes, but provides prototypes for modeling strong AI with artificial neural networks. However, modern artificial neural networks are nothing more than a way to approximate functions of many variables, there is no "mind" [7] (see also [8–13]), only reflexes. We propose to transfer the emphasis of the implementation of this hypothesis to neurographs using convolutional polycategories and their higher analogs.

In addition to the connections modeled in traditional artificial neural networks, neurons in the brain have a much richer set of possible connections, modeled by convolutional polycategories, including their higher versions, and neurographs. Our task of finding formalisms for modeling strong AI consists in the transition to networks based on convolutional polycategories, higher polycategories and their varieties. In this work, two important steps have been taken: the higher dimension categories are reduced to convolutional polycategories with a very particular type of convolutions, and a hierarchy of categorical functional systems based on higher convolutional G-polycategories is defined, which are reduced to convolutional polycategories with convolutions of a certain structure. These results direct the search for formalisms for modeling strong AI in the field of convolutional polycategories with convolutions reflecting the hierarchical structure of a neural network. If we rely on the neural network model with the help of categories in which the neurons themselves are objects and the arrows are organized through synapses on dendrites, as was done in [14], then it is necessary to move from categories to higher-dimensional categories. In category theory, strict and weak higher-dimensional categories are represented by the approaches of May, Penon, Leinster, Batanin, Joyal, Street, Baes, and other scientists [15]. Among the main constructions, the strict n-category is used here. For clarity, we first give the definition of a 2-category.

Definition. 2-category C is a set of 2-cells $x, y, \ldots$, on which functions of the source s_0, s_1 and functions of the target t_0, t_1 with ones $s_0 x, \ldots, t_1 x, x \in C$ are defined, as well as partial binary operations of composition $*_0, *_1 : C \times C \to C$, satisfying the properties of that $C^{(0)} = \{C, *_0, s_0, t_0\}$ and $C^{(1)} = \{C, *_1, s_1, t_1\}$ - categories. In this case, $C^{(0)}$ category units are $C^{(1)}$ category units, category structures and commute among themselves. The usual conditions for the applicability of compositions, properties of units, as well as associativity and replacement rules are met.

One of the main examples of 2-categories is the Cat category of small categories. The subcategory here is the simplicial category SC, which is most important in category theory, algebraic topology, and algebraic geometry (objects are finite ordinals, arrows are nondecreasing functions between ordinals). The category $C^{(0)}$ consists of objects $\overline{m}, \overline{n}, \ldots$ with arrows $\overline{m} \leq \overline{n}$ and the category $C^{(1)}$ consists of objects $f, g, \ldots$ with arrows $f \leq g$, that are natural transformations.

Let's move on to matching 2-categories to convolutional polycategories. For 2-cells and for horizontal multiplication we have.

This uses a compound convolution $\sigma = (\sigma_{st}, \sigma_t, \sigma_s)$ with the values shown in the diagram. There are also diagrams for vertical multiplication and substitution rules.

Definition. A collection of P elements is called an ***n-category*** if the following is defined and satisfied. Source and target mappings $\alpha_i, \omega_i : P \to P, i = 1, 2, \ldots, n$ are defined. Compositions $f_i^\circ g, f, g \in P$ for $\omega_i(f) = \alpha_i(g)$ are defined.

$$\omega_j \omega_k = \omega_k \omega_j, \alpha_j \alpha_k = \alpha_k \alpha_j, \ \alpha_j \omega_k = \omega_k \alpha_j (j \neq k), \ \omega_k \omega_k = \omega_k, \alpha_k \alpha_k = \alpha_k,$$
$$\alpha_j \omega_j = \omega_j, \omega_k \alpha_k = \alpha_k, \alpha_j (g \ {}_k^\circ f) = \alpha_j(g) {}_k^\circ \alpha_j(f),$$
$$\omega_j (g \ {}_k^\circ f) = \omega_j(g) {}_k^\circ \omega_j(f)(j \neq k), \alpha_j (g \ {}_j^\circ f) = \alpha_j(f), \omega_j (g \ {}_j^\circ f) = \omega_j(g),$$
$$h_j^\circ \left(g_j^\circ f \right) = \left(h_j^\circ g \right)_j^\circ f (\text{associativity}),$$
$$\left(g_k^\circ f \right)_j^\circ \left(q_k^\circ h \right) = \left(g_j^\circ q \right)_k^\circ \left(f_j^\circ h \right) (\text{replacement}),$$
$$\text{if} = \alpha_j(f) = \omega_j(f), \text{ then } g_j^\circ f = g, f \ {}_j^\circ g = g.$$

The construction given for 2-categories carries over to n-categories, the following theorem is true.

Theorem
A strong n-category is an associative compositional convolutional polycategory.

Teorem shows that a neuron-object model, links-arrows, equipped in a similar way, does not lead to a more general construction than the usual convolutional polycategory.

5 Hierarchies of Functional Systems as Hierarchies of G-polycategories

Let's return to functional systems in their categorical implementation. Brain neurons implement variants of higher polycategories [1, 2], but their interaction and properties depend on other functional systems of the body, such as the cardiovascular system (CVS), which provides nutrition, oxygen supply, and transport of hormones, and others. How legitimate is the above hypothesis about the capabilities of neural networks for modeling strong AI, taking into account only the spikes between neurons? Perhaps, taking into account the mutual influence of, for example, CVS and neuron systems will make it possible to give the necessary models for a strong AI. Here we have the highest G-polycategory.

The concept of hierarchy should be taken broader than the relationship between systems and their subsystems. In categorical systems theory, there are all the possibilities for such a broader approach. Moreover, even the first attempts to strictly construct hierarchically related subsystems of the organism lead to such more complex hierarchies. We will consider below an example of a G-hierarchy implemented by the cardiovascular system (CVS) in its interaction with neural and other subsystems of the body.

We have defined a neurograph as a set of higher polyarrows, in addition to n-polycategories, as examples of neurographs, we will consider G-neurographs defined for a fixed directed graph G as follows [2].

Let us choose one of the arrows of the graph G. This arrow is associated with some arbitrary polygraph P. The beginning of the arrow is the set of polyobjects, the end of the arrow is the set of polyarrows P.

We carry out the same procedure with all the arrows of the graph G, taking into account the following: end, contains polyobjects of the arrow polygraph, whose beginning coincides with the end of this arrow (in case the end of this arrow coincides with the beginning of the second one considered). The resulting set of polygraphs is called a G-neurograph. The theorem on the representation of higher categories by convolutional polycategories is generalized to G-neurocategories (neurographs with additional properties), according to the generalization, G-neurocategories are reduced to convolutional polycategories with suitable convolutions. For 2-categories, as the graph G we obviously have a graph $\cdot \to \cdot \to \cdot$.

Let's move on to the physiological example of CVS. The CVS includes the heart, blood vessels (veins, arteries, arterioles, venules, etc.) and the blood filling them. The system-forming factor of CVS is the result of the transfer from cells and other subsystems of oxygen, carbon dioxide, nutrients, signaling molecules, and so on to other cells and subsystems. Similar results are needed throughout the life of the organism and the CVS functions while the organism is alive. The CVS is a polyarrow no lower than the second level, since it connects the already functioning subsystems. Consider a 2-polygraph of neurons as a polyarrow with objects in the form of synapses and 2-neurons of a polyarrow connecting the first neurons, the corresponding graph G coincides with the given graph for 2-categories. Both neurons and 2-neurons receive nutrients and oxygen from the blood, releasing carbon dioxide and other metabolic products into the blood. For simplicity, we will not take into account other connections and subsystems. As a graph G, we have the following graph (0 - a vertex for synapses, 1 - a vertex for neurons, 2 - a vertex for 2-neurons) (Fig. 6).

The issue of the hierarchy of functional systems in the theory of functional systems in physiology is at the non-formalized intuitive level. The given example is the

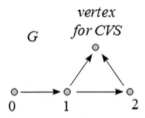

Fig. 6. Ierarchy picture

simplest, but it already shows the need to apply hierarchies associated with higher polycategories. Many of the functional systems have been studied in physiology in great detail, this allows an adequate hierarchy of systems to be carried out within the framework of the categorical theory of systems, which serves as the topic of separate works.

6 Conclusion

The paper proposes a hybrid functional system equipped with a universal AI agent that implements a strong AI variant. The system is a versatile prototype for medical prosthetics. The task that we have set and are solving is to find formalisms for modeling strong AI within neural networks based on higher categories and convolutional poly-categories. One of the main paper results is n-categories are reduced to convolutional polycategories with convolutions of a special form. The necessity of using hierarchies based on the highest G-neurocategories for the description of hierarchies of functional systems is shown, examples of such hierarchies are given. The listed results are new and original.

References

1. Tolokonnikov, G.K.: Mathematical categorical systems theory. In: Biomachsystems: Theory and Applications, vol. 2, pp. 22–114 (2016)
2. Tolokonnikov, G.K.: Mathematical foundations of the theory of biomachsystems. In: Biomachsystems: Theory and applications, vol. 1, pp. 31–213 (2016)
3. McCarthy, J.: Programs with common sense. In: Proceedings of the Symposium on Mechanisation of Thought Processes, London, vol. 1, pp. 77–84 (1958)
4. Russell, S., Norvig, P.: Artificial Intelligence. Modern Approach. Publishing House "Williams" (2006). 1408 p.
5. Tolokonnikov, G.K.: Computable and non-computable physical theories according to R. Penrose. Part 3 Appl. Math. Quantum Theory Programs 9(4), 3–294 (2012)
6. Tolokonnikov, G.K.: Universal zomby and virtual reality. In: Proceedings of the II All-Russian Conference with International Participation "Modern Systems of Artificial Intelligence and Their Applications in Science", Kazan, 14 May 2014, pp. 88–102 (2014)

7. Tolokonnikov, G.K., Chernoivanov, V.I., Sudakov, S.K., Tsoi, Y.A.: Systems theory for the digital economy. In: Advances in Intelligent Systems and Computing, vol. 1127, pp. 447–456 (2020)

8. Karande, A.M., Kalbande, D.R.: Weight assignment algorithms for designing fully connected neural network. IJISA **10**(6), 68–76 (2018)

9. Dharmajee Rao, D.T.V., Ramana, K.V.: Winograd's inequality: effectiveness for efficient training of deep neural networks. IJISA **10**(6), 49–58 (2018)

10. Hu, Z., Tereykovskiy, I.A., Tereykovska, L.O., Pogorelov, V.V.: Determination of structural parameters of multilayer perceptron designed to estimate parameters of technical systems. IJISA **10**, 57–62 (2017)

11 Awadalla, M.H.A.: Spiking neural network and bull genetic algorithm for active vibration control. IJISA **10**(2), 17–26 (2018)

12. Abuljadayel, A., Wedyan, F.: An approach for the generation of higher order mutants using genetic algorithms. IJISA **10**(1), 34–35 (2018)

13. Gurung, S., Ghose, M.K., Subedi, A.: Deep learning approach on network intrusion detection system using NSL-KDD dataset. IJCNIS **11**(3), 8–14 (2019)

14. Ramırez, J.D.G.: A New Foundation for Representation in Cognitive and Brain Science: Category Theory and the Hippocampus, Depart. de Automatica, Ingenierıa Electronica e Informatica Industrial Escuela Tecnica Superior de Ingenieros Industriales (2010). 238 p.

15. Cheng, E., Lauda, A.: Higher-Dimensional Categories: An Illustrated Guide Book. University of Cambridge (2004). 182 p.

Genetic Interpretation of Neurosemantics and Kinetic Approach for Studying Complex Nets: Theory and Experiments

Ivan V. Stepanyan[1(✉)], Michail Y. Lednev[1], and Vladimir V. Aristov[2]

[1] Mechanical Engineering Research Institute of the Russian Academy of Sciences, 4, Malyi Kharitonievsky pereulok, 101990 Moscow, Russian Federation
neurocomp.pro@gmail.com, miklesus@mail.ru
[2] Dorodnicyn Computing Center of Federal Research Center "Computer Science and Control" of RAS, Vavilov st. 40, 119333 Moscow, Russian Federation
aristovvl@yandex.ru

Abstract. The neurosemantic approach allows the formation of information management systems based on incoming information flows. This paper explores the possibilities of molecular genetic algorithms of multiscale information analysis for displaying neurosemantic networks in various parametric spaces. The practical application of such a combined approach for tasks of molecular diagnostics of genetic diseases using Walsh's orthogonal functions is proposed. Issues of interaction between semantics and semiotics in the context of the phenomenon of consciousness and brain are considered. A kinetic approach is also used to construct random graphs, which in a geometric version allows obtaining complex structures; after a percolation transition, a complex cluster can be compared with elements of consciousness. The importance of the study is due to the fact that scientific progress requires new methods to optimize the perception of the results of neural network analysis, as well as the perception of big data with the possibility of analysis at various scales of visualization. Also, of interest to science are model experiments in the field of the theory of the origin of consciousness.

Keywords: Information processing · Genetic coding · Visualization · Consciousness modeling · Kinetic approach · Complex nets

1 Introduction

The neurosemantic mechanism of auto-structuring of information [1] was proposed by V.I. Bodiakin from the Institute of Control Sciences Academician VA Trapeznikov as an algorithm for synthesis of a self-learning neural network structure. Neurosemiotics refers to sign aspects of synapses, neurons and neural networks [2, 3] and also refers to cognition and the human being and information processing systems.

The interaction of semantics and semiotics in the context of the phenomenon of consciousness and the brain has long been of interest to researchers. In neural networks, neurosemiotics can be interpreted as the structure of connections of each neuron ("pattern/sign" formalism), as well as some "signs" or "internal images" that the consciousness operates with.

Genometry – a proposed in [4, 5] class of molecular-genetic algorithms of multi-scale processing of information based on the system of orthogonal Walsh functions with the possibility of visualization in various parametric spaces. The geometric approach is applicable for processing arbitrary information presented in quadric code (nucleotide alphabet A,G,C,T/U). This fact allows us to extend the scope of this method beyond molecular genetics.

Research in this direction became possible due to the fact that experience appeared in the field of RAS developments - neurosemantics, matrix genetics, as well as the development of the kinetic approach. This also applies to neural network approach. At the same time, the analysis of visual information and the ergonomics of perception of big data are closely related to the issues of consciousness and thinking.

The aim of the study is to comprehend the combination of the possibilities of various approaches at the junction of genetics and neuroinformatics to get a new toolkit for various-scale analysis of genetic information and to compare the results obtained with methods of studying consciousness.

The goal of the research is to conduct a model experiment at the junction of neurosemantics and molecular genetics on various scales with the calculation of the frequencies of encoding neurons DNA fragments with the multilevel fractal structure of genetic coding, as well as a model experiment for studying complex nets.

The importance of the study is due to the fact that scientific progress requires new methods to optimize the perception of the results of neural network analysis, as well as the perception of big data with the possibility of analysis at various scales of visual-ization. Also of interest to science are model experiments in the field of the theory of the origin of consciousness.

In this paper the task is to search for a variant of expanding the neurosemantic approach using genometric algorithms based on the principles of genetic coding. The second part of the work, where the structures of a random graph on the basis of the kinetic-geometric approach, is methodologically consistent with this was studied.

2 Literature Review and Mechanism of Neurosemantic

Cartesian genetic programming (CGP) is a branch of genetic programming in which candidate designs are represented using directed acyclic graphs. Evolutionary circuit design is the most typical application of CGP. Paper [6] presents a new software tool developed to analyse and visualise a genetic record.

At [7] causal regulatory relationships among the genes in each module are deter-mined by a Bayesian network inference framework that is used to formally integrate genetic and gene expression information. Paper [8] presents a new algorithm for imputing missing values in medical datasets. This algorithm evolutionarily combines Information Gain and Genetic Algorithm to mutually generate imputable values. Paper [9] introduces an algorithm to locate the longest common subsequences in two different DNA sequences. In [10] a cryptographic security technique is proposed. The proposed system incorporate cryptology technique of encryption inherits the concept of DNA based encryption using a 128-bit key.

Neurosemantic graph or neurosemantic structure (NSS) is a multi-line hierarchical graph consisting of so-called N elements. The automatic construction of the semantic hypertext graph is based on the property of NSS - auto-structurization. This process results the localization of individual semantic units from the text form into individual neural-like elements of NSS: each semantic unit of the subject area corresponds to appropriate N element of NSS.

Autostructurization is a self-learning algorithm when, with sufficient input information, during the dynamic construction of the NSS structure, the process of convergence with the semantic structural structure occurs, which fully corresponds to the processes in the subject area, i.e. their topology is isomorphic (by the isomorphism of structures we mean the correspondence of N elements and their relations).

In a simplified version, the auto-structuring mechanism defines synaptic NSS relations, which display the input information in layers: the first layer contains an alphabet of symbols, the second layer contains pairs of symbols, the third layer contains triplets of symbols, etc. In general, auto-structuring implements a statistical optimization algorithm where the minimum possible structure is searched, in which layers of NSS are corresponded to identified semantic blocks: syllables, words, phrases, sentences, paragraphs, etc. As a result of the auto-structuring mechanism, when processing non-random text sequences, the information is compressed. The quality of autostructurization is the ratio of the number of localized semantic units of the subject area in individual elements of NSS to their total number in the text form [1].

With the automatic formation of the neurosemantic structure of the graph, the total resources of R_{ncc} resources spent on common vertices (N elements) and edges of the graph when displaying a symbolic information flow from the specified subject area are minimized. Links in the structure are corresponded to the numbers of elements of the previous layers. The NSS resource R_{ncc} is calculated as follows [11]:

$$R_{HCC} = \sum_{i=0}^{m} \left(n_i d_i + \sum_{j=1}^{n_i} \left(k_j + k_j d_{i-1} \right) \right),$$

where k - number of bits per number of connections of N element; d - number of bits per number of N element; i - number of the layer; j - number of the element in the layer. The resource of the text form:

$$R_{T\Phi} = \sum_{t=1}^{l} S_t,$$

where l is the length of the text form; s is the number of bits per character of the text form.

3 Genometry of Big Data

Let us recall the basic algorithm described in [4]. The sequence of characters from the set of {A,G,C,T/U} encoding the nitrogen bases is divided into fragments of equal length N, where N is a free parameter of the algorithm - the length of "semantic units". We will call the obtained length fragments N-meters or N-elements. Taking into account the system of genetic subalphabets: G = C "3 hydrogen bonds" / A = T "2 hydrogen bonds"; C = T "pyrimidines" / A = G "purines"; A = C "amino" / G = T "keto" sequence of nitrogenous bases can be represented as three binary sequences consisting of zeros and units. The choice of coding method (what to consider zero or unit) affects the transformation of symmetry of the final visualization. The resulting binary recording of the fragments is their representation in the form of three sequences of decimal or other unambiguously identifying values (code frequency, number of units in code, etc.). Thus, the conversion of binary N-elements to decimal numbers allows them to be displayed in the coordinate system (it is possible to visualize in spherical, cylindrical and other coordinate systems).

Other numerical interpretations of "semantic units" for each of the subalphabets are also possible. The obtained values specify the coordinates of points in parametric space. Such spaces can be with different dimensions depending on the number of parameters taken into account during visualization. It is also possible to apply for other systems of genetic subalphabets [5]. Therefore, we are talking about the class of so-called DNA-algorithms of visualization.

Application of the class of DNA-algorithm allows detecting a latent order of genetic structures and other texts. Examples of two-dimensional visualization of the nucleotide composition are given in Fig. 1. It is seen that genetic coding of living organisms of different species can have individual fractal character.

Fig. 1. Examples of two-dimensional imaging (abscissa axis: purines-pyrimidines, ordinate axis: amino-keto, N = 8). Left - genome of bacteria Ralstonia eutropha, center - genome of bacteria Candidatus Arthromitus, right - genome of bacteria Actinomadura madurae.

By converting any information into a quadric code (e.g., after the preliminary conversion this information into binary code), a sequence of four-digit alphabet characters can be obtained. For this type of sequence, the method of geometric analysis is also applicable. As our studies have shown [4, 5], a randomly generated long sequence of symbols encoding nucleotides gives a pattern, all points of which are scattered chaotically (short sequences do not contain enough information to show the pattern structure).

Here by information we mean a certain signal as a sequence of symbols of some alphabet with low entropy. Observation of the facts that chaotic sequences give chaotic two-dimensional visual patterns, and sequences encoding real DNA (as well as long texts in natural language) give complex fractal structures, allows us to obtain a new way of assessing entropy in the form of a geometric mosaic (symbol).

Depending on the encoding method, each mosaic has an internal interpretation. For example, each point of the pattern corresponds to the present N-elements. Thus, a possible variant of application of the genometric approach is the analysis of texts in natural language (Fig. 2). Let us consider the way of application of these properties in neurosemantics.

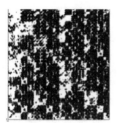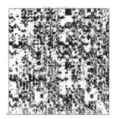

Fig. 2. Quaternary [A,G,C,T] representation of binary code encoding text in natural language (N = 12, full texts of classic literature works: left and center – Bible, right – The Twelve Chairs by Ilf and Petrov, 1928)

4   Displaying Multiscale Molecular Genetic Information in a Neurosemantic Graph

As noted earlier, the simplified automatic structuring algorithm implies working with 1-, 2-,... n-digit N-elements at each corresponding level. The N parameter in each layer of the neural network is simultaneously a free parameter of the DNA algorithm, which allows this algorithm to be applied in each layer of the neural semantic structure of the network keeping all connections between layers. N-elements can be placed on the plane of a two-dimensional genometric projection, while the possibility of constructing synaptic connections between layers is preserved from neurosemantic (Fig. 3).

Thus, the layers of the neurosemantic structure can be represented in the form of genometric mosaics while maintaining the structures of connections between the layers. The result is a class of neurosemiotic networks that combines the ideas of cognitive computer graphics [12], semiotic modeling [13], matrix genetics [14] and the neurosemantic approach [1]. The structural relationship with deep learning networks [15] can be traced.

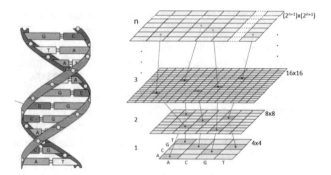

Fig. 3. The scheme of displaying the genome as a neurosemiotic network and the hierarchy of information display. Parametric visualization spaces are integrated into the layers of the semantic neural network.

An example of the application of the described approach in molecular genetics is the problem of training a neural network to identify genetic tumor markers. The neurosemiotic network is based on all connections, and each layer of the network has an internal interpretation at the level of individual fragments of genetic nucleotides. This approach will allow not only to solve the problems of training the neural network, but also to identify markers of genetic disorders and determine their position in DNA [16–18].

5   Experiments and Geometric-Kinetic Approach in Modeling the Elements of Consciousness

For the experiment, Homo sapiens chromosome 1 was examined in the sub-alphabet '2 or 3 hydrogen bonds' simultaneously at several imaging scales. The human chromosome was analysed along its length from one end to the other. As a result of the experiment, an image was obtained in Fig. 4.

As a result of the experiment was possible to visualize the fractal nesting of the structure of genetic information in the human chromosome. This confirms the applicability of the neurosemantic approach, since neurosemantics is based on multiscale. The results of the multiscale frequency analysis are shown in the table. Each neuron in the network corresponds to N-measures, and each neuron has its own activation frequency.

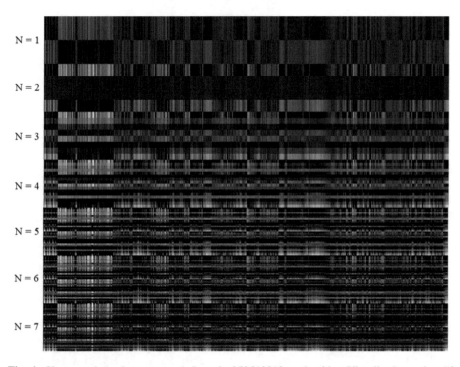

Fig. 4. Homo sapiens chromosome 1. Length: 252513013 nucleotides. Visualised propriety "2 or 3 hydrogen bonds". The abscissa axis is the serial number of the N-string, the ordinate axis is its code. The scales shown are from N = 1 to N = 7.

N	Code of N-plet	Frequency		
		Pyrin-Pyrimidin	Amino-Keto	2 or 3 hydrogen bonds
1	0	10347680	10353805	10190319
	1	10348810	10342685	10506171
2	0	2944380	2855887	2351974
	1	2230771	2322366	2744715
	2	2228149	2319665	2741656
	3	2944945	2850327	2509900
3	0	1156832	1053984	719992
	1	807581	849495	848225
	2	679508	697936	1082710
	3	806158	850341	745620
	4	805168	850155	848518
	5	681715	697679	981108
	6	804797	848661	744709
	7	1157071	1050579	927948

(*continued*)

(*continued*)

N	Code of N-plet	Frequency		
		Pyrin-Pyrimidin	Amino-Keto	2 or 3 hydrogen bonds
4	0	514326	445520	243874
	1	353067	345384	295228
	2	283126	291693	401417
	3	321379	345951	234889
	4	279735	299897	400685
	5	230373	224254	412008
	6	250580	291412	283143
	7	354883	345073	275999
	8	353064	345567	296582
	9	251500	291562	341207
	10	226860	231967	410762
	11	283095	290613	322963
	12	325357	336354	235425
	13	280260	300530	324437
	14	353063	344884	274819
	15	513454	443461	420684
5	0	244941	205703	94345
	1	166366	150176	101324
	2	129536	127969	157393
	3	153573	148239	79582
	4	128717	135864	168032
	5	97045	96654	152786
	6	103576	128492	102899
5	7	153249	148530	85540
	8	135245	132515	156246
	9	87994	108026	163773
	10	87967	81986	197841
	11	96167	97213	131217
	12	113101	125209	102733
	13	88399	107044	124099
	14	116281	125307	101172
	15	166039	150232	119163
	16	166469	150740	100936
	17	116337	126209	135681
	18	96783	105607	163520
	19	104153	128184	108720
	20	94051	104320	152324
	21	87368	81481	177434
	22	97402	105364	123749

(*continued*)

(continued)

N	Code of N-plet	Frequency		
		Pyrin-Pyrimidin	Amino-Keto	2 or 3 hydrogen bonds
	23	129225	127655	134745
	24	147970	143736	79807
	25	113316	126163	108312
	26	93806	103903	132070
	27	129613	135763	127373
	28	147779	143579	85651
	29	135323	132510	134379
	30	166412	151104	118444
	31	245095	203821	218008

The issues of perception of complex information and human interaction with the outside world are closely related to the phenomenon of consciousness. The use of the kinetic and statistical method allows one to describe rather complex systems, including modeling the properties of biological structures, see [19, 20]. In this approach, we rely on the construction of random graphs and their kinetic interpretation [21–25].

A basic statement is connected with the interpretation of consciousness as a statistical phenomenon and the identification of cycles ("vortices") of signals corresponding to its manifestation. In this consideration, an attempt is made to reduce the description of manifestations of consciousness to the study of a kinetic-statistical model capable of reproducing its essential properties. Moreover, the term "consciousness" will be in connection with the term "self-consciousness", although it will differ from it. The principal assumption is that consciousness (and various manifestations of this concept associated with it: mind, reflection, abstract thinking, etc.) should be interpreted as manifestations of the properties of the whole, i.e. there is no center of consciousness in the brain.

It may be useful here to distinguish between the consciousness ("preconsciousness") of the animal and the consciousness of man.

Having understood what the similarities and differences are, we will be able to imagine the design features of the model. We correlate the phenomenon of consciousness with the work of the brain (this is one of the characteristics of this reductive model), and both humans and animals have the concepts of "input" (input of information, receptor) and "output" (effector, system of reactions - motor or other - after information processing). Immediate and quick reactions, correlated with a set of instincts, are comparable in both animals and humans. But if in an animal a slowdown in response (or its absence) only leads to the exclusion of a response signal, then in a human some abstraction of the response is possible. This is manifested in the effect of thought, formulated verbally or visually in the form of a drawing, or even in the form of a vague image, a rudiment of an abstracted sign (word), which can be considered an element of human consciousness. The absence of a direct reaction leads to the conclusion that there are fundamental differences: in an animal there is a direct response, in humans it is also possible to slow down the reaction, which is manifested in the formation of complex systems of neurons in the brain. Such structures can be simulated using an increasingly complex neural network with random ER graphs.

The complication of the human cerebral cortex in comparison with the analogous neural system of an animal can, in our opinion, lead to more indirect paths for the passage of signals, "entangling" them in such a neural system.

We consider the formation of stable cycles, which corresponds to the formation of elements of abstract thinking. Formally, the appearance of cyclic (circulation) structures is associated with the possible transition of graph trees, primary loops and cycles to large clusters, to the creation of global structures of cycles. This giant cluster can also have localized levels, therefore, in two trees of graphs originating from the "input" and "output" (input-output) and connected by vertices (nodes) when the threshold value of complexity is crossed, local giant cluster systems can be formed, which are primary semiotic elements.

These can be potential signs, sound limited statements, visual, visual images (there is so far no connection with the well-studied differentiation of the brain, which can be included in the subsequent development of the model). The system of such trees of neural network graphs with the study of the percolation transition (analogue of the phase transition) when the system becomes more complex and the detection of the occurrence of cycles and clusters is the basic formal model of this kinetic-statistical approach. The structure of cycles and clusters could be associated with the identification of the "system of thought circulation".

For the neural system of an animal, they are constructed into a system of two graph trees connected by vertices. It is mainly the complication and diversity of the system of "stimulation" (internal or external) that goes through the input and the response that goes through the output. But there is no reflection, "abstraction of such abstraction", which is inherent exclusively in human consciousness and meaning the rudiment of self-consciousness.

It can be expected that self-consciousness is associated with the emergence of at least a couple (or more) giant clusters that have a self-sufficient meaning and in a certain self-sustaining "cyclical", "vortex" structures, but at the same time are open systems, interconnected. The statistical fullness of each of these systems means the image of self-consciousness reduced to a simple model. At the same time, the system must interact with another such system, but their possibility of constant autonomization means the reflection of self-consciousness. This evolution of random graphs can lead to the study of complex structures of two or more large cycles. Moreover, the separation-connection between them is capable, from our point of view, to convey the properties of self-reflection, self-consciousness. Consciousness is a statistical phenomenon, there are no local small areas responsible for consciousness and self-consciousness. But a large enough or central large cyclic area capable of "contemplating" other large areas can convey essential properties of consciousness. Interruption of such connections corresponds to loss of consciousness (for example, in a dream). The emergence of sporadically emerging connections between such global cycles corresponds to the return of consciousness, but in a certain probabilistic version, which can be found with random connections of images and thoughts in dreams. Thus, the model of consciousness represents a certain statistical integrity, in which it is precisely the possibility of division into parts that leads to the concept of self-contemplation, self-consciousness.

Let us consider in general terms the method of constructing and describing a random Erdëos-Renyi (ER) graph in the kinetic formalism, see [21, 22].

We can present the ER graph as a kinetic problem by starting with N isolated nodes and introducing links (edges) one by one between randomly selected node pairs. For convenience, we set to $N/2$ the rate at which links are introduced. Consequently, the total number of links at time t is $Nt/2$, so that the average degree $2L/N$ equals t. Therefore the average degree grows at rate 1.

We now investigate the time evolution of the cluster size distribution, from which we can probe the percolation transition of the ER graph. Initially the network consists of N isolated single-site clusters. As links are introduced into the network, clusters can only merge, so that the number of clusters systematically decreases and their mean size grows. The probability that two disconnected clusters of sizes i and j join to create a cluster of size $k = i + j$ equals $(i\,Ci/N) \times (j\,Cj/N)$; here Cj is the total number of clusters of size j. We can find the master equation for $ck(t) = Ck(t)/N$, the density of clusters with k nodes at time t, is.

$$\frac{dc_k}{dt} = \frac{1}{2} \sum_{i+j=k} ic_jjc_j - kc_k,$$

Since the network starts with all nodes isolated, the initial condition is $ck(0) = \delta k,1$. We can obtain the cluster size distribution.

$$c_k(t) = \frac{k^{k-2}}{k!} t^{k-1} e^{-kt},$$

These basic results can be applied to determine a variety of geometrical features of the ER graph.

Consider the construction of a random graph ER in a geometric formalism. Modeling the growth of random graphs in a certain algorithm for adding edges (links) to nodes corresponds to the kinetic approach. According to the classification from [21, 22] we can distinguish between three types of clusters. *tree, unicyclic clusters, or complex clusters*. By definition, a tree contains no closed paths (cycles), a unicyclic cluster contains a single cycle, while a complex cluster contains at least two cycles. We can describe these types of clusters using the notion of the Euler characteristics. For a graph, the Euler characteristic is defined as $E = Nlinks - Nnodes$, where the $Nlinks$ is a total number of links (edges) and $Nnodes$. is a total number of nodes. The Euler characteristic of any tree $E = -1$; for a unicyclic cluster, the Euler characteristic $E = 0$; a complex cluster $E > 0$.

An ER graph starts with N disconnected nodes, i.e. N trees of size 1. Adding a link between two trees gives rise to a larger tree. During the initial stage of evolution, the random graph is a *forest*; that is, all its clusters are trees. Eventually a link inside a tree will be added, and this tree will become a unicyclic component. Adding a link between a tree and a unicyclic cluster results in a unicyclic cluster, while a link between two unicyclic clusters leads to a complex cluster. Because there will be a spectrum of clusters of various complexities, it would seem that the classification in to trees, uni-cyclic clusters, and complex clusters is insufficient. Instead we should distinguish complex clusters more precisely and group them according to their Euler characteristic. Remarkably, this precision is unnecessary for large networks ($N \rightarrow \infty$), since there is at most one complex cluster. More precisely, below the percolation threshold, there is a macroscopic ($\sim N$) number of trees and a finite number (fluctuating from realization to realization) of unicyclic clusters, and above the percolation threshold, the situation is the same as below the threshold, except there is a single complex cluster, the giant component.

In fact, we will be engaged in the construction of complex clusters with the allo-cation of a giant component in modeling a random ER graph, since we mean that the appearance of such a giant cycle in a system of neurons, as indicated above, is the most important element of preconsciousness.

Consider the graph generation algorithm used to obtain images:

1. A set N of nodes (vertices) $N = \{1,2,3.... N\}$ is given. Each node is numbered.
2. The initial structure of the graph is set from two connected nodes.
3. Repeat N times steps 4–6.
4. The set of vertices of the graph G is recalculated (at the first step it is $\{1,2\}$).
5. One of N nodes is randomly selected and connected to an arbitrary node of the graph G.
6. Go to step 3.
7. Stop.

The next step in this general problem is to introduce dynamics into the graphs of different complexity obtained by structures. You can give a certain node (or a system of nodes) the meaning of a receptor (s), and to some other node (s) give the meaning of an effector (s). A signal is fed through the receptor (input), in the effector (output) it is transformed in some way. In a simple graph structure, the signal paths from input to output are fairly simple, although in a tree they become more complex. But if there is a giant component - a complex cluster, then the signal is "delayed" - it wanders through the cycle of a complex graph. This is also the appearance of the element of "preconsciousness".

A certain set of particles (or signals) is fed to the input - it can be described by a certain energy distribution function, at the output the distribution function is obtained in a transformed form. In a complex cluster, some circulation of motion occurs with a possible hydrodynamic (kinetic) analogy of a vortex. In principle, signals can be modeled more abstractly, more "mathematically", referring to the graph as a system of neurons and axons.

The following approach was proposed: one of the terminal nodes is taken as an input, some other (or all others) terminal nodes as an output. A particle is fed to the input, which follows a random path until it reaches any exit. Each transition from top to top increases the weight of this particle, which we register at the output. We repeat the experiment with a wandering particle a given number of times (numerical experiment). This allows not to calculate all the cycles mathematically and to do other operations to obtain the spectra of the distribution of weights at the outputs. For a statistically significant sample, $f(N, M)$ tests will suffice, where N is the number of nodes, M is the number of edges of the graph. Results of simulation are shown in Fig. 4, 5 and 6.

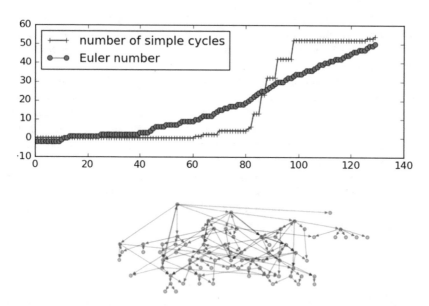

Fig. 5. The random graph (bottom) and its parameters: the number of simple cycles and Euler's number (top). The abscissa is the iteration step, the ordinate is the parameter value.

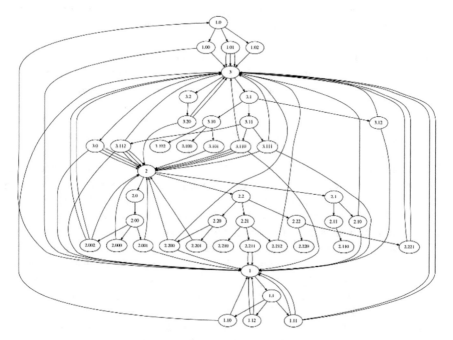

Fig. 6. An example of the structure of a random directed graph with cycles, consisting of three subtrees.

6 Discussion

There is a connection between neural networks and the genetic code. Fractal structures in neural networks are necessary to describe the fractal properties of the real world. Studying DNA methods of mathematical biology we come to a new class of algorithms for building networks of deep learning, information processing and approaching to the consciousness modeling.

The result of visualization of this or that low-entropic information is geometry (sign, symbol). Therefore, geometry allows us to move from semantics to semiotics through automatically generated cognitive graphics. Genometric and neurosemantic approaches refer to the analysis of big data. At the same time, methods of neurosemantics are able to display information in spaces of higher dimensionality, and neurosemiotics - in spaces of fractal dimensionality.

Biological sequence comparison is one of the most important and basic problems in computational biology. The problem of annotating and structuring (taxonomy) sequenced data has not been completely resolved. The proposed partial solution to the problem is annotation with genometric signs.

7 Conclusions

The proposed concept allows you to reach a sign (a genometric mosaic, often fractal in nature, can be interpreted as a sign) directly from texts and large data. The kinetic-geometric modeling of growth and functioning of random graphs, where it is possible to reveal systems of complex clusters, is also correlated with this approach, which allows to approach the problem of describing complex structures of compounding and action of neuronal systems.

The genomometric method is connected with the kinetic-geometric simulation of random graphs, where complex clusters can be observed, the next step of the study is comparison of the theoretical models with the real structures of the neuron systems in the brain.

The generalization of the new results is demonstrating the multilevel fractal structure of DNA using the example of the human chromosome. It is shown that the neurosemantic approach allows to consider such a complex object as DNA at various levels of detail with visualization. This allows to visualize a complex hierarchy of information and the results of neural network modeling. An experiment of frequency analysis of N-mers that activate neurons in a network demonstrates this principle.

One of the significant contributions made in this study is the application of the kinetic approach to the study of complex networks, since the issues of consciousness and perception of information dough are related to each other and to the phenomenon of genetic coding. Described approach is able to visualize the cluster and fractal nature of genetic information. Despite the presence of heuristics for determining the optimal scaling parameter N, multiscale increases the variety of genometric visualizations.

Here some points can be outlined for future possible research in modeling and experimentation. Let's suppose the search for an adequate neural network to reproduce the origin and development of cycles during the development of a graph tree. One can try to mark the network complexity threshold when the marked percolation transition occurs. In the case of successful network experiments, recommendations for carrying out real experiments can be developed, although this will require improving the old or developing new methods. Purpose could be the detection of the appearance of the complication of signals in neuronal systems up to the detection of cyclic structures (as elements of the manifestation of consciousness) when comparing neural signal systems in vertebrates and humans.

In the future, it is planned to develop those outlined paths that have been demonstrated in this work to obtain new scientific results and visualisation tools. This is particularly important for visualizing markers of cancer or other socially significant diseases.

References

1. Bodiakin, V.I.: Concept of self-learning information and control systems based on the neurosemantic paradigm Upravlenie razvitiem krupnomasshtabnyh sistem [Management of large-scale systems development]. In: Proceedings of the Sixth International Conference, vol. 2, pp. 289–298. Establishment of the Russian Academy of Sciences Institute of Management Problems, Moscow, V.A. Trapeznikov (2012). (in Russian)
2. Rumelhart, D.E., McClelland, J.L.: Parallel Distributed Processing. Explorations in the Microstructure of Cognition. MIT, Cambridge (1986)
3. Roepstorff, A.: Cellular neurosemiotics: outlines of an interpretive framework. University of Aarhus, Aarhus (2003)
4. Stepanyan, I.V., Petoukhov, S.V.: The matrix method of representation, analysis and classification of long genetic sequences. Information 8, 12 (2017)
5. Stepanyan, I.V.: Biomathematical system of the nucleic acids description methods. Comput. Res. Model. 12(2), 417–434 (2020). https://doi.org/10.20537/2076-7633-2020-12-2-417-434. (in Russian)
6. Kuznetsova, I., Filipovska, A., Rackham, O., Lugmayr, A., Holzinger, A.: Circularized visualisation of genetic interactions. In: Proceedings of the 26th International Conference on World Wide Web Companion, pp. 225–226, April 2017
7. Zhang, B., Tran, L., Emilsson, V., Zhu, J.: Characterization of genetic networks associated with Alzheimer's disease. In: Systems Biology of Alzheimer's Disease, pp. 459–477. Humana Press, New York (2016)
8. Elzeki, O.M., Alrahmawy, M.F., Elmougy, S.: A new hybrid genetic and information gain algorithm for imputing missing values in cancer genes datasets. Int. J. Intell. Syst. Ap. (IJISA) 11(12), 20–33 (2019). https://doi.org/10.5815/ijisa.2019.12.03
9. Khalil, M.I.: Locating all common subsequences in two DNA sequences. Int. J. Inf. Technol. Comput. Sci. (IJITCS) 8(5), 81–87 (2016). https://doi.org/10.5815/ijitcs.2016.05.09
10. Gupta, L.M., Garg, H., Samad, A.: An improved DNA based security model using reduced cipher text technique. Int. J. Comput. Netw. Inf. Secur. (IJCNIS) 11(7), 13–20 (2019). https://doi.org/10.5815/ijcnis.2019.07.03
11. Lednev, M..Yu.: An innovative neurosemantic approach in the analysis and processing of unstructured data. Intellect. Property Exch. XVI(2), 15–24 (2017)
12. Zenkin, A.A.: Cognitive computer graphics — application to decision support systems. In: Proceedings of II International Conference MORINTECH—1997, St. Petersburg, Russia, vol. 8, pp. 197–203 (1997)

13. Osipov, G.S.: Semiotic modeling: an overview. In: Strohn, R.J. (ed.) Proceedings of Workshop on Russian Situation Control and Cybernetic/Semiotic Modeling, Columbus, USA, March 1995, pp. 38–64 (1995)
14. Petukhov, S.V.: Matrix genetics, algebras, genetic codes, noise immunity. RCD (2008). 316 p.
15. Washburn, J.D., Mejia-Guerra, M.K., Ramstein, G., Kremling, K.A., Valluru, R., Buckler, E.S., Wang, H.: Evolutionarily informed deep learning methods for predicting relative transcript abundance from DNA sequence. Proc. Natl. Acad. Sci. **116**(12), 5542–5549 (2019)
16. Schumacher, M., Graf, E., Gerds, T.: How to assess prognostic models for survival data: a case study in oncology. Methods Inf. Med. **42**(05), 564–571 (2003)
17. Angermueller, C., Lee, H.J., Reik, W., Stegle, O.: DeepCpG: accurate prediction of single-cell DNA methylation states using deep learning. Genome Biol. **18**(1), 1–13 (2017)
18. Abdelhady, H.G., Allen, S., Davies, M.C., Roberts, C.J., Tendler, S.J., Williams, P.M.: Direct real-time molecular scale visualisation of the degradation of condensed DNA complexes exposed to DNase I. Nucleic Acids Res. **31**(14), 4001–4005 (2003)
19. Aristov, V.V., Ilyin, O.V.: Methods and problems of the kinetic approach for simulating biological structures. Comput. Stud. Modell. **10**, 851–866 (2018)
20. Aristov, V.V.: Biological systems as nonequilibrium structures described by kinetic methods. Results Phys. **13**, 102232 (2019)
21. Krapivsky, P.L., Redner, S., Ben-Naim, E.: A Kinetic View of Statistical Physics. Cambridge University Press, Cambridge (2010)
22. Ben-Naim, E., Krapivsky, P.L.: Kinetic theory of random graphs: from paths to cycles. Phys. Rev. E **71**, 026129 (2005)
23. Krapivsky, P.L., Redner, S.: Emergent network modularity. J. Stat. Mech. 073405 (2017)
24. Yang, W., et al.: Simultaneous multi-plane imaging of neural circuits. Neuron **89**, 269–284 (2016)
25. Severino, F.P.U., et al.: The role of dimensionality in neuronal network dynamics. Sci. Rep. **6**, 29640 (2016)

Focal Curves in the Problem of Representing Smooth Geometric Shapes

T. Rakcheeva[(⊠)]

Mechanical Engineering Research Institute of the Russian Academy of Sciences,
4, Maly Kharitonievskiy Pereulok, Moscow 101990, Russia
rta_ra@list.ru

Abstract. A focal method for approximating an empirical geometric shape represented by a smooth closed curve is developed. The analytical basis of the method is multifocal lemniscates. The finite number of foci inside the lemniscate and its radius are control parameters of this method. In contrast to the harmonic representation, focal control parameters have of the same nature as the shape itself. *In this paper we consider the more general class of multifocal curves from the point of view of using them as a basis for describing geometric shapes.* Making the most general requirements for the metric distance function and the invariant description of geometric shapes sets *gives* a parameterized set of bases in the quasilemniscates class that meet these requirements. Properties of each basis, defined by different functions distances are also reflected in their approximate possibilities. The family multifocal lemniscates is distinguished as the limiting case of parameterized set of basic quasilemniscates. The focal representation of empirical geometric shapes by multifocal lemniscates makes it possible to apply them in different applications.

Keywords: Shape · Curves · Focuses · Lemniscates · Ellipses · Quasilemniscates · Approximation · The metric · The degrees of freedom · The generation of forms

1 Introduction

"The splitting of the discrete and continuous seems to me the main result all forms of teaching."

H. Weil

The need for an approximate description of empirical geometric forms is widely represented in modem applied mathematics. It can be a contour shape of various objects for construction-production or aesthetic-design purposes; and a graphical representation of the results of complex computational procedures for the purpose of meaningful interpretation or search for solutions; and the border of an abstract set in the space of feature for identification or the diagnostics [1–6]. The latter refers to pattern recognition – the most relevant area of modern applied mathematics, such as artificial intelligence [4, 6, 7].

Z. Hu et al. (Eds.): AIMEE 2020, AISC 1315, pp. 29–41, 2021.
https://doi.org/10.1007/978-3-030-67133-4_3

Analytical description of empirical geometric forms, represented by a smooth closed curve, as the traditional problem of applied mathematics is constantly evolving, responding to the needs of newly emerging problems. For approximate description of curves, numerous methods use certain classes of approximating functions, depending on the requirements of the applied problem and the corresponding properties of the curves. The widely used class of trigonometric polynomials is adequate such properties signals, how the periodicity and smoothness.

What should also be adequate analytical representation of an empirical closed curve for the description of its shape?

Classical sources define the shape as spatially organized structure, whose geometric properties are independent. Thus, the approximate description of empirical geometric forms certain requirements for the approximation method; in this case, it is invariance with respect to the choice of the coordinate system. Mathematically, this is known to mean invariance with respect to motion transformations (congruence) and scale transformations.

Continuous approximation of the discrete description of what is the description of the empirical form is obviously a composite of continuous and discrete. The continuous component reflects the general character of the empirical form, while the discrete component reflects its individual features. All approximation methods differ by these components. Thus, a harmonic approximation represents an empirical curve as a discrete system of coefficients (control parameters) in a continuous basis of harmonics. The harmonic description has, as it is known, a number of remarkable properties that make them widely used not only for describing signals, but also for approximating descriptions static geometric shapes, solving problems of compressed representation and filtering.

This paper is devoted to the focal approximation in the class of continuous functions called lemniscates. The focal approximation represents an empirical curve via a discrete system of focus points inside the curve. In contrast to the harmonic representation, the focal representation has a discrete component (control parameters) of the same nature as the shape itself. This gives new opportunities for applying the focal approximation of the curve shape in solving applied problems [3].

2   Multifocal Lemniscates

Lemniscate – the focal curve is completely determined by the system of a finite number k of points-foci in it and a numeric parameter R radius. Invariant of the lemniscate is constant along the curve product the distances to the foci of all k:

$$\prod_{j=1}^{k} r_j = R^k, \qquad (1)$$

where r_j – Euclidean distance from any point of the lemniscate (x, y) to the j-th focus f_j with the coordinates of (a_j, b_j). Lemniscate with k foci will be called kf-lemniscate, and the focal system of k foci $f = \{f_1, ..., f_k\}$ – kf-system or kf-structure.

Lemniscates are a smooth closed plane curves without self-intersections, not necessarily simply connected. For a fixed set of k foci lemniscates with different radii form a family of embedded curves from k-connected, for small values of radius R, to simply connected, for large values. The curves with a larger radius cover the curves with a smaller radius without crossing (Fig. 1a). In a certain range of values of the radius the lemniscate have a large variation in form that may influence their use as approximating functions for a wide range of curve forms [1–3].

The family of a multifocal lemniscates forms class of approximating functions for the approximation of smooth closed curves (continuous basis). Foci of the lemniscate $f_j = \{a_j, b_j\}$, $j = 1, \ldots, k$ form discrete focal system.

3   Basis of the Lemniscates

The fact that the approximating class of the lemniscates is a family of smooth functions brings the focal method of approximation with such classical methods as an approximation of power or trigonometric polynomials. However, there is a fundamental difference of focal approximation from these methods. The power and trigonometric approximation, based on a parametric description of the curves, are adequate to the tasks of analysis of signals – one-dimensional curves. Their degrees of freedom are objects of others in relation to curve spaces and in general not invariant under coordinate transformations, that preserve the shape of the curve, such as shift, rotation, scaling. The focal method more appropriate for tasks related to the representation shape of the curve. Freedom degrees of lemniscates are invariant under these transformations. This is easily seen, since the object of approximation (points of the curve) and freedom degrees of approach (foci of the lemniscate), belong to the same space.

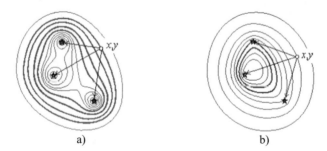

a)   b)

Fig. 1. The families of the curves with the same focal system: a) 3*f*-lemniscates and b) 3*f*-ellipses

4   Approximation by the Multifocal Lemniscates

The task of approximation of closed curves by the mulityfocal lemniscates is formulated as follows:

Let be an arbitrary closed curve C in the plane of its points $\{x_i, y_i\}$, *i = 1, ..., n and some nearness criterion of curves. Required to find the inside of the curve focal system* $\{f_j\}$, *j = 1, ..., k, such that at a certain radius R, corresponding to them lemniscate L will be close to the curve C in terms of selected criterion.*

In [1], devoted to the analysis of approximation possibilities of the multifocal lemniscates, Hilbert showed that by choosing an appropriate number of foci and their location on the plane, we can get the lemniscate, which is close to any predetermined smooth closed curve. However, the question of how to determine the radius and the corresponding system of focal points for concrete empirical curve, given by the coordinates of its points, remained open. Appropriate methods of producing the focal approximation of an arbitrary curve, determining the number, location of the focal points and the radius of the approximating lemniscate, are developed for both the real description, and for the complex. For work with the focal method two criteria are determined. One of them is universal and estimates a degree of affinity between the curve C and the lemniscate L according to this or that function of distance between points of these curves in terms of usual Euclidian metric. The other one is special and follows from lemniscate invariant property to keep constant radius value in all points. The criteria are topologically equivalent: at aspiration to zero of distance between a curve and lemniscate radius fluctuations on a curve aspire to zero and wise versa [3].

Examples of focal approximation of the empirical curve shown in Fig. 2a, 5d. For comparison the result of approximating of the same curve in the harmonic basis is shown in Fig. 2b (to achieve equal precision it is required approximately the same number of parameters: 19 parameters for the focal and 22 parameters for the harmonic method).

a)   b)

Fig. 2. Approximation of the empirical curve: *a)* foci (9), *b)* harmonics (5)

Focal representation of the form does not impose on the data of the curve any requirements on parameterization and ordering – the point can be given in any order (as opposed to harmonic representation). The result of the focal approach is also free of the parameterization and is determined only by the form of the curve.

5 Additive Invariant

A natural question arises about the possibility of focal approximations in other class of the focal curves close to lemniscates. Represents, in particular, of interest to consider a more general class of such functions in terms of the approaching problem of the shape of the curves and determine thus lemniscates place in a series of similar families of approximating functions. In other words:

"*What is singularity about the class of lemniscates in the class of other focal curves in terms of describing the shape of the curve?*"

Invariant of the multifocal lemniscate in the form of (1) allows identify it as a multiplicative invariant of Euclidean distances between each point and all its foci. Consider the wider class of functions, including class of the multifocal lemniscate as a special case.

The equation of the lemniscate (1) can convert to an additive invariant of logarithmic functions of distances as follows:

$$\sum_{j=1}^{k} \ln r_j(x, y) = S, \tag{2}$$

where $S = k \ln R$. Similarly, an additive invariant can be represented by the class of curves, which are a multifocal generalization of the ellipses for which as known the invariant is a sum of the distances to its foci inside the curve:

$$\sum_{j=1}^{k} r_j(x, y) = S \tag{3}$$

Multifocal curves with additive invariant (3) is called, by analogy with the lemniscates, multifocal ellipses with a radius of S. Parameterized family of the isofocal ellipses is shown in Fig. 1b for the same focal system, as the lemniscates family in Fig. 1a. The limiting case of large values of S is also a circle, for small values of S the connection of curves, in contrast to the lemniscates, is not broken – the curves, remaining convex, contracted to the point (in this case to a single focus at the apex of the obtuse angle) with the smallest total distance to the foci. A comparison of the two families in Fig. 1 shows that the multifocal lemniscates demonstrate much greater variety of forms than multifocal ellipses with the same foci. In particular, in the case of k-foci, forming a regular structure of located in the vertices of a regular k-gon, kf-ellipses shrink to the center of the circumscribed circle. This is well illustrated by the family of $6f$-ellipses (Fig. 3a) and by similar family $7f$-ellipses (Fig. 4a, *bottom*) in contrast to the lemniscates (Fig. 3b and Fig. 4a, *top*).

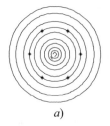 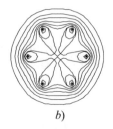 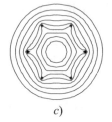

a)   *b)*   *c)*

Fig. 3. Families of isofocal 6*f*-quasi-lemniscates with hexagonal focal structure: *a)* $f(r) = r^{1/3}$; *b)* $f(r) = \ln(r)$ (lemniscates); *c)* $f(r) = r$ (ellipses)

6   Quasilemniscates

Bearing in mind, the common representations of both families: multifocal lemniscates (2) and ellipses (3), it is interesting to consider the equation:

$$\sum_{j=1}^{k} f(r_j(x,y)) = S, \tag{4}$$

where an arbitrary function of *f* is a generalization of them. Defined by Eq. (4) curves is called *quasilemniscates*. Ordinary multifocal lemniscates are obtained by setting *f* (*r*) = ln*r*, and multifocal ellipse, if we set $f(r) = r$ [8].

This paper devoted to the analysis the possibilities of using quasilemniscates as a class of approximating curves. The main conclusion is that the lemniscate in some sense are the most appropriate for the approximation of arbitrary form of the curves among all quasilemniscates under certain requirements to the function *f*. And namely:

- should be monotonically increasing, because it is in some sense an analogue of the distance;
- should provide an approximation complexity of the method, sufficient for the approximation of smooth curves of arbitrary shape;
- must be invariant under transformations that preserve the shape of the curve: shift, rotation and scaling.

The last requirement is divided into two parts: the invariance under movement of rigid body: *shifts and rotations*, and *scaling* invariance.

7   Invariance of the Lemniscates Basis

It is easily seen; the lemniscates satisfy the formulated requirements to function *f*. Indeed:

Lemniscate function $f = \ln r$ satisfies obviously demands of monotone increase and complexity;

Euclidean distance r is invariant under transformations of movement, so that all the functionals of type (4) are invariant under transformations of shift and rotation;

The scale transformation, as is not difficult to show, converts the kf-lemniscate in kf-lemniscate but with a different radius.

Therefore, up to the radius values, family of the kf-lemniscates is invariant under scale transformations, which corresponds to the definition of the shape. Thus, the realistic requirements for the class of functions that are useful for approximating the shapes of the curves are to be fulfilled for the class multifocal lemniscates.

And as these requirements are satisfied quasilemniscates?

8 Convex Functions

The statement is that, if the function f convex and the corresponding quasilemniscates given by Eq. (4) are convex curves. This refers to the convexity of its limited set. The proof of this fact, obviously, should be of the following statements simply provable: 1) if f – a convex monotonically increasing function, then the function $f(r(x,y))$ of points (x,y) is also a convex; 2) the sum of convex functions is a convex function; 3) level line of a convex function is a convex curve.

Thus, if the distance is given by convex functions, the functional (4) generates only convex families of functions that cannot be generally used as the class of approximating functions in the development of an approximation method. The convex functions of distance determine and multifocal ellipses (3), so their use is only possible if a priori knowledge by that approximated curve itself can be a convex function.

9 Scale Invariance

The following statement concerns the invariance under the transformations of the movement. It is provided the invariance under these transformations of the Euclidean distance is in the argument of the functional (4).

As for the scale transformation, the lemniscates, as shown above, satisfied to this requirement up to the radius R. Let us consider the quasilemniscates. The assertion consists in the fact that:

If

$$\sum_{j=1}^{k} f(r_j(x,y)) = const,$$

then and

$$\sum_{j=1}^{k} f(\alpha(r_j(x,y))) = const,$$

which may, generally speaking, have a different meaning; α – scale coefficient.

Assume that the function $f(r)$ defines the class of quasilemniscates invariant under scale transformation. In order to obtain the necessary conditions, we can consider the case of the two foci. The assertion is that:

if

$$f(r_1) + f(r_2) = const,$$

then

$$f(\alpha r_1) + f(\alpha r_2) = const.$$

To prove it, we fix foci and define the function $F(\alpha, S)$, as follows: the value of this function will be assumed radius of quasi-lemniscate obtained from the quasi-lemniscate with radius S as a result of stretching with a coefficient of α. Thus, for any point on the quasi-lemniscates with radius S, the following equality holds:

$$f(\alpha r_1) + f(\alpha r_2) = F(\alpha S) = F(\alpha, f(r_1) + f(r_2)).$$

Differentiating of r_1 and r_2 the extremes of this equality, we obtain:

$$\alpha f'(\alpha r_1) = \frac{\partial F}{\partial S}(\alpha, f(r_1) + f(r_2)) f'(r_1), \quad \alpha f'(\alpha r_2) = \frac{\partial F}{\partial S}(\alpha, f(r_1) + f(r_2)) f'(r_2).$$

The equations can be rewritten as:

$$\frac{f'(\alpha r_1)}{f'(r_1)} = \frac{1}{\alpha} \frac{\partial F}{\partial S}, \quad \frac{f'(\alpha r_2)}{f'(r_2)} = \frac{1}{\alpha} \frac{\partial F}{\partial S}.$$

Equating the left-hand sides, we find that the ratio of the derivatives on the new curve and on the original at the same point does not depend on this point, but only on α. Denote this ratio by $g(\alpha)$:

$$\frac{f'(ar_1)}{f'(r_1)} = \frac{f'(ar_2)}{f'(r_2)} = g(a).$$

Differentiating further of α, we get:

$$\frac{rf''(\alpha r)}{f'(r)} = g'(\alpha)$$

For $\alpha = 1$ the right side of the equation becomes a constant $q = g(1)$. As a result, we obtain the following equation:

$$rf''(r) = qf'(r).$$

By adopting the notation $h = f'$, we obtain the first order differential equation:

$$rh'(r) = qh(r),$$

solution of which is known to be:

$$h(r) = C_1 r^q.$$

Returning to the function f, we finally obtain:

$$f(r) = C_1 \int r^q dr + C_2 = \left\{ \frac{C}{q+1} r^{q+1} + C_2, \ q \neq -1; \ C_1 \ \ln r + C_2, \ q = -1 \right\}$$

Thus, we have the class of functions for which specified above functional is invariant under a scale transformation:

$$f(r) = c \ln r \tag{5}$$

or

$$f(r) = cr^\alpha, \tag{6}$$

the first of which is the lemniscates (2).

Direct substitution verifies that the found functions really define classes of quasi-lemniscates invariant under scaling, already for any number of foci.

Thus, the interest for the purposes of approximation, apart from the usual lemniscate given by the solution of (5), may represent quasilemniscates defined by power functions of the form (6), where only the values of α in the range $0 < \alpha < 1$. Indeed, at $\alpha \geq 1$ the function $f(r)$ turns out to be convex. It is not difficult to show results in convexity of the entire class of approximating functions, severely restricting the class of the approximated functions. When $\alpha < 0$ the function $f(r)$ is decreasing, what contradicts the requirement of monotonic increasing of the function with the meaning of the distance (see Sect. 6).

10 Families of Quasilemniscates

On materials of this work it was carried out comparative computer experiment. For the same foci system the families of quasilemniscates were built for different f-distance functions and for a wide range of radius S (Fig. 4).

Figure 4a (*top*) shows a family of curves with a logarithmic function of distance $f(r) = \ln r$ – the family of lemniscates (5). Here, as in Fig. 1a also revealed considered above properties, in particular, holding all foci inside lemniscates. Figure 4a (*bottom*) shows another family of a quasilemniscates given by the solution of (6) for the $\alpha = 1$: $f(r) = r$ – family of multifocal ellipses. All curves in this family, as in Fig. 1b, are convex, their form is poor and with decreasing radius they "lose" their foci, leaving them outside of the curve (and the more $\alpha > 1$, the stronger these properties are manifested).

The other two pictures (Fig. 4b) are another two families of quasilemniscates generated by the solution of (6), with the functions of the distance:

$f(r) = \sqrt{r}$ (Fig. 4b, *bottom*); $f(r) = \sqrt{4r}$ (Fig. 4b, *top*).

The curves of these families have not all convex, in contrast to ellipses, but their forms poorer than the lemniscates form, and they, like ellipses, "lose" their foci, though not as "easy". The comparison of the same families of curves with power functions of distance (Fig. 4b) together shows that the manifestation of such characteristics as the holding of foci and variability of forms, from a family with a lower value of the degree (Fig. 4b, *top*), is more than in the family with the higher value (Fig. 4b, *bottom*).

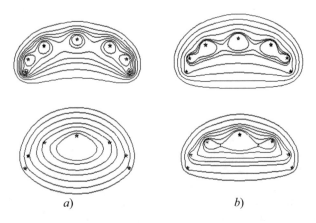

$$a) \qquad\qquad\qquad\qquad b)$$

Fig. 4. The families of a multifocal quasilemniscates with the different f-function of the distance: *a*) the lemniscates (top), ellipses (bottom) and *b*) $f(r) = r^{1/4}$ (top), $f(r) = r^{1/2}$ (bottom)

It can be concluded that with increasing degree of the root n in a function of distance $f(r) = \sqrt[n]{r}$ corresponding family of quasilemniscates will move away in its properties from the family of ellipses, and close to the family of lemniscates. It is natural, having in mind a relation between the logarithm and the roots:

$$n(\sqrt[n]{r} - 1) \to \ln r, \quad n \to \infty.$$

11   Computer Experiment in the Approximation of Curves

The observed regularity is vividly expressed in the task of approximating the same empirical curve in the four discussed bases (Fig. 5). The above properties of quasilemniscates given by the different functions of distance are reflected in their approximation properties. As an illustration, Fig. 5 shows examples of approximations

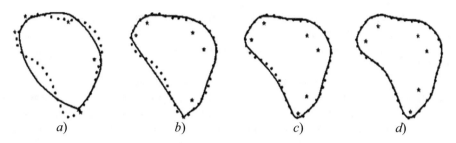

Fig. 5. Approximation of the empirical curve by quasilemniscates $f(r) = r^{\alpha}$: a) $\alpha = 1$, b) $\alpha = 1/2$, c) $\alpha = 1/4$ and lemniscates: d) $f(r) = \ln r$

of the same empirical curve by the six-focus quasilemniscates: ellipses with $f(r) = r$, $f(r) = \sqrt{r}$, $f(r) = \sqrt[4]{r}$ $(a - c)$; lemniscates with $f(r) = \ln r$ (d).

As can be clearly seen by comparing the figures, quasilemniscates are not able to approximate a given curve (the result of approximation by ellipses remains convex), while the approach, carried out with the same number of foci lemniscates can be considered quite satisfactory [3].

It should be noted that the higher the root of α the closer approximation to the approximation by the lemniscates (and to required).

Thus, analysis of kf-family of quasilemniscates identifies family of lemniscates as the most satisfying the general requirements for the distance function and description of geometric forms and their invariants.

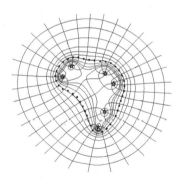

Fig. 6. Poly-polar own system coordinates of the empirical form

12 Conclusion

Focal system has a representation in the same coordinate system as the curve, and is subject to transformations, preserving the form of the curve. Preserving the compressed information about the curve form, the focal system inherits also the symmetries of the curve form, unlike the harmonic system. It is shown that the lemniscate, and hence the basis of lemniscates, have the same symmetries group as its focal system. Thus, the focal representation that is invariant with respect to the coordinate system, allows separating the invariants of the components of identity and individuality of geometric form [3, 9].

Focal representation of the form of the curves allows entering the plane polypolar coordinate system. Lemniscates family and the family of gradient curves form two mutually orthogonal families of coordinate curves $\rho(x, y) = $ const and $\varphi(x, y) = $ const, where the metric component can be any shape, for example, the shape of a given curve [10]. Figure 6 shows the grid polypolar lemniscate coordinate system generated by the focal approximated empirical curve. Thus, the shape of the object image can be related to *its own, individual, polypolar coordinate system* by implementing a focal approximation. Metric component can be in this case rather complex, manually adjusted or automatically, and for any form of the metric component the angular component is obtained as orthogonal.

The pattern recognition problem in the focal paradigm makes it possible to apply the focal model both at the learning stage and at the identification stage. Focal representation of curves and surfaces is using for description of classes and their boundary forms. An important advantage is that the focal classification space also forms a continuous affiliation function, which marks each point of the space as belonging to a certain class, but it also preserves the quantitative evaluation of the degree of this belonging. Thus, the multidimensional classification problem reduces to decision making in one-dimensional space [4].

The analysis of the properties of focus bases for describing empirical geometric forms has shown that quasilemniscates, unlike lemniscates, cannot be used in these and most other applied problems. Finally, this material can also be useful in the educational process, since it is important for students not only to know abstract mathematical objects, but also to feel the possibility of their practical application [9, 11, 12].

References

1. Hilbert, D.: Gessamelte Abhandlungen, vol. 3, p. 435. Springer, Berlin (1935)
2. Markushevich, A.I.: The theory of analytical functions, p. 486. T.1 – M., Nauka (1967)
3. Rakcheeva, T.A.: Multifocus lemniscates: approximation of curves. Comput. Math. Math. Phys. **50**(11), 1956–1967 (2010)
4. Rakcheeva, T.A.: Focal model in the pattern recognition problem. In: Advances in Intelligent Systems and Computing, vol. 902, pp. 127–138 (2019). https://doi.org/10.1007/978-3-030-12082-5_12
5. Al-Jubouri, H.A.: Integration colour and texture features for content-based image retrieval. Int. J. Mod. Educ. Comput. Sci. (IJMECS) **12**(2), 10–18 (2020). https://doi.org/10.5815/ijmecs.2020.02.02
6. Mahmoud, S.M., Habeeb, R.S.: Analysis of large set of images using mapreduce framework. Int. J. Mod. Educ. Comput. Sci. (IJMECS), **11**(12), 47–52 (2019). https://doi.org/10.5815/ijmecs.2019.12.05
7. Fahim, A.: A clustering algorithm based on local density of points. Int. J. Mod. Educ. Comput. Sci. (IJMECS) **9**(12), 9–16, (2017). https://doi.org/10.5815/ijmecs.2017.12.02

8. Rakcheeva, T.A.: Quasilemniscates in the task of approximation of the curve forms. Intellect. Syst. **13**(1–4), 79–96 (2009)
9. Rakcheeva, T.A.: Symmetries of polypolar coordination. Vestnik MGOU. Ser. "Phys. Math." **1**, 10–20 (2011)
10. Rakcheeva, T.A.: Polypolar lemniscate coordinate system. Comput. Res. Model. **1**(3), 256–263 (2009)
11. Gourav, T.S., Singh, H.: Computational approach to image segmentation analysis. Int. J. Mod. Educ. Comput. Sci. (IJMECS) **9**(7), 30–37 (2017). https://doi.org/10.5815/ijmecs.2017.07.04
12. Osaci, M., Cuntan, C.D.: Graphical programming environment for performing physical experiments. Int. J. Mod. Educ. Comput. Sci. (IJMECS) **12**(1), 11–17 (2020). https://doi.org/10.5815/ijmecs.2020.01.02

Method of Fuzzy Agreed Alternative Selection in Multi-agent Systems

N. Yu. Mutovkina[✉]

Tver State Technical University, Tver, Russia
letter-boxNM@yandex.ru

Abstract. The article considers the method of fuzzy coordinated alternative selection in multi-agent systems. Alternatives may include strategies for the behavior of intelligent agents, options for their actions to solve the problem, current problems, and sets of resources to solve them. In this study, the method of fuzzy agreed selection is proposing an example of choosing the optimal way to solve complex problems in multi-agent systems in the presence of both quantitative and qualitative selection criteria. The methodology is basing on evaluating alternatives based on the importance of parameters and trust levels for agents. Estimates are values of membership functions that characterize the compatibility of each variant with the term of "best". The theoretical analysis allowed us to identify the conditions for coordinated decision-making in multi-agent systems, as well as the reasons that prevent this. They can eliminate by agreed optimization and control methods. Methods of statistics are using to determine the consistency of expert opinions and the reliability of expert assessments.

Keywords: A multi-agent system · Intelligent agent · Psycho-behavioral type · Fuzzy logic · Agreed choice · Confidence level

1 Introduction

Multi-agent systems (MAS) are designing to solve complex, non-standard problems, the solution of which is onerously due to their unclear formulation, the presence of many both quantitative and qualitative characteristics, and the lack of known solutions. One such task is the selection of the optimal variant of the plan or course of action (best alternative), which is complicated by the diversity of intelligent agents, as well as quality characteristics that are difficult to formalize and to evaluate objectively. This led to the importance and practical significance of this study. Also, when evaluating alternatives and then selecting the optimal alternative, it is necessary not only to form a representative system of evaluation criteria but also to measure the level of trust in agents. Due to the anthropomorphic definition of an intelligent agent and the practice of applying MAS in various fields, it is establishing that not all agents can be trusted completely, equally. The main reason for this is that each agent has their views, opinions, and interests [1]. At the same time, the question of the knowledge unsuitability of intelligent agents for solving poorly formalized problems is not considering.

Z. Hu et al. (Eds.): AIMEE 2020, AISC 1315, pp. 42–54, 2021.
https://doi.org/10.1007/978-3-030-67133-4_4

It is assuming that all agents are well aware of the problem area and have the necessary knowledge and skills to solve the problem.

In this study, we consider the idea of the consistency of agents' actions in negotiations regarding the choice of the optimal way to solve a global problem (Ω). Each agent's opinion of any of the evaluated alternatives is unclear (multiple, ambiguous) [2]. The vagueness in the assessment is due to a variety of factors, including the psycho-behavioral type of agent currently in specific conditions. There is a high degree of subjectivity in evaluating and choosing the optimal alternative, but there are the following predefined conditions:

1. All agents know exactly their target functions and MAS target functions.
2. All agents are qualified to participate in the task (Ω).
3. Each agent has enough resources to participate in the task (Ω).
4. The sets of possible ways to solve the problem (Ω) and the criteria for evaluating them are strictly limited.
5. The amount of time allocated to solve the problem (Ω) is clearly defined.
6. The values of the membership functions are obtaining directly from the agents following the proposed scale $L = [0, 1]$.

The purpose of this work is to develop a methodology that allows implementing a fuzzy agreed choice of the optimal way to solve a complex problem in the MAS in the presence of both quantitative and qualitative selection criteria. The coordinated choice is considering as a decision-making stage, following the phase of obtaining coordinated expert assessments. To achieve this goal, it is necessary to review the literature on the problem of optimal choice, formulate a theoretical basis for the functioning of the MAS, formalize the task of choice, synthesize the method and conduct its practical testing.

2 Literature Review

The problem of a coordinated choice of alternatives is certainly not new and has been considering in many works. However, all research has been limited, as a rule, to the use of well-known methods, which include: Bayesian approach [3–5], decision trees [6–8], method of analysis of hierarchies [9–11], production rules [12–14], fuzzy cognitive maps [15–17]. Analysis of these approaches allows us to conclude that fuzzy cognitive maps are the most appropriate way to describe the tree of choosing the optimal alternative to solving a complex weakly formalized problem. However, they also do not take into account the degree of trust between intelligent agents in evaluating alternatives.

In this study, MAS is considering as an anthropocentric, realistic dynamic system characterized by incomplete and partial reliability of the information, in which behavioral models of agents are representing using fuzzy logic [18–20]. A set of fuzzy "IF-THEN" rules and motivation as one of the input variables are using to model agent behavior. It is assuming that the agents that make up the MAS have different levels of rationality and can take into account the interests of other agents and the MAS as a whole along with their interests. The basis of motivation is generally accepted norms, laws of social life, projected on the agents who make up the MAS, agents are in an equilibrium state only when they manage to align their views and interests with generally accepted laws and principles. Agents usually reject the prospect of small gains if there is a high probability of losses [21].

3 Method

Developing a joint solution to the problem (Ω) first involves generating N alternatives to the solution ($i = \overline{1,N}$). At the same time, M agents are in the MAS, ($j = \overline{1,M}$). Based on the specifics Ω, the agents jointly form a representative set of evaluation criteria ($k = \overline{1,K}$), and agent j gives its assessment α_{jk} of alternative i for each of the parameters. Since the evaluation criteria can be both quantitative and qualitative, to be able to compare estimates, it is necessary to move from the values of different types of parameters to their fuzzy assessments, measured on the same quantitative scale. The measurement scale is representing as a numerical segment [0, 1], and the evaluation $\alpha_{jk} \rightarrow \mu_k^j(i) \in [0, 1]$ characterizes how much the method i of solving the problem (Ω) corresponds to the concept of "best by criterion k".

As a result, each alternative i will be represented not by a set of parameter values α_{jk}, but by a set $\{\mu_1^j(i), \mu_2^j(i), ..., \mu_k^j(i), ..., \mu_K^j(i)\}$ of corresponding numerical estimates. Thus, for each criterion $k \in K$, there is a set $\{\mu_k^j(1), \mu_k^j(2), ..., \mu_k^j(i), ..., \mu_k^j(N)\}$, in which each element characterizes the correspondence of alternative i to the concept of "best" according to this criterion. This definition can represent by a fuzzy set defined on the universal set of alternatives N:

$$\tilde{k} = \left\{ \frac{\mu_k^j(1)}{1}, \frac{\mu_k^j(2)}{2}, ..., \frac{\mu_k^j(N)}{N} \right\} \tag{1}$$

with the membership function $\mu_k^j(i)$ that characterizes the compatibility of alternative i with the concept of "best".

The following options are available for representing evaluation matrices:

I) the evaluation of alternative i by $k = \overline{1,K}$ criteria by M agents, $j = \overline{1,M}$ (Table 1);

II) the evaluation of N, ($i = \overline{1,N}$) alternatives according to $k = \overline{1,K}$ criteria by an expert j (Table 2).

Table 1. Evaluation of alternative i by MAS agents

k ⧹ j	1	2	3	...	K-1	K
1	$\mu_1^1(i)$	$\mu_2^1(i)$	$\mu_3^1(i)$	...	$\mu_{K-1}^1(i)$	$\mu_K^1(i)$
2	$\mu_1^2(i)$	$\mu_2^2(i)$	$\mu_3^2(i)$	...	$\mu_{K-1}^2(i)$	$\mu_K^2(i)$
3	$\mu_1^3(i)$	$\mu_2^3(i)$	$\mu_3^3(i)$	...	$\mu_{K-1}^3(i)$	$\mu_K^3(i)$
...	...	...	...	...	...	...
M-1	$\mu_1^{M-1}(i)$	$\mu_2^{M-1}(i)$	$\mu_3^{M-1}(i)$	...	$\mu_{K-1}^{M-1}(i)$	$\mu_K^{M-1}(i)$
M	$\mu_1^M(i)$	$\mu_2^M(i)$	$\mu_3^M(i)$	...	$\mu_{K-1}^M(i)$	$\mu_K^M(i)$

Table 2. Evaluation of N alternatives by agent j MAS

k ⧹ i	1	2	3	...	K-1	K
1	$\mu_1^j(1)$	$\mu_2^j(1)$	$\mu_3^j(1)$	...	$\mu_{K-1}^j(1)$	$\mu_K^j(1)$
2	$\mu_1^j(2)$	$\mu_2^j(2)$	$\mu_3^j(2)$	...	$\mu_{K-1}^j(2)$	$\mu_K^j(2)$
3	$\mu_1^j(3)$	$\mu_2^j(3)$	$\mu_3^j(3)$	...	$\mu_{K-1}^j(3)$	$\mu_K^j(3)$
...	...	...	...	...	...	...
N-1	$\mu_1^j(N-1)$	$\mu_2^j(N-1)$	$\mu_3^j(N-1)$	...	$\mu_{K-1}^j(N-1)$	$\mu_K^j(N-1)$
N	$\mu_1^j(N)$	$\mu_2^j(N)$	$\mu_3^j(N)$	...	$\mu_{K-1}^j(N)$	$\mu_K^j(N)$

Each of the criteria can have specific importance (v_k, $k = \overline{1,K}$), determined by a specially designed scale, for example, $v_k \in (0, 1)$, $\sum_{k=1}^{K} v_k = 1$. Taking into account the values of v_k, the adjusted estimates of alternative i, $i = \overline{1,N}$ are calculated, for example, as shown in Table 3.

Table 3. Adjusted estimates of alternative i

$\diagdown$ k j	1	2	3	...	K
1	$v_1\mu_1^1(i)$	$v_2\mu_2^1(i)$	$v_3\mu_3^1(i)$	...	$v_K\mu_K^1(i)$
2	$v_1\mu_1^2(i)$	$v_2\mu_2^2(i)$	$v_3\mu_3^2(i)$	...	$v_K\mu_K^2(i)$
3	$v_1\mu_1^3(i)$	$v_2\mu_2^3(i)$	$v_3\mu_3^3(i)$	...	$v_K\mu_K^3(i)$
...	...	...	...	...	...
M-1	$v_1\mu_1^{M-1}(i)$	$v_2\mu_2^{M-1}(i)$	$v_3\mu_3^{M-1}(i)$	...	$v_K\mu_K^{M-1}(i)$
M	$v_1\mu_1^M(i)$	$v_2\mu_2^M(i)$	$v_3\mu_3^M(i)$	...	$v_K\mu_K^M(i)$

To determine the integral estimate (α_i) for each matrix (Table 3) Maximin rule of the composition of fuzzy sets is used. For the first version of the estimation matrix representation, it looks like this:

$$\alpha_i = v_1\mu_1^j(i) \circ v_2\mu_2^j(i) \circ ... \circ v_K\mu_K^j(i) = \max_k\left\{\min_j\left[v_1\mu_1^j(i), v_2\mu_2^j(i), ..., v_K\mu_K^j(i)\right]\right\} \tag{2}$$

for the second version:

$$\alpha_i = v_k\mu_k^j(1) \circ v_k\mu_k^j(2) \circ ... \circ v_k\mu_k^j(N) = \max_i\left\{\min_k\left[v_k\mu_k^j(1), v_k\mu_k^j(2), ..., v_k\mu_k^j(N)\right]\right\}. \tag{3}$$

The final decision is making by a majority vote.

To increase the objectivity of the assessment (to minimize uncertainty), "weight" coefficients can also be assigned to each of the M agents $w_j, j = \overline{1,M}$. They characterize the level of trust in agent j, depend on its psycho-behavioral type $r_j(t), t = \overline{1,T}$ [1] and the average square deviation of the change in $r_j(t)$ over time. The values of w_j are calculating using the formula:

$$w_j = \left(w_{\sigma_j} + w_{\bar{r}_j}\right)/2, \tag{4}$$

$$w_{\sigma_j} = 1/e^{7\sigma_j}, \sigma_j \in [0,01; 0,5], \tag{5}$$

$$w_{\bar{r}_j} = \frac{0,73}{\sigma_{\bar{r}_j}\sqrt{2\pi}} \cdot e^{-\frac{\tau^2}{0,73}}, \tau = \frac{r_j - \bar{r}_j}{\sigma_{\bar{r}_j}}, \bar{r}_j \in (0; 1) \tag{6}$$

where σ_j is the mean square deviation calculated by the formula:

$$\sigma_j = \sqrt{\frac{\sum\limits_{t=1}^{T}\left(r_j(t) - \bar{r}_j\right)^2}{T}},$$

where T is the number of time cycles;

$\bar{r}_j$ is a psycho-behavioral type of agent j, which can be determined by the formula:

$$\bar{r}_j = \frac{1}{T}\sum\limits_{t=1}^{T}r_j(t);$$

e; π; $0,7$; $0,73$ are constants.

The smaller value σ_j, the more stable the agent type, so the agent is more trustworthy; however, if $\bar{r}_j \in (0,0;0,4)$ or $\bar{r}_j \in (0,7;1,0)$, the trust in such an agent is small. The mean square deviation cannot take a zero value or a value greater than $0,5$ [22]. Minor (within the established norm) changes in the psycho-behavioral type of the agent over time indicate its activity in the MAS; $\sigma_j = 0,5$ is the maximum deviation of the psycho-behavioral type of the agent from the average value. With this deviation, there is no confidence in the agent, so he can exclude from the MAS.

The uniformity of the states of agent j is checking using the coefficient of variation:

$$V_j = \left(\sigma_j/\bar{r}_j\right) \cdot 100\% < V_j^{crit.} \tag{7}$$

where $V_j^{crit.}$ is the critical value of the coefficient of variation, exceeding which indicates the instability of the agent's states in the MAS, therefore, such an agent cannot enjoy a high level of trust. In the practice of statistics, for example, in [23, p. 83] $V_j^{crit.} = 33\%$ for distributions close to normal. If $V_j > V_j^{crit.}$, the agent is removing from the MAS.

Taking into account w_j, the matrices of estimates of alternative i are calculated for the representation of I) by the formula:

$$\mu_{K^* \cup W}^j(i) = \mu_{K^*}^j(i) + \mu_W^j(i) - \mu_{K^*}^j(i) \cdot \mu_W^j(i) \tag{8}$$

where $\mu_{K^*}^j(i)$ is the values of the membership functions from Table 3;

K^* is set of adjusted estimates of alternative i;

$\mu_W^j(i)$ is the values of the membership functions calculated by the formula (4);

W is a set of "weight" coefficients that characterize the trust level to the j-th agent.

Then the final score is calculated:

$$\alpha_i = \bigcap_{k=1}^{K}\left(\bigcup_{j=1}^{M}\mu_{K^* \cup W}^j(i)\right) = \min_k\left\{\max_j\left[\mu_{1 \cup W}^j(i), \mu_{2 \cup W}^j(i), ..., \mu_{K^* \cup W}^j(i)\right]\right\}. \tag{9}$$

If the second version of the evaluation matrix representation is using, then the formula (3) is applied to find the final evaluation, which is converted with w_j in mind by the rule of multiplying the fuzzy set by a number:

$$\alpha_i = \max_i \left\{ \min_k \left[w_j v_k \mu_k^j(1), ..., w_j v_k \mu_k^j(N) \right] \right\}. \tag{10}$$

Next, the best way to solve the problem (Ω) is chosen based on the majority of votes received by alternative i.

In both cases, the difficulty lies in establishing the truth of the values of v_k, $\mu_k^j(i)$ due to the vagueness and subjectivity of the reasonings of agents acting as experts. There are two most obvious options for fuzzy reasonings that lead to the creation of conditions of vague uncertainty:

1) the best are several alternatives out of N; either each expert "welcomes" several options approximately equally, or $\exists \alpha_i^{optim.}$, $i \geq 2$, where $\alpha_i^{optim.}$ is the estimate of the optimal variant calculated by the formula (9) or (10);

2) Experts can't find any alternatives that match their preferences; there are several alternatives, but none is suitable – so most experts believe.

Thus, the optimal decision in the MAS is hindering by the crisis of evaluation and the crisis of choice. The evaluation crisis is that it is often difficult for an agent to objectively evaluate the proposed alternatives, especially if one of them (or several) belongs to the same agent, and they are interested in choosing it in the end. The choice crisis occurs when it is hard to choose one alternative after the evaluation since several of the N alternatives received the same high rating, or none of them received an estimate sufficient for the choice.

There are different ways to determine the consistency of expert opinions and the reliability of expert assessments, such as those proposed in [24, 25], and others. These include statistical methods for estimating variance, estimating probability for a given range of changes in estimates, estimating the Kendall, Spearman rank correlation, and the concordance coefficient. Methods for improving the consistency of expert assessments using techniques of statistics for processing the results of an expert survey are widespread. In this case, to analyze the consistency of estimates $\mu_k^j(i)$ using the coefficient of variation:

$$V_k = \left(\sigma_{v_k \mu_k^j(i)} \Big/ \overline{v_k \mu_k^j(i)} \right) \cdot 100\% < V_k^{crit.} \tag{11}$$

where $\sigma_{v_k \mu_k^j(i)}$ is the mean square deviation of the agent's estimates by the k criterion;

$\overline{v_k \mu_k^j(i)}$ is an average rating of agents according to the criterion k of alternative i;

$V_k^{crit.}$ is the critical value of the coefficient of variation, the excess of which indicates complete inconsistency of the agents' estimates according to the criterion k.

The lower the value of V_k, the more consistent the estimates are considered.

4   Results and Discussion

Let the problem (Ω) be solved by five agents who have knowledge and understanding of the four possible ways to solve it. Each of these ways is characterizing by three main parameters, the "weight" coefficients of which the agents also determine together:

1) cost of the problem-solving method (Ω),$v_1 = 0{,}5$;
2) the speed of solving the problem (Ω) using this method, $v_2 = 0{,}2$;
3) accuracy of the results obtained, $v_3 = 0{,}3$.

These parameters are considering as linguistic variables, each of which has a base term set T = {"low" "medium" "high"}. Then the aggregated criterion for choosing an alternative can be formulated as follows: IF the COST of using this method is low OR medium and the SPEED of solving the problem by this method is high OR medium, and the ACCURACY of the results is high, THEN this method is accepting for execution.

Accordingly, the values of the membership functions $\mu_k^j(i)$ characterize the compatibility of alternative i with the concept of "best" according to three criteria: an inexpensive method, a quick solution, and accurate results.

During their activity in the MAS, agents have established themselves in a certain way, which is reflecting in the level of trust w_j in the opinion of each of them. The confidence level is calculating using the formula (4). Over 18 clock cycles, the values r_j changed, as shown in Fig. 1.

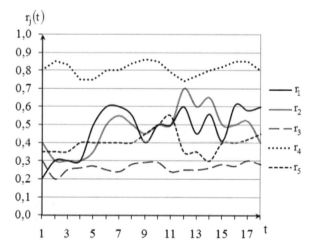

Fig.1. Dynamics of the behavioral type of agents in the MAS

The results of calculating levels of trust and checking the uniformity of agent States are presenting in Table 4.

Table 4. Characteristics of MAS agents

	σ_j	$\bar{r}_j$	w_{σ_j}	$w_{\bar{r}_j}$	w_j	V_j, %
r_1	0,13	0,48	0,40	0,99	0,70	26,43
r_2	0,11	0,47	0,46	0,98	0,72	23,76
r_3	0,02	0,26	0,87	0,39	0,63	9,28
r_4	0,04	0,81	0,76	0,21	0,49	4,60
r_5	0,06	0,40	0,66	0,85	0,76	14,44

Thus, the first, second, and fifth agents have the compromise type, and the fifth agent has the greatest stability of its type since its σ_5 is small compared to the σ_j of the first and second agents. The third agent is of the evasive type, and the fourth is of the coercive type, also with a high degree of stability. Therefore, the fifth agent deserves the most confidence, which is confirmed by the calculations.

Agents evaluate each alternative based on the selected criteria. The following estimation matrices D_i were obtained (according to Table 1):

$$D_1 = \begin{bmatrix} 0,7 & 0,7 & 0,6 \\ 0,8 & 0,6 & 0,7 \\ 0,5 & 0,6 & 0,6 \\ 0,6 & 0,8 & 0,5 \\ 0,5 & 0,7 & 0,6 \end{bmatrix}; D_2 = \begin{bmatrix} 0,3 & 0,5 & 0,5 \\ 0,4 & 0,3 & 0,5 \\ 0,3 & 0,3 & 0,6 \\ 0,5 & 0,4 & 0,5 \\ 0,4 & 0,4 & 0,4 \end{bmatrix}; D_3 = \begin{bmatrix} 0,6 & 0,9 & 0,9 \\ 0,7 & 0,8 & 0,9 \\ 0,6 & 0,9 & 0,8 \\ 0,7 & 0,7 & 0,9 \\ 0,6 & 0,8 & 0,8 \end{bmatrix};$$

$$D_4 = \begin{bmatrix} 0,7 & 0,6 & 0,6 \\ 0,8 & 0,7 & 0,7 \\ 0,7 & 0,5 & 0,9 \\ 0,8 & 0,7 & 0,7 \\ 0,8 & 0,6 & 0,8 \end{bmatrix}.$$

According to Table 3, adjusted estimates of alternatives were obtaining, taking into account the importance coefficients v_k:

$$D_1^{v_k} = \begin{bmatrix} 0,35 & 0,14 & 0,18 \\ 0,40 & 0,12 & 0,21 \\ 0,25 & 0,12 & 0,18 \\ 0,30 & 0,16 & 0,15 \\ 0,25 & 0,14 & 0,18 \end{bmatrix}; D_2^{v_k} = \begin{bmatrix} 0,15 & 0,10 & 0,15 \\ 0,20 & 0,06 & 0,15 \\ 0,15 & 0,06 & 0,18 \\ 0,25 & 0,08 & 0,15 \\ 0,20 & 0,08 & 0,12 \end{bmatrix};$$

$$D_3^{v_k} = \begin{bmatrix} 0,30 & 0,18 & 0,27 \\ 0,35 & 0,16 & 0,27 \\ 0,30 & 0,18 & 0,24 \\ 0,35 & 0,14 & 0,27 \\ 0,30 & 0,16 & 0,24 \end{bmatrix}; D_4^{v_k} = \begin{bmatrix} 0,35 & 0,12 & 0,18 \\ 0,40 & 0,14 & 0,21 \\ 0,35 & 0,10 & 0,27 \\ 0,40 & 0,14 & 0,21 \\ 0,40 & 0,12 & 0,24 \end{bmatrix}.$$

Checking the consistency of estimates in each matrix $D_i^{v_k}$ using the coefficient of variation gave the following results (Table 5):

Table 5. Checking the consistency of ratings, %

k / V_k	Inexpensive method	Quick decision	Accurate results	Average percentage of variation
V_{D_1}	18,8	11,0	10,5	13,5
V_{D_2}	19,7	19,7	12,6	17,3
V_{D_3}	7,7	9,1	5,7	7,5
V_{D_4}	6,4	12,1	13,8	10,8

Rule (2) determines the final scores α_i: $\alpha_1 = 0,25$, $\alpha_2 = 0,15$, $\alpha_3 = 0,30$, $\alpha_4 = 0,35$. Thus, if we assume that all agents without exception enjoy the same trust, then the fourth way to solve the problem (Ω) is a priority. However, there are some "vague" conditions for making a decision, since the difference between α_3 and α_4 is small, and the greatest agreement was reached in the evaluation of the third alternative. To increase the objectivity of the alternative evaluation procedure, the levels of trust (w_j) in agents are further applying. The estimates adjusted for w_j according to rule (8) are representing by the following evaluation matrices:

$$D_1^* = \begin{bmatrix} 0,80 & 0,74 & 0,75 \\ 0,83 & 0,75 & 0,78 \\ 0,72 & 0,67 & 0,70 \\ 0,64 & 0,57 & 0,56 \\ 0,82 & 0,79 & 0,80 \end{bmatrix} ; D_2^* = \begin{bmatrix} 0,74 & 0,73 & 0,74 \\ 0,78 & 0,74 & 0,76 \\ 0,69 & 0,65 & 0,70 \\ 0,61 & 0,53 & 0,56 \\ 0,80 & 0,77 & 0,78 \end{bmatrix} ;$$

$$D_3^* = \begin{bmatrix} 0,79 & 0,75 & 0,78 \\ 0,82 & 0,76 & 0,80 \\ 0,74 & 0,70 & 0,72 \\ 0,67 & 0,56 & 0,62 \\ 0,83 & 0,79 & 0,81 \end{bmatrix} ; D_4^* = \begin{bmatrix} 0,80 & 0,73 & 0,75 \\ 0,83 & 0,76 & 0,78 \\ 0,76 & 0,67 & 0,73 \\ 0,69 & 0,56 & 0,59 \\ 0,85 & 0,78 & 0,81 \end{bmatrix} .$$

According to the rule (9) determines the final grade α_i: $\alpha_1 = 0,79$, $\alpha_2 = 0,77$, $\alpha_3 = 0,79$, $\alpha_4 = 0,78$. The consistent selection rule applies: since the variation in the estimates of the third alternative is lower than the first (Table 6), and according to the initial final estimates the third alternative is better than the first, and it is acceptable for execution.

Table 6. The final consistency check of the estimates, %

k \ V_i	Inexpensive method	Quick decision	Accurate results	Average percentage of variation
V_{D_1}	9,5	11,1	11,8	10,8
V_{D_2}	9,4	12,9	11,1	11,1
V_{D_3}	7,8	11,8	9,2	9,6
V_{D_4}	7,3	11,6	10,3	9,7

For the case of representation of evaluation matrices in the form of Table 2, five matrices D_j were formed, of which in four matrices also chose the third alternative. She was chosen by the first, second, third, and fifth agents. According to the accepting rules and according to the data presenting in Table 4, the opinion of these agents can be trusted.

5   Conclusions

The fuzzy agreed choice of an alternative in the MAS is a process that involves a group of agents who have the necessary information about the problem area. The result of an agreed choice is the acceptance of one alternative from the generated set of alternatives. However, if several alternatives received a high rating, or none of them received a high enough rating, then a choice crisis occurs. Also, because of the anthropomorphic definition of an agent adopted in this study, agents involved in evaluating alternatives and choices have their own opinions and preferences. They can act in their interests, ignoring the goals of the MAS. Each of the agents is characterizing by a psycho-behavioral type, determined by the intensity of changes in its states in the MAS. It is establishing that the level of trust in this agent depends on the characteristic behavioral type of the agent, as well as on the frequency and intensity of his changes. At the same time, the level of trust is an objective value for this reason justifiably that an agent with a high level of trust is unlikely to report false estimates. As practice shows, in a situation where several "good" variants there are in MAS, the choice is made in favor of the one chosen by the agent with the highest level of trust. If the opposite situation exists, that there is no suitable option out of all possible options, the evaluation of alternatives should be repeated, expanding the selection criteria base.

References

1. Tarassov, V.B.: Information granulation, cognitive logic, and natural pragmatics for intelligent agents. Open semantic technologies for intelligent systems. In: Proceedings of the 8th International Conference, BSUIR Editions, Minsk, pp. 45−55 (2018)
2. Mitra, M., Chowdhury, A.: A modernized voting system using fuzzy logic and blockchain technology. Int. J. Mod. Educ. Comput. Sci. (IJMECS) **12**(3), 17–25 (2020). https://doi.org/10.5815/ijmecs.2020.03.03
3. Berger, J.O., Pericchi, L.: Objective Bayesian methods for model selection: introduction and comparison (with discussion), pp. 135–207. In: Model Selection. IMS, Beachwood, OH (2001)
4. Chipman, H.A., George, E.I., Mcculloch, R.E.: The practical implementation of Bayesian model selection (with discussion), pp. 65–134. In: Model Selection. IMS, Beachwood, OH (2001)
5. Clyde, M., George, E.I.: Model uncertainty. Stat. Sci. **19**(1), 81–94 (2001)
6. Larichev, O.I.: Theory and methods of decision-making, and chronicle of events in magic kingdoms, Logos, p. 392 (2003)
7. Yadav, K., Thareja, R.: Comparing the performance of naive bayes and decision tree classification using R. Int. J. Int. Syst. Appl. (IJISA) **11**(12), 11−19 (2019). https://doi.org/10.5815/ijisa.2019.12.02
8. Kolo, K.D., Adepoju, S.A., Alhassan, J.K.: A decision tree approach for predicting students academic performance. Int. J. Educ. Manage. Eng. (IJEME) **5**(5), 12−19 (2015). https://doi.org/10.5815/ijeme.2015.05.02
9. Saati, T.L.: Decision-making at the dependencies and feedback: analytic network: per. s angl. Izd. 3th.-M., p. 360 (2011). LIBROKOM
10. Tutygin, A., Korobov, V.: Advantages and disadvantages of the analytic hierarchy process. Proc. Herzen Russ. State Pedagogical Univ. **122**, 108–115 (2010)
11. Diabagaté, A., Azmani, A., El Harzli, M.: The choice of the best proposal in tendering with AHP method: case of procurement of IT master plan's realization. Int. J. Inf. Tech. Comput. Sci. (IJITCS) **7**(12), 1−11 (2015). https://doi.org/10.5815/ijitcs.2015.12.01
12. Leonenkov, A.: Fuzzy modeling in MATLAB and FuzzyTech. BHV-Peterburg, SPb, p. 736 (2005)
13. Kosarev, N.I.: Production model of knowledge representation in the decision-making support systems. Bull. Siberian Law Inst. Ministry Internal Affairs of Russ. **2**(13), 136–140 (2013)
14. Davis, R., Shrobe, H., Szolovits, P.: What is a knowledge representation? AI Mag. **14**(1), 17–33 (1993)
15. Kosko, B.: Fuzzy cognitive maps. Int. J. Man Mach. Stud. **24**(1), 65–75 (1986)
16. Taber, W.R.: Knowledge processing with fuzzy cognitive maps. Expert Syst. Appl. **2**, 83–87 (1991)
17. Craiger, J., Coovert, M.D.: Modeling dyanamic social and psychological processes with fuzzy cognitive maps. In: Proceeding of the 3rd IEEE Conference on Fuzzy Systems v3, pp. 1873−1877 (1994)
18. Zadeh, L.A.: Fuzzy sets. Inf. Control **8**, 338–353 (1965)
19. Bellman, R., Zadeh, L.: Decision-making in vague conditions. In: Questions of analysis and decision-making procedures, Mir, vol. 230, pp. 172–215 (1976)
20. Bazmara, M.: A novel fuzzy approach for determining best position of soccer players. Int. J. Int. Syst. Appl. (IJISA) **6**(9), 62–67 (2014). https://doi.org/10.5815/ijisa.2014.09.08

21. Kahneman, D., Tversky, A.: Prospect theory: an analysis of decision under risk. Econometrica **47**, 263–291 (1979)
22. Das, S., Roy, K., Saha, C.K.: Establishment of automated technique of FHR baseline and variability detection using CTG: statistical comparison with expert's analysis. Int. J. Inf. Eng. Elect. Business (IJIEEB) **11**(1), 27–35 (2019). https://doi.org/10.5815/ijieeb.2019. 01.04
23. Efimova, M.R., Ganchenko O.I., Petrova E.V.: Workshop on the general theory of statistics. Finance and statistics, p. 368 (2008)
24. Litvak, B.G.: Expert information: methods of obtaining and analyzing. Radio and communications, p. 184 (1982)
25. Makarov, I.M., Vinogradskaya, T.M., Rubchinsky, A.A.: Theory of choice and decision-making. Nauka, p. 328 (1982)

Evaluation of the Effect of Preprocessing Data on Network Traffic Classifier Based on ML Methods for Qos Predication in Real-Time

Vladimir A. Mankov[1], Vladimir Yu. Deart[2],
and Irina A. Krasnova[2(✉)]

[1] Training Center Nokia, 8A, Aviamotornaya, 111024 Moscow,
Russian Federation
vladimir.mankov@gmail.com
[2] Moscow Technical University of Communications and Informatics,
8A, Aviamotornaya, 111024 Moscow, Russian Federation
vdeart@mail.ru, irina_krasnova-angel@mail.ru

Abstract. The paper proposes a traffic classification model based on Machine Learning methods for definition of QoS, which contains a new element - data preprocessing block. In earlier works data preprocessing is understood only as forming a feature matrix for classification. This paper explores nine main methods based on normalization, standardization, scaling, generating additional features and working with outliers. Our experiment on a data set obtained from a real network showed that the most important procedure was to replace outliers with median values. In General, the most successful preprocessing methods for classifying traffic were Quantile and Power transformation, since they transform different scales and distributions similar to Gaussian distribution and it is more convenient to work with many ML methods. Using of preprocessing methods can increase the classification accuracy by more than 50%.

Keywords: Preprocessing · Machine learning · ML · QoS · Standardization · Scaling · Normalization · Network traffic

1 Introduction

Recently, telecommunications are increasingly using data mining methods, especially Machine Learning (ML) methods to solve a wide range of problems [1–3]. One of such tasks is real-time traffic classification, which would allow identifying flows that cannot be allocated at the stage of network design and providing measures to support of QoS for detected flows. The use of machine learning methods is becoming more and more urgent because early solution methods such as port-based classification may not always be applicable in modern networks because ports change dynamically, traffic is encrypted and new applications appear regularly. We described in more detail this problem and the general algorithm for solving it in [4].

The problem of the ML methods is that they are sensitive to outliers or large data scatter; therefore, various methods of traffic preprocessing are often used. There are a number of works that give an idea of preprocessing network traffic data, but mainly these works are devoted to traffic classification for network security purposes.

The purpose of this work is to study the influence of data preprocessing methods on the results of network traffic classification for supporting QoS. To achieve this goal, the paper develops a traffic classification model, forms a set of methods for preprocessing data for research, prepares a dataset for the experimental part of the research and evaluates and discusses the results of experiments.

The material of the paper is described as follows: in Sect. 2 provides an analysis of the most promising works on the research topic, in Sect. 3 - methodology of our research: in Subsect. 3.1 the stages of constructing our model for classifying traffic in general terms are described, in Subsect. 3.2 - the preprocessing methods used in the work are presented, Subsect. 3.3 gives a brief description of the ML methods, the Subsect. 3.4 describes the dataset under study and the feature matrix. Section 4 includes the experimental results of traffic classification and discussion. Section 5 presents general conclusions on the work and plans for future work.

2   Related Work

In [5] preprocessing is defined as a pipeline of several stages: processing missing values, encoding port numbers and protocol, determining correlating features and detecting outliers. But the work does not present the results of applying various methods or comparing and analyzing any other approaches for preprocessing, the result of the work is a dataset prepared for classification.

In [6], the Kohonen Networks method and outlier detection and removal methods are used as the main preprocessing tool. In [7], normalization is considered according to four schemes: Frequency, Maximize, Mean Range and Rational using Naïve Bayes and J.48 methods (a variation of Decision Tree). In [8], the standardization of data for classifiers built on the basis of kNN, PCA and SVM turned out to be the best approach, but preprocessing is used exclusively for one of the features and not for all.

The classification of traffic for determination of QoS and classification of traffic for detection of intrusions has a significant difference in many details, such as: class marking - QoS class or a specific application and the presence/absence of intrusion, the use of outliers to detect anomalies for network security purposes, differences in the feature matrix - for example, for QoS purposes, it is not recommended to use the port numbers used as indicators, as many applications use dynamic ports or encryption, etc.

In this work, we analyzed the main areas of data preprocessing for classifying network traffic in order to determine QoS for models operating in real-time and presented comparative accuracy results before and after applying these methods.

3   Methodology

3.1   Traffic Classification Model

Figure 1 depicts the process of constructing a traffic classification model using Supervised ML methods for our model. The first step is to collect statistics about network traffic packets, such as packet size and arrival time between packets or Interarrival Time (IT). Next, the packets are combined into flows defined by 5-tuple, i.e. having the same set of characteristics, consisting of IP source and destination, protocol type, ports source and destination. At the second stage, for each of the flows, based on statistical information, a matrix of attributes is calculated and the classes to which these flows belong are determined. Further, the available sample is divided into training and test sequences. The third stage is the subject of analysis of this paper. Here, based on the training sequence, data preprocessing is performed, such as outlier removal, normalization, standardization, scaling, etc. The preprocessing methods discussed are described in more detail below. At the fourth stage, the classifier itself is built by the training sequence. For the test sequence, at the third stage, data are preprocessed according to the templates obtained by processing the training sequence (maximum and minimum boundaries, average value, etc.). At the fifth stage, the test sequence without class marks is fed to the trained classifier and its forecasts are compared with the true class values for these flows - thus, according to the test results, various evaluations of the classifier's efficiency are given.

Fig. 1. Building a traffic classification model using ML methods

3.2   Methods of Data Preprocessing

Below shows the notation and brief characteristics of the preliminary data processing methods used in this work. Methods 3–9 are implemented using the Python Sklearn library [9].

1. **Outlier removal (Out):** Data that deviates from the median by a value greater than STD changes to the median. The calculation is carried out for each class separately.
2. **Outlier removal (Out, 3STD):** Similar to Out, but outliers are data that deviate from the median by 3STD.
3. **Scaling Standart:** Converts the data to a standard form, so that the average value is 0 and the standard deviation is 1.
4. **Scaling MinMax:** Scales data so that it is in the range of [0, 1].
5. **Scaling Robust:** Scaling data between the 25th and 75th quantile. Centering is relative to the median.

6. **Power Transformation:** Map data to a normal distribution using power transforms. Power transforms are a family of parametric transformations by which data are mapped to a distribution that is as similar as possible to a Gaussian distribution.
7. **Quantile Transformation:** A nonlinear transformation based on the mapping of the cumulative distribution function onto a uniform distribution and then onto a Gaussian distribution. When using this method, the most frequent values are propagated, which reduces the effect of emissions. If the transformation is non-linear, then the relationships between linear correlations between variables may be distorted, but the data measured at different scales are more comparable.
8. **Normalizer:** Each sample is normalized independently of the others to a norm of 1.
9. **Poly:** Generation of additional features based on polynomials of the 2nd degree. For example, [a, b], a polynomial transformation of the 2nd degree has the form: [1, a, b, a2, ab, b2]. It is important that along with an increase in the number of features, the duration of the model construction increases linearly.

3.3   Supervised ML Methods

ML methods used in the paper:

1. **Decision Tree, DT (CART)** is a logical model consisting of nodes, leaves and branches. Each node contains some conditional function built on the basis of one of the signs. This function divides the data into different directions, which ultimately come to one of the leaves associated with certain classes.
2. **Random Forest, RF** represents a certain number of such Decision Tree, which are built and make decisions independently of each other and the final classification result is determined by the results of voting of most trees.
3. **XGBoost, XGB** also consists of a set of a certain amount of Decision Tree, but unlike RF, they do not work in parallel, but sequentially, one after another, correcting errors of previous classifiers.
4. **Gaussian Naive Bayes, GNB** is a probabilistic model based on Bayes theorem. In the case of GaussianNB, the distribution of characters is considered normal.
5. **Logistic Regression, LR** is a linear classification method using a logistic curve that describes the probabilities for a given flow.
6. **k Nearest Neighbours, kNN** assigns specific coordinates to each sample object, depending on its feature vector. For the predicted class, k neighbours are selected with a minimum distance to them. The final class is determined by the most common class among k neighbours.
7. **Neural network: Multi-layer Perceptron, MLP** implemented using input, output and several intermediate hidden layers. The input layer consists of a set of neurons based on the signs of flow. Each neuron of subsequent hidden layers uses a weighted linear summation of features with a nonlinear activation function. The classification result is obtained on the output layer.

3.4 Dataset and Internet Traffic Classes

We propose an effective approach to collect the statistics in a real network that does not introduce additional delays and is described in detail in our paper [10]. But for the purposes of research and obtaining a diverse sample, we used an open database compiled by the MAWI research team as part of the WIDE project [11] at checkpoint G, which is a 10 Gb/s connection with DIX-IE in Tokyo. For the relevance of the data collected, 15 min routes were collected in January–February 2020, with a total duration of 1,5 h and a volume of 200,25 GB. The classification was carried out on unidirectional TCP and UDP flows identified by 5-tuple tuples and data markup was carried out using the nDPI tool. For building TCP dataset 1500 flows were selected from 13 different applications, the most common in the sample: SSL, HTTP_Proxy, DNS, Apple, IMAPS, HTTP, Skype, SSH, SMTP, RTMP, Telnet, POPS and IMAP. For UDP data set 250 flows were selected, 250 flows from 8 applications were used: DNS, NTP, Quic, IPsec, SNMP, NetBIOS, STUN, UPnP. Since the developed model is a real-time algorithm, only the first 15 packets were analyzed for each flow. The matrix consists of 45 signs, including the sizes of the first 15 packets (15), the IT (14), their total, minimum, maximum, mean, median, standard deviations, variance of values (14), as well as a packet and byte bit rate (2). This approach to the formation of characters was tested by us in [12] and proved its effectiveness. As an estimate of the classification parameters in the test sample, the Accuracy parameter was used, which is equal to the ratio of the number of correctly classified flows to the total number of test flows.

4 Experiment Results and Discussion

For the obtained datasets, the data preprocessing methods 1–8 listed in 3.2 were applied. The ML algorithms chosen for analysis responded differently to the proposed data preprocessing methods. The classification results for TCP and UDP datasets are shown in Fig. 2, 3 and Table 1. The top three results for each ML method are highlighted in color.

As can be seen from the results of the tables, replacing the emissions by the median increases the classification results for all methods, for the TCP dataset to 22,15% (MLP) and for the UDP dataset 19,5% (Gaussian NB). Determining the outlier using the out method as a value that is not more than the STD behind the median shows better results than the out with 3STD value, in which the outlier is considered to be more than 3 STD behind the median. Therefore, for further analysis, the out with 3STD method is not further considered and the out method is used in conjunction with various scaling, normalization and standardization methods, increasing the classification results.

Random Forest, Decision Tree and XGB methods, as algorithms based on building a random tree, are less prone to outliers and sparseness of data, so the use of data preprocessing methods compared to other methods slightly improved classification results. Nevertheless, replacing outliers with a median (out) allows increasing the classification results by 4,3–5,9% for a TCP dataset and by 1,25–6,5% for a UDP dataset.

Table 1. The accuracy results by ML classification with data preprocessing methods

	TCP dataset (Fig. 2)							UDP dataset (Fig. 3)						
Method	RF	DT	GNB	LR	XGB	kNN	MLP	RF	DT	GNB	LR	XGB	kNN	MLP
base data	0,83	0,76	0,14	0,18	0,84	0,67	0,31	0,89	0,90	0,36	0,33	0,96	0,86	0,19
out	0,87	0,81	0,21	0,20	0,90	0,74	0,53	0,96	0,93	0,56	0,43	0,97	0,92	0,27
out, 3 STD	0,84	0,77	0,19	0,20	0,86	0,70	0,34	0,92	0,92	0,42	0,37	0,96	0,86	0,20
standart	0,83	0,75	0,20	0,45	0,85	0,62	0,53	0,90	0,91	0,60	0,64	0,96	0,65	0,64
standart+out	0,87	0,81	0,50	0,55	0,90	0,70	0,63	0,95	0,93	0,82	0,74	0,97	0,79	0,76
minmax	0,83	0,76	0,24	0,39	0,84	0,65	0,47	0,90	0,90	0,60	0,54	0,96	0,67	0,64
minmax+out	0,88	0,81	0,50	0,50	0,89	0,73	0,58	0,95	0,95	0,82	0,69	0,96	0,81	0,81
robust	0,83	0,76	0,14	0,19	0,84	0,63	0,50	0,89	0,91	0,18	0,33	0,96	0,70	0,63
robust+out	0,87	0,81	0,39	0,42	0,90	0,71	0,65	0,95	0,93	0,57	0,71	0,96	0,90	0,81
power	0,83	0,75	0,39	0,60	0,84	0,71	0,74	0,92	0,89	0,54	0,69	0,95	0,79	0,72
power+out	0,87	0,81	0,50	0,67	0,90	0,79	0,80	0,95	0,92	0,79	0,83	0,97	0,83	0,86
quantile	0,83	0,75	0,40	0,62	0,85	0,74	0,73	0,90	0,90	0,67	0,69	0,96	0,82	0,78
quantile+out	0,88	0,81	0,51	0,71	0,91	0,80	0,80	0,96	0,94	0,81	0,82	0,97	0,89	0,88
norm	0,80	0,71	0,30	0,32	0,80	0,71	0,54	0,91	0,86	0,31	0,31	0,91	0,90	0,32
norm+out	0,85	0,77	0,32	0,36	0,86	0,75	0,61	0,94	0,94	0,59	0,46	0,94	0,94	0,42

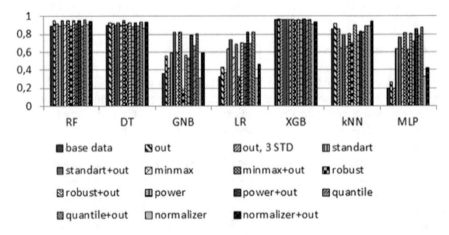

Fig. 2. The accuracy results by ML classification of the TCP dataset with data preprocessing methods

The accuracy of the GaussianNB method was increased by 37,8% for a TCP dataset while using the Quantile transformation and outlier removal and for a UDP dataset by 45,75% when scaling using Minmax or Standart methods and removing outliers.

For logistic regression, accuracy is increased by 52,9% using the Quantile transform for a TCP dataset and by 49,75% using a Power transform for a UDP dataset.

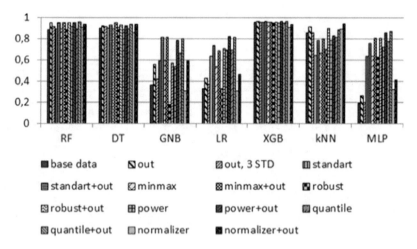

Fig. 3. The accuracy results by ML classification of the UDP dataset with data preprocessing methods

In the case of the kNN method, Quantile transformation increases the classification accuracy by 13% and thus reaches 80,4% in the case of the TCP dataset. For a UDP dataset Normalizer improves accuracy by 8,25% and as a result, the kNN method becomes a competitive classifier with an accuracy of 94,25%. Unlike other ML, kNN algorithms, accuracy decreases when using some scaling methods (in cases with UDP datasets).

For the MLP neural network, accuracy increases by 49,2% in cases with a TCP dataset and for UDP by 68,5%. Thus, 68,5% is the largest increase in accuracy obtained using data preprocessing methods, which allows you to change the accuracy from 19,25% to 87,75%.

In general, the best preprocessing methods for the assembled TCP and UDP dataset turned out to be Power and Quantile transformations together with outlier removal.

Figure 4, 5 and Table 2 show the results of adding polynomial dependencies of the 2nd degree as additional features. In this case, the classification results are considered before and after applying the most effective data preprocessing methods listed in 3.2. As can be seen from the presented results, the addition of nonlinear dependences for the results together with the removal of outliers not only does not improve the classification accuracy, but even worsens it. For algorithms built on the basis of a decision tree (Random Forest, Decision Tree and XGB), accuracy also decreases, regardless of what additional preprocessing methods are used. This is due to the features of the construction of a mathematical model. For Power and Quantile transformation the addition of nonlinear dependencies increases the classification accuracy. The accuracy of neural network classification is increased by 2–3%, kNN by 1–7% and logistic regression - up to 13%. For Naive Bayes, accuracy is practically unchanged.

In [6] with the help of Kohonen Networks, it was possible to achieve a classification accuracy of 75,6% on the route of the MAWI project for 2012 using the Decision Tree method. We also used MAWI data and but for 2020 we achieved

Table 2. The accuracy results by ML classification with data preprocessing methods and additional nonlinear dependencies

Method	TCP dataset (Fig. 4)							UDP dataset (Fig. 5)						
	RF	DT	GNB	LR	XGB	kNN	MLP	RF	DT	GNB	LR	XGB	kNN	MLP
base data	0,83	0,76	0,14	0,18	0,84	0,67	0,31	0,89	0,90	0,36	0,33	0,96	0,86	0,19
out	0,87	0,81	0,21	0,20	0,90	0,74	0,53	0,96	0,93	0,56	0,43	0,97	0,92	0,27
poly+out	0,73	0,79	0,16	0,15	0,88	0,73	0,37	0,95	0,92	0,53	0,38	0,96	0,91	0,26
power+out	0,87	0,81	0,50	0,67	0,90	0,79	0,80	0,95	0,92	0,79	0,83	0,97	0,83	0,86
poly+power+ +out	0,73	0,78	0,51	0,81	0,88	0,80	0,83	0,95	0,93	0,79	0,91	0,96	0,90	0,89
quantile+out	0,88	0,81	0,51	0,71	0,91	0,80	0,80	0,96	0,94	0,81	0,82	0,97	0,89	0,88
poly+out+ +quantile	0,87	0,80	0,51	0,80	0,90	0,80	0,82	0,95	0,93	0,80	0,91	0,96	0,92	0,91

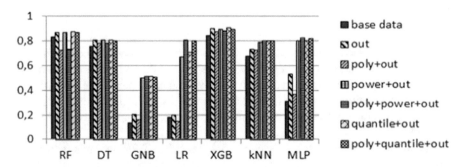

Fig. 4. The accuracy results by ML classification of the TCP dataset with data preprocessing methods and additional nonlinear dependencies

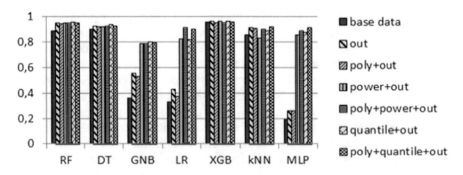

Fig. 5. The accuracy results by ML classification of the UDP dataset with data preprocessing methods and additional nonlinear dependencies

maximum results for DT 81,1% for TCP and 95,3% for UDP datasets by combining the out and minmax methods.

In [8], an assumption was made about the inefficiency of preprocessing for the Decision Tree model, because initial features are used by conditional functions in the nodes of the tree and models working in real-time, because the average, minimum and maximum values cannot be calculated from a small part of the flow. In our work, to overcome such difficulties, Decision Tree, like other models, were built on the basis of already processed data, therefore this fact did not prevent classification. For calculation the average, minimum, maximum values and the accuracy of these statistical estimates within the confidence intervals, we used strictly 15 first packets from each flow. Also, for objective rationing, scaling, etc. we used the boundaries obtained by processing the training sequence. Improving the accuracy of the classification proves the effectiveness of our solutions.

5   Conclusion

Main results of our work:

- We have developed an effective traffic classification model based on Machine Learning methods that differ from other models in the presence of a unique component - the data preprocessing block and universal feature matrix and custom collection method statistical information. The proposed model is suitable for classification of both TCP and UDP flows, while the other works are devoted only to one protocol and its structure cannot be used for other goals. A unique feature matrix that includes information only from the first 15 packets allows for real-time classification time and apply QoS policies to active network flows.
- Based on the results of our experiments, we found the dependence of the results accuracy of traffic classification from data preprocessing methods. We chose the accuracy metric as a criterion for evaluating the impact of preprocessing methods on traffic classification results. The best result for the studied dataset - 90% for the TCP Protocol and 97% for the UDP Protocol (XGB method). Highest accuracy gain −68,5% (MLP with Quantile and outlier removal).
- We came to the conclusion that in the data preprocessing block for the studied ML algorithms need to work with outliers, which allows you to increase the accuracy by 12% (for MLP). In addition, in models using methods of logistic regression, naive Bayes, kNN and MLP are recommended for the Power and Quantile transformations. For Logistic Regression and Neural Network methods can also be used generating additional nonlinear features, while for others however, generating nonlinear features reduces the accuracy of classification.
- We have proposed a research methodology that includes the most effective methods of data preprocessing. Despite the fact that in our model, some of them are low or even negative as a result, this methodology can be applied in the study of other situations, for example, other ML methods, where possible, will show a different accuracy.

Restriction the application of the model is its scope-for maintenance purposes QoS, for other purposes, additional tests are required research - for example, if intrusions are detected by emissions, they can some types of attacks can be identified.

In future work, we plan to add the detection capability to the model and classification of new flows that are not represented in the training sample.

References

1. Singh, K., Agrawal, S., Sohi, B.S.: A near real-time IP traffic classification using machine learning. Int. J. Intell. Syst. Appl. (IJISA) **5**, 83–93 (2013). https://doi.org/10.5815/ijisa. 2013.03.09
2. Kaur, J., Agrawal, S., Sohi, B.S.: Internet traffic classification for educational institutions using machine learning. Int. J. Intell. Syst. Appl. (IJISA) **4**, 36–45 (2012). https://doi.org/10. 5815/ijisa.2012.08.05
3. Almubayed, A., Hadi, A., Atoum, J.: A model for detecting tor encrypted traffic using supervised machine learning. Int. J. Comput. Netw. Inf. Secur. (IJCNIS) **7**, 10–23 (2015). https://doi.org/10.5815/ijcnis.2015.07.02
4. Mankov, V.A., Krasnova, I.A.: Algorithm for dynamic classification of flows in a multiservice software defined network. T-Comm **11**(12), 37–42 (2017). (in Russian)
5. Alothman, B.: Raw network traffic data preprocessing and preparation for automatic analysis. In: 2019 International Conference on Cyber Security and Protection of Digital Services (Cyber Security), pp. 1–5 (2019). https://doi.org/10.1109/CyberSecPODS.2019. 8885333
6. Kahya-Özyirmidokuz, E., Gezer, A., Çiflikli, C.: Characterization of network traffic data: a data preprocessing and data mining application (2012)
7. Ashok Kumar, D., Rajan Venugopalan, S.: The effect of normalization on intrusion detection classifiers (Naïve Bayes and J48). Int. J. Future Revolution Comput. Sci. Commun. Eng. **3**, 60–64 (2017)
8. Wang, W., Zhang, X., Gombault, S., Knapskog, S.J.: Attribute normalization in network intrusion detection. In: 2009 10th International Symposium on Pervasive Systems, Algorithms, and Networks, pp. 448–453 (2009)
9. Buitinck, L., Louppe, G., Blondel, M., Pedregosa, F., Mueller, A., Grisel, O., Niculae, V., Prettenhofer, P., Gramfort, A., Grobler, J., Layton, R., VanderPlas, J., Joly, A., Holt, B., Varoquaux, G.: API design for machine learning software: experiences from the scikit-learn project. ArXiv, abs/1309.0238 (2013)
10. Mankov, V.A., Krasnova, I.A.: Collection of individual packet statistical information in a flow based on P4-switch. In: Hu, Z., Petoukhov, S., He, M. (eds.) Advances in Intelligent Systems, Computer Science and Digital Economics. CSDEIS 2019. Advances in Intelligent Systems and Computing, vol. 1127. Springer, Cham (2020). https://doi.org/10.1007/978-3-030-39216-1_11
11. Cho, K., Mitsuya, K., Kato, A.: Traffic data repository at the WIDE project. In: USENIX Annual Technical Conference, FREENIX Track (2000)
12. Mankov, V.A., Krasnova, I.A.: Klassifikatsiya potokov trafika SDN-setei metodami mashinnogo obucheniya v rezhime real"nogo vremeni. In: Informatsionnye Tekhnologii I Matematicheskoe Modelirovanie Sistem 2019, Odintsovo (2019). https://doi.org/10.36581/citp.2019.31.51.016. (in Russian)

The Task of Improving the University Ranking Based on the Statistical Analysis Methods

A. Mikryukov$^{(\boxtimes)}$ (iD) and M. Mazurov (iD)

Plekhanov Russian University of Economics, Moscow, Russia
Mikrukov.aa@rea.ru, Mazurov37@mail.ru

Abstract. The task of ranking the university in international rating systems is urgent. An approach to solving the problem is proposed to ensure the required values of the basic indicators of the university's activity in the international institutional ranking QS using models developed on the basis of methods of correlation-regression analysis and factor analysis. Estimates of basic indicators and ratings were obtained based on the methods of correlation and regression analysis. A comparative analysis of the results obtained for universities in the reference group Interpretation of the results of factor analysis revealed a set of latent factors that have a significant impact on the baseline indicators. It is shown that measures to achieve the specified indicators must be carried out considering the identified correlation dependences of latent exponential factors and basic indicators, as well as the results of interpretation of the developed factor model. The novelty of the developed proposals lies in the assessment of the significance of latent factors affecting the basic indicators of the analysis of the university's activities, based on the use of correlation-regression methods and methods of factor analysis. The developed factorial model made it possible to group and structure the obtained data, as well as to reduce the dimension of the problem being solved. The results obtained allow us to solve the problem of identifying latent factors and to substantiate the conditions for achieving the required indicators of the university ranking in the international financial ranking QS.

Keywords: Correlation-regression analysis · Factor analysis · Basic indicators · Institutional rating

1 Introduction

As you know, the Ministry of Education and Science has launched Project 5-100, which is a state program to support the largest Russian universities [1]. The goal of the project is to increase the prestige of Russian higher education and to bring at least five universities out of the project participants to the top 100 universities in three authoritative world rankings: Quacquarelli Symonds (QS), Times Higher Education (THE) and Academic Ranking of World Universities.

Currently, the QS institutional ranking includes 25 Russian universities [2, 3]. In the first place among Russian universities is the Moscow State University, who entered the top 100 of the institutional ranking at 84th position, in second place - Novosibirsk

Z. Hu et al. (Eds.): AIMEE 2020, AISC 1315, pp. 65–75, 2021.
https://doi.org/10.1007/978-3-030-67133-4_6

National Research State University (231st position), participating in the "Project 5-100", in third place - St. Petersburg State University (234th position). The Higher School of Economics (HSE), as of 2020, is ranked 322nd, National Research Technological University "MISIS" - 451st, Plekhanov Russian University of Economics - 776th position.

Over the past 5 years, Russian universities have shown noticeable dynamics in entering the top 500 of the QS institutional ranking, having increased their representation by one and a half times (from 9 to 16), mainly due to the participants of the "Project 5-100". This indicator is one of the main guidelines of the federal project "Young Professionals".

In view of the above, the leadership of the Plekhanov Russian University of Economics, the task was formulated to move in the world ranking of universities QS by 2025. to the position currently occupied by MISIS. For this purpose, an analysis of the conditions for achieving a given position was carried out and, based on the developed models, proposals were justified to ensure the fulfillment of the task.

The purpose of the research is to develop scientifically grounded proposals for increasing the target performance indicators of the university in the international institutional ranking QS to the required values, considering the totality of latent factors. The degree of achievement of the preset values of the basic indications and, consequently, the level of the university rating depends on changes in latent factors.

To achieve this goal, the methods of correlation - regression and factor analysis are used, which make it possible to identify the degree of influence of latent factors on the basic indicators and the main indicator (rating) functional.

The task was solved in 3 stages. At the first stage, the analysis of the correlation of basic indicators that ensure the promotion of Plekhanov Russian University in the QS World University Ranking institutional ranking: academic reputation, reputation with the employer, the ratio of the number of students to the number of research and teaching staff, citations per teacher, international teachers, international students. The listed indicators are included in the university ranking system and are presented in the information and analytical system QS - analytics [4]. Based on the methods of correlation and regression analysis in the environment of the analytical platform Deductor 5.3, the pairwise correlation coefficients of the functional values and basic indicators for the Plekhanov Russian University and MISIS University. Based on the obtained values, the analysis of the correlation of indicators providing the promotion of Plekhanov Russian University in the QS World University Ranking.

The results of the calculations made it possible to reveal the tightness of the relationship between the base indicators and the rating functionality. At the next stage of the study, latent factors influencing the basic indicators of the university's activity were identified using the methods of factor analysis. An assessment of their significance and degree of influence on the basic indicators, as well as their grouping, was carried out. The use of the mathematical apparatus of factor analysis made it possible to reduce the dimension of the problem being solved and to ensure the grouping and structuring of the data obtained.

Interpretation of the results of factor analysis made it possible to identify latent factors that provide the main contribution to obtaining the result. At the third stage of the study, a set of measures was substantiated to achieve the planned indicators to increase the institutional ranking of the QS University.

Thus, a new approach to solving the problem of providing conditions for achieving the required values of the university performance indicators in the international institutional ranking QS using models developed based on statistical analysis methods is proposed. The novelty of the approach is determined by obtaining estimates of the strength of the relationship between the basic indicators and their relationship with the rating functional based on the methods of correlation-regression analysis, solving the problem of identifying latent factors based on the application of the method of principal components of the developed factor model, reducing the dimension of the problem being solved, since a large number of interconnected (dependent, correlated) variables significantly complicates the analysis and interpretation of the results, and a reasonable assessment of the degree of influence of latent factors on the basic indicators, which made it possible to formulate a list of necessary measures to solve the problem of increasing the university's ranking.

Section 2 provides a literature review on the research topic, Sect. 3 presents the results of assessing the correlation of university performance indicators, Sect. 4 based on the factor analysis method provides identification of latent factors and an assessment of their significance, Sect. 5 substantiates measures to achieve the planned performance indicators of the university.

2 Literature Review

The problem of applying the methods of correlation-regression and factor analysis is considered in a fairly large number of works by domestic and foreign scientists [5–16]. In works [5–8] theoretical issues of statistical analysis are considered, in works [8–11] the features of the application of methods of correlation-regression and factor analysis in the socio-economic sphere are considered, in works [12–16] the features of constructing applied statistical models are considered.

The analysis of the sources showed that in the presented formulation, the problem of justifying the conditions for achieving the required values of the university performance indicators in the international institutional ranking QS using models developed on the basis of the methods of correlation-regression and factor analysis was not solved.

3 Methodology. Correlation-Regression Analysis of University Performance Indicators

In the institutional ranking of the best universities in the world, QS World University Ranking, universities are assessed by the value of the rating functionality, calculated based on the following six basic indicators [4]:

- academic reputation (AR), contributes 40% to the rating value;
- reputation with the employer (PP) - 10% contribution to the rating value;

- the ratio of the number of students to the number of scientific and pedagogical workers (SPD) (SP), - 20% contribution to the value of the rating;
- citations per teacher (CP) - 5% contribution to the rating value;
- international teachers (MT) - 20% contribution to the rating value;
- international students (MS) - 5% contribution to the rating value.

Each indicator has a weight equal to the degree of contribution to the value of the rating functionality. The value of the rating functional R is calculated using the formula (1):

$$R = \sum_{i=1}^{6} w_i x_i, \tag{1}$$

where w_i is the weight of the corresponding indicator; x_i is its value.

The following initial data on the university for the period 2013–2019 were taken as a basis for calculations: rating functionality, basic indicators - academic reputation; reputation with the employer; the ratio of the number of students to the number of teaching staff; citations per teacher; international teachers; international students.

Based on the methods of correlation-regression analysis in the environment of the analytical platform Deductor 5.3 using the module "Correlation analysis", the coefficients of pairwise correlation of the values of the functional and basic indicators were calculated for the Plekhanov Russian University and for "MISIS" (Table 1) using the Pearson test (allows you to assess the significance of differences between the actual and theoretical number of characteristics of the sample).

Table 1. Matrix of correlation of rating functional with basic indicators using Pearson criteria

Basic indicators	Rating functional, Plekhanov Russian University of Economics	Rating functional, MISIS
AR	0,152	0,854
ER	0,726	0,607
RS/T	0,939	0,511
CT	0,141	0,883
IT	0,182	0,494
IS	0,604	0,667

Coefficients of pairwise correlation between the basic indicators were calculated in a similar way. In accordance with the Chaddock scale (Table 2), an assessment of the tightness of the correlation coefficients was carried out [4].

The calculations made it possible to draw the following conclusions.

It was revealed that the rating functional is strongly correlated with the basic indicators: "The ratio of the number of students to the number of teaching staff" (r = 0.939), "employer's reputation" (r = 0.726) and "International students" (r = 0.604). The strength of the link between the rating functionality and other indicators is practically absent.

Table 2. Chaddock scale

Pairwise correlation coefficient	Bond strength
up to 0,3	Practically absent
0,3–0,5	Weak
0,5–0,7	Noticeable
0,7–0,9	Strong

For the MISIS rating functional the greatest closeness of connection was found for the basic indicators "Academic reputation" (0.854) and "Citations per teacher" (0.883), the smallest - for the indicator "International teachers".

A more reliable criterion for assessing the tightness of relationships is a statistical assessment of the coefficients of pair correlation by comparing its absolute value with the table value r_{crit}, which is selected from a special table [5].

If the inequality $\left| r_{calc} \geq r_{crit} \right|$ is satisfied, then with a given degree of probability (usually 95%) it can be argued that there is a significant linear relationship between the numerical populations under consideration. That is, the hypothesis about the significance of the linear relationship is not rejected. In the case of the opposite relation, i.e., for $\left| r_{calc} < r_{crit} \right|$, a conclusion is made about the absence of a significant connection.

In accordance with the table "Critical values of the correlation τ_{crit} for the significance level $\alpha = 0.05$, the probability of an admissible error in the forecast 0.95 and the degree of freedom $f = n\text{-}k = 4$ (for a given number of measurements $n = 6$, the number of calculated constants $k = 2$, in the formula for calculating r involves two constants — x and — y), the tabular value $r_{crit} = 0.811$ is found.

The calculation results (hypothesis testing) are presented in Table 3 practically confirmed the grades obtained on the Chaddock scale, except for the indicator "International students".

Table 3. Strength of connection between functionality and indicators

$R = r^2$	r_{calc}	r_{crit}	Bond strength
AR	0,140	0,811	Insignificant
IT	0,194	0,811	Insignificant
IS	0,636	0,811	Insignificant
RS/T	0,952	0,811	Significant
ER	0,854	0,811	Significant
CT	0,174	0,811	Insignificant

The calculation of the coefficients of determination ($R = r^2$), which is a measure of the variability of the result y (the value of the rating functional) as a percentage of the change in the factor (basic indicator) x showed that for the basic RS/T indicator

$r^2 = 0.9063 = 90.3\%$ means that 90.3% of the variation of the functional is determined by the basic RS/T indicator.

At the next stage of the study, the identification and interpretation of the hidden (latent factors) that affect the baseline indicators were carried out using the methods of factor analysis, which is a class of multivariate statistical analysis procedures aimed at identifying the latent factors responsible for the presence of linear statistical relationships (correlations) between the observed variables [6–8].

4 Results. Identification of Latent Factors and Assessment of Their Significance

Factors are groups of certain variables that correlate with each other more than with variables included in another factor. Thus, the meaningful meaning of the factors can be identified by examining the correlation matrix of the initial data.

To assess the influence of latent factors on basic indicators, one of the most common methods of factor analysis was used - the method of principal components, which makes it possible to reduce a large number of interconnected (dependent, correlated) variables, since a large number of variables significantly complicates the analysis and interpretation of the obtained results [9].

The mathematical model of factor analysis is a set of linear equations in which each observed variable xi is expressed as a linear combination of common factors: F_1, F_2, ..., F_n and a unique factor U_i [10]:

$$x_i = \sum_{k=0}^{n} a_{ik}F + U_i \qquad (2)$$

where x_i is a variable, $i = 1$, m, (m is the number of variables); n is the number of factors; $n < m$, a_{ik} - factor load; F_k - common factor, $k = 1$, n; U_i is a particular factor.

The factor analysis procedure includes the following stages [11–13].

Stage 1. Construction of the correlation matrix of the system of variables by calculating the Pearson's linear correlation coefficients.

Stage 2. Extraction of factors and calculation of factor loadings a_{ik}, which are the main subject of interpretation. At this stage, methods of component analysis (principal component analysis), principal factors and maximum likelihood are used. When solving the problem, the most common method was used, which is the method of principal components. It made it possible to identify groups of closely correlated variables in a multidimensional space and replace them with the main components without loss of information content.

The mathematical model of the principal component method is represented by the formula (3).

$$y_i = \sum_{i=1}^{k} \alpha_{ij}z_i, \qquad (3)$$

where: y_j is the main component; α_{ij} is the coefficient reflecting the contribution of the variable z_i to the principal component y_i; z_i - standardized initial variable, $z_i = (x_i - \bar{x_i})/s_i$, s_i - variance, $i = 1, k$.

The calculation of the principal components is reduced to the calculation of the eigenvectors and eigenvalues ($\lambda_1, \lambda_2, ..., \lambda_k$) of the correlation matrix of the initial data. The α_{ij} values are factor loadings. They represent the correlation coefficients between the original variables and the principal components. Factors include those variables for which $|\alpha_{ij}| > 0,7$.

To reduce the dimension of the space $Y = (y_1, y_2, ..., y_k)$ by cutting off non-informative variables, the Kaiser criterion is used, which is associated with eigenvalues: the number of principal components includes variables that correspond to the eigenvalues $\lambda_i > 1$, since their informative value is higher.

Stage 3. Rotation of the factorial solution, which is used if the selected factors cannot be interpreted clearly enough.

For the analysis and interpretation of the results obtained, the varimax method and the quartimax method are used [14–16]. Varimax is a method most often used in practice, the purpose of which is to minimize the number of variables that have high loads on a given factor (which helps to simplify the description of a factor by grouping around it only those variables that are more associated with it than with the rest), cannot be used, since in the problem being solved, the variables (basic indicators) cannot be reduced, since they are all significant. Considering the above, to interpret the results of factor analysis, we used quartimax, a method that ensures the reduction (minimization) of the number of factors necessary to explain the variation of a variable.

The mathematical apparatus of factor analysis made it possible to solve the following two problems [17–20]:

1) reducing the dimension of the number of variables used due to their explanation by a smaller number of factors.
2) grouping and structuring of the received data.

Based on the principal component method, the eigenvalues of the factors were calculated and a matrix of factor loadings was constructed (Table 4), which represents the correlation coefficients between the initial variables (basic indicators) and the principal components (factors).

Table 4. Factor loading matrix

Basic indicators	Final factors (Quartimax method)			
	Factor 1	Factor 2	Factor 3	Factor 4
AR	0,2466		0,684	
ER	0,2938	0,8769	0,1026	0,3664
RS/T	0,1357	0,9455	0,1416	0.2490
CT	0,9754		0.1113	0,1470
IT	0,9906			0,1090
IS	0,7938	0,5435	0,2365	

The eigenvalue of the factor λ_i reflects its contribution to the variance of variables, explained by the influence of general factors. According to the Kaiser criterion, it is considered that those factors for which this indicator is significantly less than 1.0 do not make a significant contribution to the explanation of the result. The second indicator is the percentage of variance explained in the variables.

It is generally accepted that with a well-grounded factorial solution, so many factors are chosen so that they together explain at least 70–75% of the variance. In some cases, this figure can reach 85–90%. The factor loadings matrix illustrates the strength of the relationship between a variable and a factor. The higher the factor load in absolute terms, the higher the bond strength.

Thus, the main subject of interpretation of the results of the performed factor analysis was the extraction of significant factors and the calculation of factor loads.

Table 5 presents the results of identifying latent factors associated with baseline indicators.

Below is the interpretation of the results of factor analysis. In accordance with the results of the identification of factors presented in Table 4, the most significant factors influencing the baseline are:

- Factor 1 (includes a set of latent factors: the number of teaching staff, the level of their qualifications and the presence of close collaboration (the number of joint research projects) with foreign universities and research organizations) affects the basic indicators of the AR, ER, CT, IT, IS;
- Factor 2 (includes a set of latent factors: number of teaching staff, level of payment for teaching staff) influences on the basic RS/T indicator.
- Factor 3 (includes a set of particular factors: the presence of well-known scientific schools and dissertation councils, the presence of close collaboration (the number of joint scientific projects) with foreign universities and scientific organizations, the number of teaching staff, the level of their qualifications) affects the basic indicator of AR;
- Factor 4 (includes a set of latent factors: the level of training (competencies) of students, the demand for graduates from the employer, the presence of basic departments in enterprises) affects the basic indicator of the ER.

Thus, the results of the interpretation showed that the factors 1,2,3 have the greatest influence on the basic indicators. Those the task of increasing the values of the basic indicators is directly related to the increase in particular indicators characterizing: The number of teaching staff and the level of their qualifications; Close collaboration (number of joint research projects) with foreign universities and research organizations; The level of training (competencies) of students, The demand for graduates from the employer, The presence of basic departments at enterprises, The presence of well-known scientific schools and dissertation councils.

Table 5. Relationships between latent factors and baseline

Basic indicators	Factors				
	Notable scientific schools and dissertation councils	Close collaboration (number of joint research projects) with foreign universities and research organizations	Востребованные направления и профили подготовки	Base departments at enterprises	Level of teacher's qualification
AR	The level of training (competencies) of students	The demand for graduates from the employer		The presence of basic departments at the enterprises	The qualification level of the teaching staff
ER	Number of teaching staff	Payment level of teaching staff			
RS/T	Number of publications In DB Scopus		Stimulating factors		Qualification level of teaching staff
CT		Close collaboration (number of joint research projects) with foreign universities and research organizations			The qualification level of the teaching staff
IT; IS	Availability of places in the hostel for foreign students. Pay level of foreign teachers	Additional areas for education	Number of teaching staff with language training		Proficiency level of teaching staff

5 Discussion Justification of Measures to Achieve the Planned

The results obtained in paragraphs 3–4 made it possible to substantiate a set of measures to increase the values of particular indicators (factors) necessary to solve the problem of achieving university performance indicators by 2025, corresponding to the level of MISIS indicators in 2019.

Correlation dependences between functional and basic indicators are obtained. The presence of a strong connection between the functional and the following indicators was revealed: "The ratio of the number of students to the number of teaching staff" ($r = 0.952$), "Reputation with employers" ($r = 0.854$) and "International students" ($r = 0.636$). The strength of the relationship between the functional and the rest of the basic indicators is insignificant.

The largest contribution (98.9%) to the final result (the value of the rating functional and the corresponding place in the QS rating) is made by the following particular indicators: The number of teaching staff and the level of their qualifications; Close collaboration (number of joint research projects) with foreign universities and research organizations; The level of training (competencies) of students, The demand for graduates from the employer, The presence of basic departments at enterprises, The presence of well-known scientific schools and dissertation councils.

Measures to increase the values of latent indicators should be carried out considering the obtained correlation dependences of the most significant factors affecting the basic indicators.

6 Conclusion

An approach to solving the problem of providing conditions for achieving the required values of the university performance indicators in the international institutional ranking QS using models based on the methods of correlation-regression and factor analysis has been developed.

The developed correlation-regression model made it possible to calculate the pairwise correlation coefficients of the values of the functional and basic indicators for the Plekhanov Russian University and the MISIS University, the rating indicators of which are taken as a basis, as well as to carry out a comparative analysis of the results obtained for the universities of the reference group, to reveal the strength of the relationships between the basic indicators and from the links with the rating functionality.

The procedures of multivariate statistical analysis using the method of principal components of the developed factor model made it possible to solve the problem of identifying and interpreting latent factors affecting the basic indicators, to identify the most significant factors, to ensure their grouping and structuring, as well as to reduce the dimension of the problem being solved, which made it possible to analyze its results.

The results obtained on the basis of the developed models made it possible to formulate a list of activities and substantiate the expediency of their implementation in order to solve the problem of achieving the specified indicators of the university's activity.

The proposed approach is new. The obtained estimates of the correlation dependences between latent factors and basic indicators, the results of identifying latent factors formed the basis for building a cognitive model of scenario forecasting of measures to achieve the required values of the target indicators of the university's activity in the international institutional ranking QS.

Acknowledgments. The article was prepared with the support of the Russian Foundation for Basic Research, grants No. 18-07-00918, 19-07-01137 and 20-07-00926.

References

1. Decree of the President of the Russian Federation, On measures to implement state policy in the field of education and science. 7 May 2012. https://www.pravo.gov.ru
2. Information and analytical system QS – analytics. https://analytics.qs.com/#/signin
3. The project to increase the competitiveness of leading Russian universities among the world's leading scientific and educational centers. https://www.5top100.ru/
4. Ratings of world universities. https://www.educationindex.ru
5. Baraz, V.R.: Correlation and regression analysis of the relationship between indicators of commercial activity using the Excel program: a training manual. Yekaterinburg: State Educational Institution of Higher Professional Education "USTU – UPI", p. 102 (2005)
6. Draper, N., Williams, I.D.: Applied regression analysis, p. 912 (2019)
7. Sokolov, G.A., Sagiotov, R.V.: Introduction to regression analysis and planning of regression experiments in the economy: Textbook, Infra-M, p. 352 (2016)
8. Ayvazyan, S.A., Buchstaber, V.M., Enyukov, I.S., Meshalkin, L.D.: Applied statistics. Classification and reduction of dimension. Finance and Statistics, p. 607 (1989)
9. Rao, S.R.: Linear statistical methods and their applications. Science (Fizmatlit), p. 548 (1968)
10. Fomina, E.E.: Factor analysis and categorical method of the main components: a comparative analysis and practical application for processing the results of the survey. Humanitarian Bulletin, no. 10 (2017). http://dx.doi.org/10.18698/2306-8477-2017-10-473
11. Wu, J.F.: Reproducing an alternative model of intercultural sensitivity for undergraduates from Taiwan and Hainan. Int. J. Mod. Educ. Comp. Sci. (IJMECS) **8**(1), 1–7 (2016). https://doi.org/10.5815/ijmecs.2016.01.01
12. Olabiyisi, S.O., Adetunji, A.B.: An evaluation of the critical factors affecting the efficiency of some sorting technique. Int. J. Mod. Educ. Comput. Sci. (IJMECS) **8**(2), 25–33 (2013). https://doi.org/10.5815/ijmecs.2013.02.04
13. Meher, J., Barik, R.C., Panigrahi, M.R.: Cascaded factor analysis and wavelet transform method for tumor classification using gene expression data. Int. J. Mod. Educ. Comput. Sci. (IJMECS) **9**(2), 73–79 (2012). https://doi.org/10.5815/ijitcs.2012.09.10
14. Seagal, E.: Practical business statistics. Publishing house "Williams", p. 1056 (2002)
15. Godin, A.M.: Statistics: textbook. Publishing and Trading Corporation "Dashkov and Co.", p. 368 (2002)
16. Bureeva, N.N.: Multivariate statistical analysis using the STATISTICA IFR. Nizhny Novgorod, Novosibirsk State University N.I. Lobachevsky, p. 112 (2007)
17. Jolliffe, I.T.: Principal Component Analysis, Springer Series in Statistics, 2nd ed., Springer, New York, XXIX, 487 p. 28 illus (2002). ISBN 978-0-387-95442-4
18. Gorban, A.N., Kegl, B., Wunsch, D., Zinovyev, A.Y. (eds.): Principal Manifolds for Data Visualization and Dimension Reduction, Lecture Notes in Computational Science and Engineering 58, Springer, Berlin, XXIV, 340 p. 82 illus (2008). ISBN 978-3-540-73749-0
19. Bolch, B., Huan, K.D.: Multidimensional statistical methods for business and economics. Statistics, 1979, p. 317 (1974)
20. Nivorozhkina, L.I., Arzhenovsky, S.V.: Multidimensional statistical methods in economics: a textbook. Moscow: RIOR: INFRA-M, p. 203 (2018)

Deep-Learned Artificial Intelligence for Consciousness – Thinking Objectization to Rationalize a Person

Nicolay Vasilyev[1]([⊠]), Vladimir Gromyko[2], and Stanislav Anosov[3]

[1] Fundamental Sciences, Bauman Moscow State Technical University,
2-d Bauman Street, 5, b. 1, 105005 Moscow, Russia
nik8519@yandex.ru
[2] Computational Mathematics and Cybernetics, Lomonosov Moscow State
University, Leninskye Gory, 1-52, 119991 Moscow, Russia
gromyko.vladimir@gmail.com
[3] Public Company Vozrozhdenie Bank, Luchnikov per., 7/4, b. 1,
101000 Moscow, Russia
sanosov@cs.msu.su

Abstract. System-informational culture (SIC) is full of science big data anthropogenic environment of artificial intelligence (I_A) applications. Mankind has to live in networks of virtual worlds. Cultural evolution extends scientific thinking to everyone in the boundaries of synthetic presentations of systems. Traditional education has become overweighted problem. Because of that it is necessary to learn a person to learn oneself. Achieving level of cognogenesis educational process in SIC is to be directed on consciousness – thinking objectization. Personal self – building leans on axiomatic method and mathematical universalities. For the objective of auto-poiesis, a person come untwisted as universal rational one possessing trans – semantic consciousness. Gender phenomenology in SIC presents thinking – knowledge by I_A tools needing consonant partnership with man. The latter is based on epistemology to extend hermeneutic circle of SIC. Like that up-to-date noosphere poses objectization problem to attain Lamarck's human evolution on the ground of Leibnitz's mathesis universalis in the form of categories language. It can be solved only by means of deep – learned and natural intelligences adaptive partnership.

Keywords: System-informational culture · Deep-learned artificial
intelligence · System axiomatic method · Consciousness – thinking ·
Objectization · Language of categories · Cogno - ontological knowledge base

1 Introduction

Up-to-date system-informational culture (SIC) impacts essentially man's evolution [1]. Informatics successes and improvement of communication are prerequisites of DL I_A building as natural intelligence (I_N) partner in affaire of constitutive phenomenology [2, 3]. Noosphere requires from anybody the third world objectization and scientific thinking skills [4, 5]. On-line learning weakens social bonds creating possibility of personally social interdisciplinary activity but suffers from shortcomings of traditional teaching [6]. Objective subjectization can be achieved only by means of adequate universal education assisted by DL I_A [3, 6, 7]. It is discussed in Sects. 1.1–1.3. It is innovative and promising approach to achieve needed semantic education under conditions of SIC directed on person's rational sophistication. Scientific meanings understanding is necessary to become personality living in system world.

The paper is focused on perfection of emerging technology of man's deep-learning. Reduction method is contributed, see Sect. 1.1. It is applied to human mind development. In Sects. 2, 3 it is displayed on examples.

Cognogenesis is based on gradual conceptualization of presentations up to their expression by means of concise semantic tools of categories language. The formalization allows using deep-learned artificial intelligence (DL I_A) by trained natural one because mind rational evolution leans on universalities proper to thinking – consciousness processes. It serves to knowledge meanings understanding.

Knowledge structure must be well-founded by means of system axiomatic method application, see Sects. 2, 3. Proto-phenomenon of meanings is present in metaphors of natural language, discovered in natural sciences, and, in full measure, displayed in mathematics [8–10]. Consciousness – thinking molding corresponds to notions and theories formation in mathematics. DL I_A must function keeping in mind cogno-ontological knowledge base (CogOnt) presenting their dynamics.

Semantic and object - oriented category approach is widely used in various I_A systems [11–13]. Contributed technology realization can lean on the tools. Existing ontological bases embrace multidisciplinary electronic resources. They have to be formalized by means of axiomatic method AM) with the help of language of categories. Their including in CogOnt will support person's semantic work under cognogenesis.

1.1 Knowledge Personal Objectization Problem

Consciousness objectization answers to SIC challenge to transform cognitive intellectual processes of a person (I_N). It allows achieving trans-semantic state of mind to be able to carry out needed trans-disciplinary activity in SIC. It has become practical problem in frames of constitutive phenomenology [2]. Mind evolution occurs on the unity of art and science. Neurophenomenology ascertains that ontogenesis repeating philogenesis is modified now on anthropogenesis attaining cognogenesis. Big data of I_A applications are to be understood [4, 5]. Subjective objectization is attainable only with the help of life-long deep-learned tutoring partner – DL I_A [15]. For the purpose, mind study was undertaken through mathematics because it deals with thinking in meanings [8, 9]. Mathematical universalities give ground to application of knowledge without premises principle [2]. Artificial intelligence now essentially influences ergonomics [16–18]. On the agenda, there are semantic communication, data protection, and universal education assisted by DL I_A. Transcendental SIC personality appears due to the third world subjectization. Idea image (*eidos*) is primary for thinking. It is base of intuition. Expressed semantically in language of categories (L_C) such intuition implants meaning - thinking in rang of life.

1.2   Education in SIC

Objectization of thinking – consciousness is to be supported by DL I_A on the basis of cogno-ontological knowledge base (CogOnt). DL I_A is to be used in real time on everybody's working place. In difference from other investigations, I_A is to be applied to man's universal development, see Fig. 1. On-line education reinforces inconsistency of traditional teaching that pays insufficient attention to labor with meanings. Adequate rational education in SIC must pursue the aim to enrich Komensky's didactics by concern to present and identify knowledge achieving the level of mathematical universalities [6, 7]. Student is to be provided with adaptive help of information expression – processing to extract his benefit from the synthesis, see Fig. 1.

$$learn\ to\ know\,(calculate) \rightarrow understand\ meaning \rightarrow auto\,(self) - develop$$

Fig. 1. Evolution of the reasonable as inheritances chain in education

Substantiation plays an important role in discovering and explaining knowledge structure using adequate tools and methods. In this case it is impossible to do without deep-learned adaptive assistance as guarantee of successful personal auto-poiesis (Maturana) [14–17]. In the process of rational consciousness sophistication – building semantic languages are to be used. Similar to metaphors in natural languages mathematical tools significantly compress information selecting semantic and presenting it in universal form making objectization problem solvable by means of subjective objectization [6, 10, 16].

1.3   Contribution in DL I_A Building

DL I_A technology of man's universal rational self-building is enriched by new results. It is studied how to achieve knowledge objectivity with the help of meta-mathematical means, description in category language and tools of reasoning. They are to be included in CogOnt [14–17]. To apply system axiomatic method and formal reasoning whole theories and their models are to be substantiated in language of categories (L_C) [19, 20]. Substantiation (proof) makes knowledge structure self-obvious. DL I_A can support the process. Objectization needs well-foundation. The approach is applied here to the category of algebraic theories. Formal reasoning is displayed in Sect. 3 on example of Poincare – Birkhof – Witt theorem generalization due to its category proof.

1.4   Approbation of the Approach

Consciousness transformation is life-long process. According to contributed technology, mind is to be gradually trained. For meanings identification and highlighting studied "reality" must be modeled using different levels of AM. Significance of symbolical forms for intellectual development was investigated by Kassirer in [20].

Mathematical tools serve to discover properties of the reality. It is similar to texts understanding [21]. That is why DL I_A must use CogOnt containing linguistic and mathematical means to run through theories genesis. On the basis DL I_A can assist man to understand *eidos* presented by abstractions [2]. It gives birth to universalities expressed mathematically. For labor with them, man applies clear "real" tools.

Only universally prepared person will be able to interact with DL I_A on semantic level. For the objective, L_C is to be mastered. Highlighting corresponding commutative diagrams DL I_A can help man to study L_C itself.

Contributed technology is approbated by means of observing how our mind overcomes arising intellectual difficulties of scientific study. Additional results are obtained teaching secondary and high school students. For the purpose, work of not yet created DL I_A can be simulated by a teacher applying the principles of T_U. In the experiments students were to solve specially selected mathematical tasks which some notions are to be generalized in. Their needed solution lies on the way of AM application. Underlying meanings extraction happens by means of constructed theory.

Example 1. (*Secondary school.*) Knowing area of square how can one to measure area of a rectangle? Main property of the area notion is additivity. Area S of a square $(a+b) \times (a+b)$ is equal to $(a+b)^2 = a^2 + ab + ba + b^2$. Symmetry of the right part of the equality means that the square can be partitioned in two squares $a \times a, b \times b$ and two equal rectangles $a \times b, b \times a$. Due to the additive property of area function and areas equality of equal figures, one concludes that area $S(a \times b) = ab$.

(*High school*). Find area of a circle with radius r and perimeter $2p$ knowing that $S(a \times b) = ab$. Circle symmetry allows its transformation in another symmetric figure near to a rectangle that conserves value of initial area. Precisely, one can slit circle along radiuses in equal sectors $i = 1, 2, \ldots, 2n$. Figure 2 describes half circles $1 \le i \le n, n+1 \le i \le 2n$ developed in two equal "combs", inserted one into another, and stuck together along radiuses. (Sector $i = 2n$ is also to be split in two to replace one part in suitable way to obtain Fig. 2.)

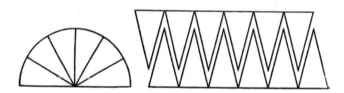

Fig. 2. Proof of the formula $S = rp$ drawn in Hindu temple (10 c. B. C.) [23]

Parallel straight sides of the figure have length r. Remained two parts of its boundary consists of small arcs of the circle put together. They have length p. If $n \to \infty$ then its limit figure is rectangle $r \times p$. So as area is continuous function of figures then formula rp is searched for.

People feel symmetry of an object. Its meaning is clarified by the notion of group that can be defined as one object category with iso - arrows.

Neurophenomenology happens by means of synthesis of all scientific presentations. In Ex. 1 students have to connect notions from "different" parts of mathematics: geometry, algebra, analysis. In the given case, synthesis is based on measurement, continuous maps, product, and symmetry universalities. Axiomatic approach guarantees the notions succession.

In the process of teaching observations confirmed that discursion on the base of universalities strengthens thinking and helps to meanings formation.

2   Fundamentality of Intuitive and Discursive Functions of Consciousness Differentiation

Natural intelligence (I_N) evolution happens due to the existing differentiation of thinking (discursion) and consciousness (intuition). Knowledge objectivity is achievable only on the level of systems, theories, and formal models. In Fig. 3 natural and artificial intelligences duality $I_N \sim DL\,I_A$ is presented. Intuition and discursive reasoning of I_N are supported by formal analogues of DL I_A. Structures in Fig. 3 are interconnected categories describing knowledge. Inheritance relation $\hookrightarrow$ transforms knowledge into unit system. The structures display meanings in mathematical form admitting investigation and comparison. Besides, formalization is needed for DL I_A functioning. It enables personal presentations with objectivity.

$$\left\langle \begin{array}{c} intuition \\ \updownarrow \\ discursion \end{array} \right\rangle \sim \left\langle \begin{array}{c} structures\ inheritance \\ \updownarrow \\ formal\ reasoning \end{array} \right\rangle \hookrightarrow \langle knowledge\ objectivity \rangle$$

Fig. 3. Objectivity as inheritance of natural and artificial intelligences partnership

Corresponding tools of meanings formal description and study are discovered in mathematics.

First of all, it is language of categories (L_C) [19, 20]. Having second order L_C is predestined for system comparison supporting meta mathematical level of knowledge presentation. It is convenient for reasoning due to its geometrical diagrams. Mentioned in Fig. 3 structures consist of protograms, universal commutative diagrams, and functors. They express definitions of general notions and theorems descript in L_C, for instance, theorem on homomorphisms and lemma on squares as inheritance of commutative property [19]. Universalities are to be used on meta-mathematical level. Syntaxes of all algebraic theories can be expressed in compressed category form [24].

Example 2. Consider the category V of all algebraic theories. It can be described as system $\{N; \varphi_{n,m}\}$ which objects are natural numbers $n, m \in N$ and morphisms $\varphi_{n,m}$: $n \rightarrow m$ are presented by sets of *n-ary* algebraic operations. All its constituents are regarded on as products of units or projections $\pi_j : n \rightarrow 1,\ j = 1, 2, \ldots, m$:

$$n = 1 \times \ldots \times 1,\ \ \varphi_{n,m} = \pi_1 \times \ldots \times \pi_m$$

Thus, intuitive presentation about algebraic structures obtained strict meaning by means of this description.

Algebraic semantic can also be expressed in L_C with the help of functors $\chi : T \rightarrow E$ acting from category of algebras $T \in V_\Omega(\Sigma)$ into any category of values E at least it had finite products. Algebras manifolds $V_\Omega(\Sigma)$ are defined by means of system Σ of identities depicted in a signature Ω with the help of variables set X. So, algebraic models are obtained as functors χ, see Fig. 3.

In its turn, formal algebraic theory $A(\Sigma, R)$ is defined with the help of a set R of inference rules. An identity can be proved by formal deduction starting from Σ and applying rules R to them. According to completeness theorem, an identity is true if it is fulfilled in all interpretations $f : X \rightarrow A,\ A \in A(\Sigma, R)$. It is sufficient to verify it for universal object that is free algebra $F_\Sigma \in V_\Omega(\Sigma)$ with free generators system X. Formal reasoning implantation is needed for DL I_A functioning, see Fig. 3.

Intuitive and discursive differentiation allows mind to overcome arising intellectual difficulties due to mathematical theories, linguistic tools formation and thinking – reasoning usage. Enriching double helix of consciousness, they guarantee man's evolution [15].

Meanings can be grasped by system axiomatic method (AM_S) application to studied models [1, 3, 6, 7, 15–18]. By the method DL I_A supports process of knowledge objectization – understanding. Proof plays an important role in thinking – consciousness objectization process. Convincing knowledge is attainable by means of its substantiation.

3 Objectization on the Level of Thinking-Consciousness

Convincing knowledge is needed for objectization. Proof helps to understand knowledge structure. It enables it with self-obviousness [7]. Universal mathematical constructions (meanings) are used in CogOnt for knowledge presentation and substantiation. Meanings are displayed on the level of whole mathematical theories and models and strictly expressed in L_C. Consciousness – thinking differentiation is to be integrated by knowledge ideation on the base of system axiomatic method (AM_S) and language of categories.

3.1 Knowledge Substantiation with the Help of DL I_A

Universal commutative proto-diagram (*protogram*) is tool to establish correlation among morphisms from different categories. For instance, free object $F \in V$ in a category V having free generators system $X \in V' = SET$ is defined by means of

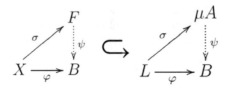

Fig. 4. Object $A \in V_1$ universality is inherited from free object F

protogram, see the left part of Fig. 4. Object X and morphisms $X \to F$, $X \to B$ belong to category V' and $\psi \in V \neq V'$. Applying forgetting functor $\Phi : V \to SET$ the protogram becomes commutative diagram in the category SET thus defining function $\psi\Phi$. The protogram meaning is existence of unique continuation of the map up to morphism $\psi \in V$. Let's consider inheritance of free object universality in case of two categories interconnected by means of a functor $\mu : V_1 \to V_2$:

Here, $A \in V_1$; L, μA, B, σ, φ, $\psi \in V_2$. Let V_1, V_2 be sub-categories of the category V of all algebraic theories containing free and factor objects. All the requirements take place for categories of associative algebras with neutral element and Lee algebras correspondingly. There is also functor μ replacing multiplication in algebra $A \in V_1$ by multipliers commutator regarded on as multiplication operation in Lee algebra $A^{(-)} \equiv \mu A$ [25, 26].

Object $A \in V_1$ is called *universal enveloping* one for given object $L \in V_2$ if the right diagram from Fig. 4 is fulfilled i.e. for any monomorphism φ there is unique morphism ψ making it commutative one.

Now DL I_A support of formal reasoning, see Fig. 3, can be illustrated on example of Poincare – Birkhof – Witt (PBW) theorem proof [25].

Example 3. *Let L be any Lee algebra over commutative ring with unit. Then there exists A – universal enveloping associative algebra with unit over the same ring.*

Category Proof. Let base X of algebra L be generators system of free objects $F \in V_1$, $G \in V_2$. Let show that universal enveloping algebra A is inheritance of free objects F, G, see Fig. 4. PBW theorem proof is sequence of the next diagrams – intermediate inheritances, see Fig. 3, 4. According to the right diagram from Fig. 4 there exist unique morphisms $\pi : F \to L$, $\sigma' : F \to G^{(-)}$. Ideal $K \subset G$ generated by image $\sigma'(\ker \pi)$ of inverse image π allows defining factor algebra $A = G/K$, natural homomorphism $\tau' : G \to A$, and functor image $\tau = \mu\tau'$. Besides, there is natural homomorphism $\theta : F \to \tilde{F} = F/\ker \pi$. Functor μ application to constituents of category V_1 and universality of factor objects [19, 20] allow commutative diagram in category V_2 to grow from the top of Fig. 5 to its bottom. Theorem on homo (iso) - morphisms can be applied to π and $\sigma' \circ \tau$ to construct arrow $\sigma : L \to A^{(-)}$ as composition $\sigma = \sigma_1 \circ \sigma_2$ of morphisms $\sigma_1 : L \tilde{\to} \tilde{F}$, $\sigma_2 : \tilde{F} \to A^{(-)}$. Factor object $A^{(-)}$ universality [19, 20] use accomplishes construction of the next commutative diagram.

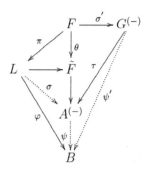

Fig. 5. Category proof of PBW theorem

Phenomenology of triangular sub diagram in Fig. 5 is needed inheritance substantiation from Fig. 4. It proves that A is universal enveloping algebra for L. In particular, quaternions body $H \simeq (R^3)^{(-)}$ gives universal representation of vector algebra R^3 with vector multiplication.

Category proof is similar to solution of geometrical problems on construction with new permitting tools [27] – universal diagrams, see Fig. 3, 4 and 5. Thus, well-founded knowledge is a result of geometrical constructions. Clear visual form of the theorem proof enables knowledge with self-obviousness. It is due to understanding of underlying structures [6]. Return to geometry origins helps to it [6, 28, 29].

Due to the category proof, PBW theorem admits generalization on case of other functors μ. For instance, it can be applied to categories of groups and associative algebras with unit correspondingly. They are compared with the help of functor μ transforming any group into group algebra. The specialization of the theorem applied to quaternions body gives enveloping group $G = \{\pm 1, \pm i, \pm j, \pm k\}$, $\mu G = H$. Specification of general constructions always maintains stability of general presentations.

4 Conclusions

Thinking   consciousness objectization is answer to SIC challenges to strengthen intellectual cognitive processes and create trans-semantic mind for trans - disciplinary activity in SIC. Co-evolution with anthropogenic environment has to be supported by universal tutoring on the ground of mathesis universalis of categories language and adaptive partnership with DL I_A. It is proved that neurophenomenology happens on the base of knowledge synthesis. Super natural limit of scientific knowledge in SIC requires super sensual speculation. To achieve it, in practice of education is introduced eidetic reduction on the basis of transcendental one. So as AM has attained formalization level notions succession and knowledge substantiation can be supported by DL I_A using CogOnt. Contributed technology has been approbated and allows avoiding didactic mistakes of education of the kind "down with Euclid!". On the agenda stands problem of DL I_A building. It will require efforts of all members of scientific society.

References

1. Gromyko, V.I., Vasilyev, N.S.: Mathematical modeling of deep-learned artificial intelligence and axiomatic for system-informational culture. Int. J. Robot. Autom. **4**(4), 245–246 (2018)
2. Husserl, A.: From idea to pure phenomenology and phenomenological philosophy. In: General Introduction in Pure Phenomenology, Book 1. Acad. Project, Moscow (2009)
3. Gromyko, V.I., Kazaryan, V.P., Vasilyev, N.S., Simakin, A.G., Anosov, S.S.: Artificial intelligence as tutoring partner for human intellect. J. Adv. Intell. Syst. Comp. **658**, 238–247 (2018)
4. Popper, K.: Objective Knowledge. Evolutional Approach. URSS, Moscow (2002)
5. Popper, K.R.: Suppositions and Refutations. Scientific Knowledge Growth. Ermak, Moscow (2004)
6. Vasilyev, N.S., Gromyko, V.I., Anosov, S.S.: Deep-Learned artificial intelligence as educational paradigm of system-informational culture. In: Proceedings of 4-th International Conference on Social Sciences and Interdisciplinary Studies, vol. 20, pp. 136–142, Palermo, Italy (2018)
7. Vasilyev, N.S., Gromyko, V.I., Anosov, S.S.: On inverse problem of artificial intelligence in system-informational culture. J. Adv. Intell. Syst. Comp. **876**, 627–633 (2019)
8. Manin, J.I.: Mathematics as Metaphor. MCNMO, Moscow (2008)
9. Gromov, M.: Circle of Mysteries: Universe, Mathematics, Thinking. Moscow Center of Cont. Math. Educ, Moscow (2017)
10. Pinker, S.: Thinking Substance. Language as Window in Human Nature. Libricom, Moscow (2013)
11. Aminu, E.F., Oyefolahan, I.O., Abdullahi, M.B., Salaudeen, M.T.: A review on ontology development methodologies for developing ontological knowledge representation systems for various domains. Int. J. Inf. Eng. Electron. Bus. **12**(2), 28–39 (2020). https://doi.org/10.5815/ijieeb.2020.02.05
12. Rahman, M.M., Akter, F.: An efficient approach for web mining using semantic web. Int. J. Educ. Manag. Eng. **8**(5), 31–39 (2018). https://doi.org/10.5815/ijeme.2018.05.04
13. Chowdhury, S., Hashan, T., Rahman, A.A., Saif, S.: Category specific prediction modules for visual relation recognition. Int. J. Math. Sci. Comput. **5**(2), 19–29 (2019). https://doi.org/10.5815/ijmsc.2019.02.02
14. Vasilyev, N.S., Gromyko, V.I., Anosov, S.S.: Emerging technology of man's life-long partnership with artificial intelligence. J. Adv. Intell. Syst. Comp. **1069**, 230–238 (2019)
15. Maturana, H., Varela, F.: The Tree of Knowledge: The Biological Roots of Human Understanding. Progress-Tradition, Moscow (2001)
16. Vasilyev, N.S., Gromyko, V.I., Anosov, S.S.: Deep-learned artificial intelligence and system-informational culture ergonomics. J. Adv. Intell. Syst. Comp. **965**, 142–153 (2020)
17. Vasilyev, N.S., Gromyko, V.I., Anosov, S.S.: Deep-learned artificial intelligence for semantic communication and data co-processing. J. Adv. Intell. Syst. Comp. **1130**, 2, 916–926 (2020)
18. Vasilyev, N.S., Gromyko, V.I., Anosov, S.S.: Artificial intelligence as answer to cognitive revolution challenges. J. Adv. Intell. Syst. Comp. **1152**, 161–167 (2020)
19. McLane, S.: Categories for working mathematician. Phys. Math. Ed., Moscow (2004)
20. Goldblatt, R.: The Categorical Analysis of Logic. North Holland Publ. Comp, Amsterdam, New York, Oxford (1979)
21. Kassirer, E.: Philosophy of Symbolical Forms. Language. Univ. book, Moscow, St.-Pet. 1 (2000)
22. Hadamer, G.G.: Actuality of Beautiful. Art. Moscow (1991)

23. Kagan, V. F.: Geometry origins. Tech. – Theor. Lit. Moscow, Leningrad (1949)
24. Handbook of Mathematical Logic. Barwize, J. (ed.), vol. 1, North Holland Publ. Comp., Amsterdam, New York, Oxford (1977)
25. Skorniakov, L.A.: Elements of General Algebra. Nauka, Moscow (1983)
26. Shafarevich, I. R.: Main notions of algebra. Regular and chaos dynamics, Izhevsk (2001)
27. Engeler, E.: Metamathematik der Elementarmathematik. Springer, Heidelberg (1983)
28. Hilbert, D.: Grounds of Geometry. Tech.-Teor. Lit, Moscow-Leningrad (1948)
29. Euclid: Elements. GosTechIzd, Moscow-Leningrad (1949–1951)

Algorithmization of Computational Experiment Planning Based on Sobol Sequences in the Tasks of Dynamic Systems Research

I. N. Statnikov and G. I. Firsov[✉]

Blagonravov Mechanical Engineering Research Institute of the Russian Academy of Sciences, 4, Malyi Kharitonievsky Pereulok, 101990 Moscow, Russian Federation
firsovgi@mail.ru

Abstract. The method of the planned computational experiment is proposed, which allows not only to reduce the volume of information obtained by modeling, but also to interpret the results of the computational experiment. The article shows the algorithmic possibilities of using the Sobol sequences (LP sequences) when constructing matrices of planned experiments for carrying out computational experiments. It is shown that with a small number of computational experiments it is possible to take into account the exact values of the boundaries of the variation intervals or to approach them from the inside as well as from the outside if necessary. New matrices of planned experiments are formed using vector shift translators calculated without regard to the above requirements. It is important that statistical estimates of the first and second moments of the new and old planned experiment matrices have remained practically unchanged.

Keywords: Algorithm · Sobol sequences · LP_τ - sequence · Experiment planning · Computational experiment · Shift translator · Randomization

1 Introduction

Let's recall the definition of the algorithm given in [1]: "Algorithm is a prescription that defines a computational process (called algorithmic in this case), which begins with an arbitrary source given (from some set of possible source data for this algorithm) and is aimed at obtaining a result fully determined by this source data". Such definition corresponded (and corresponds) to the rapid development of computer technology that began in the second half of the XX century, and as a result there appeared possibilities of solving tasks from different fields of science and technology which could not be solved by analytical methods before [2–5]. The development of methods of mathematical modeling, methods of constructing mathematical models, most adequately describing the projected device, has significantly reduced the acuteness of the problem of obtaining the required information on the influence of deviations in the values of the device parameters from the calculated ones on its.

Z. Hu et al. (Eds.): AIMEE 2020, AISC 1315, pp. 86–93, 2021.
https://doi.org/10.1007/978-3-030-67133-4_8

However, new difficulties have naturally appeared - physical restrictions on the growth of both physical and computational experiments, and the problems of interpretation of the obtained results from the point of view of the set goals (criteria). To overcome these problems there appear theories and algorithms of experiment planning, starting with physical ones, for example, for agricultural needs. Theories of the planned experiment [5] are further developed when the object of experimentation is a mathematical model describing (roughly) the work of the projected dynamic.

In the last quarter of the XX century there appeared algorithms of planning LP_τ - sequences [6], intended for solving the problems of research and design of dynamic systems, when the complexity of a mathematical model is determined (in the understanding of the authors; of course, the real concept of "complexity" can be more voluminous) by a large number of parameters $\alpha_j, j = \overline{1,J}$ in the description of the mathematical model (multiparametricity), and dozens of criteria $\Phi_k, k = \overline{1,K}$ (multicriteria), required for the functioning of the system. We are talking about the PLP-search algorithms (LP_τ Sequence Planning [7–13].

LP_τ - sequences are numerical sequences of binary rational numbers q $(0 < q < 1)$ used as pseudo-random sequences in the statistical study of mathematical models, for example, for the calculation of multidimensional integrals [6]. Their important characteristic is the length of the period L of such a sequence, when the repetition of numbers begins. In [6, 14] $L < 2^{10}$ and $L < 2^{20}$ correspondingly.

Such property of these sequences that all of them (up to the 51st [6, 14]) consist of sections (subsets) of binary-rational numbers q, and these subsets in each of these sequences represent random permutations with respect to each other, also proved to be important for the stated material. This property has allowed in the development of [7–11] to offer new algorithms of numerical research of dynamic systems and the choice of rational values of parameters of these systems, to find their new instrumental capabilities.

2   The Idea of Algorithms

The idea of algorithms consists in that on the basis of construction of matrixes of planned experiments it appears possible at the established (set) probability P_3 of received estimations to carry out the analysis of properties of the investigated (projected) dynamic system to the analysis of the same properties, but depending "on average" $\tilde{\Phi}_k (\alpha_j)$ on each varying parameter, i.e. to the analysis of "one-dimensional" dependences that psychologically and practically becomes very useful as "curse of dimensionality" on Bellman is essentially overcome.

The algorithms are based on the randomization of the vectors $\bar{\alpha}$ location in the source area $G(\bar{\alpha})$ defined by type inequalities and calculated with the help of LP_τ - sequences using the standard linear formula.

$$\alpha_{ijh} = \alpha_{j*} + q_{ihj_{\beta h}} \times \Delta\alpha_j, \tag{1}$$

The randomization consists in the fact that for each h series of experiments ($h == 1$, ...,$H(i,j)$), where $H(i,j)$ - the sample volume from elements Φ_{ijh} for one criterion in i section j variable parameter, the vector $\vec{j} = (j_{1h}, j_{2h}, ..., j_{\beta h})$ of random numbers of rows in the table of guiding numerators [6, 14] is calculated by the formula:

$$j_{\beta h} = [R \times q] + 1, \tag{2}$$

Rand values α_{ij} in h series are calculated by formula (1), where $\Delta\alpha_j = \alpha_{j^{**}} - \alpha_{j^*}$, $\alpha_{j^{**}}, \alpha_{j^*}$ - respectively, the upper and lower boundaries of the area $G(\bar{\alpha})$; $\beta = 1, ...,$ J; R - any integer number (in [6, 14] $R = 51$); j- fixed number of the variable parameter; $i = 1, ..., M(j)$ - number of the level j mountain parameter in h series; $M(j)$ - number of levels, by which divided into sections j this parameter; in general case $j_{\beta h} \neq j$ (which is one of the purposes of randomization). It was proved with the help of Romanovsky's criterion [15] that the numbers $j_{\beta h}$ produced by the formula (2) appear to be a set of integers equally distributed in probability, realized in one series. If $M(j) = M = const$ and $H(i,j) = H = const$, in this case the parameters N_0, M and H are related by a simple ratio:

$$N_0 = M \times H, \tag{3}$$

where N_0 is the total number of computational experiments after all series of experiments; at that the sample length (volume) from Φ_{ijh} exactly equal to H. But in the general case when $M(j) = var$, then both $H(i,j) = var$, and then the formula (3) for one criterion (and for all Φ_k it is the same) will take this form $N_0 = \sum_{i=1}^{M(j)} H(i,j)$.

It should be noted that the quality of the matrix of planned experiments strongly depends on the number of identical vectors $\bar{\alpha}$ in it (in common sense: the smaller the better, i.e. closer in terms of properties to the matrix of planned experiments, which have the property of orthogonality). For quantitative estimation of this quality, the parameter æ is introduced, calculated as follows: $\text{æ} = N_0/M^J$, if $M_j = const$ and $\text{æ} = N_0/\prod_{j=1}^{J} M_j$, if $M_j = var$. In [7] it was shown that the aspiration $\varkappa \to 0$ for each J of the calculated matrix of planned experiments while keeping the minimum values H_{ij} ($H_{ij} \geq 2$) may be more effective at increasing the values M_j than simply increasing the values N_0.

3   Consideration of Boundary and Boundary Values of the Area of Variation of Parameter Values

The PLP - search method, which is probabilistic in nature, fully complies with the principles of heuristic programming, allowing for the use of various algorithmic possibilities of the method in each specific practical task of designing (research) a dynamic system. In particular, when before the beginning of solving either the task of selecting rational values of the system's variable parameters or simply the task of searching for the function's extremum, based on preliminary considerations, does not attach importance to the region's boundaries $G(\bar{\alpha})$, then we deal with the usual situation of using the PLP search [7–11]. But if the values of the studied function at the boundaries of the region $G(\bar{\alpha})$ or at the points $\bar{\alpha}_i$ as close as possible to these boundaries are of interest to the researcher, there are algorithms of implementation of such interest when the number of computational experiments is not increased. Let's proceed to consideration of possible variants of use of algorithms of PLP-search for $M = const$. Note that when $M_j = var$, then all the below considered algorithms can be implemented and in this case for each j parameter.

I. It is necessary to take into account the exact limit values, i.e. $\alpha_j \in [\alpha_{j*}, \alpha_{j**}]$. Input the shift T translator for vectors by formula $T = (\alpha_{j**} - \alpha_{j*})/M$ and then we recalculate the new parameter values α_{ijH} using the formula

$$\alpha_{ijH} = \alpha_{ij} - T, \tag{4}$$

to get (N_0/M) the values of the lower boundary, and to get (N_0/M) the values of the upper boundary we take

$$\alpha_{ijH} = \alpha_{ij} + T. \tag{5}$$

where α_{ij} they come from (1). In this algorithm, the boundary values will be repeated twice as less than the intermediate values α_{ij}. For example, if $N_0 = 32$, $J = 3$, $M_j = 4, j = 1,2,3$, $\alpha_j \in [0, 1]$, then intermediate values α_{ijH} will meet in the matrix of planned experiments 8 times (in different combinations due to randomization), and values $\alpha_{ijH} = 0$ and $\alpha_{ijH} = 1$ only 4 times and also in different combinations.

II. Proximity to borders from inside and outside the interval. The intervals $\Delta\alpha_j$ for each j are the same and equal Δ. Of course, the trivial solution at the usual probing of the investigated area of acceptable solutions consists in a significant increase in the number of points N_0. For example, if $N_0 = 512$, we will come nearer to borders of unit area $(0;1)$ on the left and right only on $\varepsilon \approx 0.001$ at purely casual (not planned) choice of points α_{ij}, and outside of borders we will not get in any way (see formula (1)). In the PLP search both variants are realized by means of bilateral inequality;

$$\frac{\Delta - 2M\varepsilon}{2(M-1)} \leq T \leq \frac{\Delta + 2M\varepsilon}{2(M-1)} \tag{6}$$

And since the PLP search is always $M > 1$ and almost $M \geq 4$, the inequality (6) can be written as follows:

$$\frac{\Delta}{M} - \varepsilon \leq T \leq \frac{\Delta}{M} + \varepsilon. \tag{7}$$

In inequalities (6) and (7) T - shift translator of the points α_{ij} calculated by formula (1); $\varepsilon(0 \leq \varepsilon < \Delta)$ - permissible (desirable) error of approaching the boundaries. If we put

$\varepsilon = 0$ in (7), we obtain $T = \Delta/2M$, which gives the maximum error of approximation of the parameter values to two boundaries simultaneously inside the interval, but does not allow to go beyond the boundaries. Of course, with increasing M values, regardless of J values, this maximum error decreases rapidly. Let us consider two examples. In these examples, there were the following parameters: 1) $J = 3; H = 2; \alpha_j \in (0; 1);$ $M_j = 4$; 2) $J = 5; H = 2; \alpha_j \in (0; 1); M_j = 8$. According to formula (3), the values $N_0 = 12$ and $N_0 = 16$ were taken. For both examples from (7), at value $\varepsilon = 0$ got $T = 0.125$. When constructing the matrix of planned experiments on the specified parameters received the following internal boundary values: 0.03125 and 0.96875 in the first example and 0.0078125 and 0.9921875 - in the second with small values of computational experiments. When $\varepsilon \neq 0$ choosing such a value of T, which in accordance with formulas (4) and (5) maximizes the proximity to the interval boundaries and at the same time does not violate the physical conditions of the system under study, such as a ban on exceeding the interval boundaries.

III. Proximity to borders from inside and outside the interval when $\Delta_j = \mathrm{var}$. Introducing a new shift translator $\tilde{T} = \Delta/2(M - 1)$ and we will always assume the parameter $\varepsilon < < \Delta$. Then at simultaneous shift from initial borders to the left and to the right by the value $t_1 = \tilde{T} - \varepsilon$ we will receive matrixes of planned experiments with as close values as possible to borders inside the interval, and at similar shift by the value $t_2 = \tilde{T} + \varepsilon$ we will receive matrixes of planned experiments with as close values as possible to borders of the interval from the outside. At $\Delta_j = \mathrm{var}$ the value $\tilde{T}_j$, t_{1j} and t_{2j} we calculate by the following formulas $\tilde{T}_j = \Delta_j/2(M_j - 1)$, $t_{1j} = \tilde{T}_j - \varepsilon_j$, $t_{2j} = \tilde{T}_j + \varepsilon_j$.

Let's consider two cases when we achieve maximum approximation to the boundaries of the area $\alpha_1 \in (0; 1), \alpha_2 \in (0; 2), \alpha_3 \in (0; 3)$. inside and out. In both cases, $M = [4\ 4\ 4]$, $N_0 = 8$, $H = 2$ and $\varepsilon = 0.002$.

For the case inside: $t_{11} = 0.164(6)$, $t_{12} = 0.331(3)$, $t_{13} = 0.498$ and we get this set of boundary values: (0.0015025; 0.998495) for the first parameter, (0.0015025; 1.9984975) for the second parameter and (0.0015025; 2.9985) for the third parameter.

For a case from the outside: $t_{21} = 0.168(6)$, $t_{22} = 0.335(3)$, $t_{23} = 0.502$ and we receive such set of boundary values: (−0.0014975; 1.001495) for the first parameter, (−0.0014975; 2.0014975) for the second parameter and (−0.0014975; 2.0014975) for the third parameter. As we can see, in both cases, with small values of N_0, the error of approaching the limit values in absolute value is less ε.

Let's answer one question: do the statistical properties of the matrices of planned experiments change significantly when implementing variants I, II and III? To answer this question, we will consider that the first and second moments for the matrix of planned experiments are equal to $E(Q_i)$ and $D(Q_i)$, respectively, in the case of the usual implementation of the PLP - search. Let the implementation of I variant $Q_{i1} = Q - T$, when shifted to the left, and $Q_{i2} = Q + T$, when shifted to the right. Then according to the well-known rules, we find: $D(Q_{i1}) = D(Q_{i2}) = D(Q_i \pm T) = D(Q_i)$, and $E(Q_{i1}) = E(Q_i - T)) = E(Q_i) - T$ and $E(Q_{i2}) = E(Q_i + T) = E(Q_i) + T$. Obviously, the dispersions of random variables have not changed compared to the usual matrix of planned experiments, but two mathematical expectations $E(Q_{i1})$ and $E(Q_{i2})$ have appeared in the new matrix of planned experiments. But we can see that if we make $T \ll 1$ from formula (1) for large M (or M_j), then we can still assume that the equality $E(Q_i) \approx E(Q_{i1}) \approx E(Q_{i2})$ is fair. All the above reasoning is also true for variants II and III.

4 Example

The above described algorithms are part of a set of PLP - search algorithms, and therefore do not reveal all its capabilities. Let's consider briefly one of examples of application of the PLP-search for the decision of a problem of parametrical synthesis of such essentially nonlinear dynamic system as transmission of the main drive of a working stand of rolling mill. Dynamic properties of the system were determined by the amplitudes of vibrations arising in the transmission elements of this drive. The mathematical model of system functioning was described by a set of 5 nonlinear differential equations of the second order in which 5 parameters varied $\alpha_j (j = 1,2,...,5)$; the initial region of parameters variation was set by J-dimensional hyperparallelepiped ($J = 5$). The purpose of the solution was determined by the requirement to reduce the amplitudes of impact loads, the values of which determined the wear of the stack elements. omputational experiments were carried out in two stages, and at each stage a matrix of planned experiments was built with the following parameters: $N_0 = 256$ - the total number of computational experiments (the number of rows of the matrix); $M_j = M = 16$ - the number of sections into which each variable parameter was divided; H-sample size in each section ($N_0 = M * H = 16 * 16 = 256$) Therefore, for the analysis of the results of computational experiments, obtained on the mathematical model, was formulated 5 quality criteria $\Phi_k(\bar{\alpha}), k = \overline{1,5}$ (five coefficients of dynamism at the junction points of the stick elements), which, in physical sense, should be minimized, and preferably at the same time: $\Phi_1(\bar{\alpha})$ - coefficient of dynamism in the gearbox anchor bolts of the drive, $\Phi_2(\bar{\alpha})$ - coefficient of dynamism on the gearbox wheel support, $\Phi_3(\bar{\alpha})$ - coefficient of dynamism on the spindles, $\Phi_4(\bar{\alpha})$ - coefficient of dynamism on the clutch, $\Phi_5(\bar{\alpha})$ - coefficient of dynamism on the clutch connecting the motor to the gearbox. According to the results of computational experiments in the area $G_0(\bar{\alpha})$ has been identified [6, 16] a new area of solution search $G_{new.}(\bar{\alpha})$, which was again carried out PLP-search. This area proved to be successful for selection to the rank of the area containing compromise solutions, because, firstly, it was shown with probability $P \geq 0,95$ of strong influence of the gap values $\Delta_k \neq 0$ on the values of criteria (for some gaps $P \geq 0,96 \div 0,98$); secondly, all the gaps $\Delta_k \neq 0$ were simultaneously minimized $\Phi_k(\bar{\alpha})$ ((so, for example, if in $G_0(\bar{\alpha})$ at $\Delta_2 \neq 0$ was received $\Phi_2(\bar{\alpha}) = 2,46 \pm 1,10$, then in $G_{new.}(\bar{\alpha})$ at $\Delta_2 \neq 0$ it turned out $\Phi_2(\bar{\alpha}) = 1,40 \pm 0,29$; if in the "idealized" variant, but improving the understanding of physical processes, we $\Delta_2 = 0$ get $\Phi_2(\bar{\alpha}) = 1,20 \pm 0,16$; these results for all $\Phi_k(\bar{\alpha})$ obtained with the same values N_0); thirdly, the "volume" of the region $G_{new.}(\bar{\alpha})$ was $\sim 3.5\%$ of the region $G_0(\bar{\alpha})$.

5   Conclusion

The separate algorithmic possibilities of PLP-search noted in the article, and even more described in [7–13] and other sources, allow to draw the following conclusions:

1) At the probabilistic approach to the task of research (and designing) of the complex dynamic system at its preliminary stage, PLP-search reduces psychological tension of the process of analysis of results of computational experiments, as the problems of search of rational decisions are reduced to the analysis of the averaged one-dimensional dependences of quality criteria from the varied parameters.

2) At the given probability P of possibility of exact account of borders of the area changes of varying parameters or approached to them by the size of border values $|\varepsilon| < <1$ inside or outside at small quantity of computing experiments essentially remove the problem of this quantity.

3) The possibility of estimating the average impact of each j-th variable on each k-th quality criterion by a probabilistic method allows to select subareas of "concentration" of required solutions for each criterion and thus to search for a compromise area (whether it is a search for a Pareto area or a given compromise scheme).

4) Allocation of subareas of "concentration" of required solutions for each criterion allows for the considered dynamic system to replace its mathematical model with more simple regression dependences of criteria on parameters, and thus to make more operative the choice of rational solutions for similar systems.

The prospects of the described method in all areas of science and industry, using mathematical models, are not yet limited, since they are associated with an increase in the technical capabilities of computers in terms of memory volume and speed of computing. The reliability of the statistical conclusions of the method is based on the synthesis of the classical theory of probability and the theory of planning mathematical experiments.

The required reliability in the method can also be increased by increasing the number of computational experiments due to the repeated use of the same matrices of planned experiments within the changed boundaries of the ranges of values of the variable parameters; in this case, the total number of computational experiments will be minimal or close to this value in comparison with other methods using statistical procedures. The novelty of the article also consists in confirming the possibility, developed by the authors, of taking into account the influence of the exact boundary and ε-boundary values of the varied parameters on the criteria, if this does not contradict the physics of the system under study.

References

1. Vinogradov, I.M. (ed.): Mathematical Encyclopedia, vol. 1, p. 1152. Soviet Encyclopedia, Moscow (1977). (In Russian)
2. LinLin, W., Yunfang, C.: Diversity based on entropy: a novel evaluation criterion in multi-objective optimization algorithm. Int. J. Intell. Syst. Appl. **10**, 113–124 (2012). https://doi.org/10.5815/ijisa.2012.10.12

3. Hou, N., Wang, Z.: A growing evolutionary algorithm and its application for data mining. Int. J. Intell. Syst. Appl. **4**, 8–16 (2011). https://doi.org/10.5815/ijisa.2011.04.02
4. Roy, Ch., Rautaray, S.S., Pandey, M.: Big data optimization techniques: a survey. Int. J. Inf. Eng. Electron. Bus. **4**(4), 41–48 (2018). https://doi.org/10.5815/ijieeb2018.04.06
5. Montgomery, D.C.: Design and Analysis of Experiments, p. 684. Wiley, New York (2001)
6. Sobol, I.M.: Multidimensional Quadrature Formulas and Haar Functions, p. 288. Nauka, Moscow (1969). (In Russian)
7. Statnikov, I.N., Firsov, G.I.: Using sobol sequences for planning experiments. J. Phys. Conf. Ser. **937**(012050), 1–3 (2017)
8. Statnikov, I.N., Firsov, G.I.: Processing of the results of the planned mathematical experiment in solving problems of research of dynamics machines and mechanisms. In: 2018 International Multi-Conference on Industrial Engineering and Modern Technologies (FarEastCon), pp. 1–6. IEEE Xplore, (2018). https://doi.org/10.1109/fareastcon.2018.8602481. https://ieeexplore.ieee.org/document/86024
9. Statnikov, I.N., Firsov, G.I.: Regression analysis of the results of planned computer experiments in machine mechanics. J. Phys. Conf. Ser. **1205**(012054), 1–6 (2019)
10. Statnikov, I., Firsov, G.: Numerical approach to the solution of the problem of rational choice of dynamical systems. In: Hu, Z., Petoukhov, S., He, M. (eds.) Advances in Artificial Systems for Medicine and Education II, pp. 69–79. Springer, Heidelberg (2020)
11. Statnikov, I., Firsov, G.: Estimation of the number of calculations for solving the tasks of optimization synthesis of dynamic systems by the method of a planned experiment. In: Misyurin, S.Y., Arakelian, V., Avetisyan, A.I. (eds.) Advanced Technologies in Robotics and Intelligent Systems, AG 2020. Proceedings of ITR 2019, pp. 149–156. Springer, Cham (2020)
12. Statnikov, I.N., Firsov, G.I.: Application of information technologies for rational selection of parameters of mechanical systems by the method of a planned computational experiment. In: Kravets, O.Ja. (eds.) Modern Informatization Problems in Economics and Safety: Proceedings of the XXIV-th International Open Science Conference, Yelm, WA, USA, January 2019, pp. 114–119. Science Book Publishing House, Yelm (2019)
13. Statnikov, I.N., Firsov, G.I.: Planning a computational experiment in the problem of multi-criterial synthesis of a planetary geared reducer. In: Kravets, O.Ja. (eds.) Modern Informatization Problems in Simulation and Social Technologies MIP-2020 SCT: Proceedings of the XXV-th International Open Science Conference, Yelm, WA, USA, January 2020, pp. 182–187. Science Book Publishing House, Yelm (2020)
14. Sobol, I.M., Statnikov, R.B.: Selecting Optimal Parameters in Multicriteria Problems, 2nd edn., p. 176. Drofa, Moscow (2006). (in Russian)
15. Mitropolsky, A.K.: The Technique of Statistical Computations, p. 576. Nauka, Moscow (1971). (In Russian)
16. Bondarenko, A.M., Zhitomirskiy, B.E., Sergeev, V.I., Statnikov, I.N.: The problem of parametric synthesis of the transmission of the main drive of the working stand of the rolling mill. Mech. Eng. (2), 11–14 (1979)

Modification of the Hausdorff Metric in Combination with the Nearest Point Algorithm ICP in Point Cloud Construction Problems

Vera V. Izvozchikova[1], Alexander V. Mezhenin[2],
Vladimir M. Shardakov[1(✉)], and Polina S. Loseva[2]

[1] Orenburg State University, Ave. Pobedy 13, Orenburg, Russia
viza-8.11@mail.ru, werovulv@inbox.ru
[2] ITMO University, Kronverksky Ave. 49, St. Petersburg, Russia
mejenin@mail.ru

Abstract. The paper deals with the issues of three-dimensional modeling and visualization of objects using point Cloud construction technology. The problem of comparing point clouds - calculating some measure of their similarity/ difference-is considered in detail. The paper studies existing metrics for calculating the distance between point clouds, and suggests a way to improve the quality of the ICP (Interactive Closest Points) algorithm by modifying the Hausdorff metric. The approach proposed by the authors is to calculate the forward and reverse distance between cloud points, and, unlike the usual Hausdorff metric, the average distance value is calculated instead of the maximum. To improve the accuracy of calculating the metric, we suggest using the averaging method-calculating the weighted average value of the normal vectors formed by pairs of neighboring triangles. The proposed approach was tested: the authors developed an application in the MATLAB environment that forms a point cloud based on images, and the Application was developed based on existing functions. An application for camera calibration in the system was tested, and an m-function was developed that implements the modified Hausdorff metric. The proposed metric was tested and compared with the Manhattan metric. The metric values were compared with the data obtained in the CloudCompare program (taken as reference data). Tests on the performance of metrics were performed on a sample of combinations of pairs of point clouds. The results of research and testing allow us to speak about the effectiveness of the proposed approaches. The research data can be used to evaluate the quality of simplification of a polygon network, to evaluate the accuracy of reconstruction of three-dimensional models in photogrammetry problems, and to identify objects in computer vision systems.

Keywords: Hausdorff dimensions · 3D reconstruction · Photogrammetry · Object recognition · Reconstruction errors · ICP algorithm · Point cloud

Z. Hu et al. (Eds.): AIMEE 2020, AISC 1315, pp. 94–100, 2021.
https://doi.org/10.1007/978-3-030-67133-4_9

1   Introduction

Currently, various methods are used to synthesize the virtual environment, such as modeling in computer graphics programs and modeling using photogrammetry. However, the method of representing space objects in the form of point clouds of different densities is the most promising [1].

Data for building a point cloud can be obtained by scanning 3D objects with special devices, as well as by processing optical scanning data.

The tools listed above for the synthesis and dynamic visualization of three-dimensional images and scenes of virtual reality environments must support the full technological cycle of editing three-dimensional images and three-dimensional scenes, provide the ability to describe the behavior of objects in time and space, modeling the movement of objects, and visualization of static and controlled dynamic objects and scenes of virtual reality.

The most important aspect is the problem of combining high realism of the displayed scenes with the required speed of visualization in real time. For example, for the aviation and space industries, this problem is most acute, due to the high technical complexity of the simulated objects and systems.

2   Overview of Existing Methods for Registering Point Clouds

The iterative nearest points algorithm (ICP) is the most widely used method needed to minimize the distance between points. This algorithm is able to perform transformations by finding the rotation matrix R and the displacement vector T, which in turn align the point clouds X and Y [2, 3].

An orthogonal geometric transformation is used to translate each individual point of cloud X into the coordinate system of cloud Y:

$$y_i = Rx_i + T, \tag{1}$$

where T is the transfer vector, R is the rotation matrix, x_i is an arbitrary point of the cloud X, and y_i is an arbitrary point of the cloud Y.

Basic algorithm for determining the nearest points has a number of disadvantages, such as the need for a common overlap area of point clouds, as well as the complexity of searching for the nearest point at each iteration [4, 5].

To eliminate these shortcomings, several modifications of this algorithm have been proposed, as shown below.

The Generalized-ICP algorithm uses a probabilistic structure for determining the error function. This method selects points from solid sections and combines them, thereby improving the performance and accuracy of the ICP algorithm [6, 7].

The algorithm based on orientation histograms allows you to solve the problem without using initial approximations for the rotation matrix R and the transfer vector T (as defined in the basic ICP algorithm). In the course of work, orientation histograms are obtained for clouds of points X and y. Then the maximum of the correlation function is found for all values of the rotation matrix, which is the desired angular orientation. The advantage of this modification is the efficiency of the algorithm at large initial angles between the input point clouds [8–10].

To reduce computational complexity, various metrics are also used (instead of Euclidean) to measure the distance between points. One of these metrics is the Manhattan metric (or city block distance) $(x, y) = \sum_{i=1}^{n} |x_i - y_i|$.

Metrics has better performance and noise tolerance, and uses less computational cost and less execution time compared to the basic algorithm [11–15].

$$f = max\{d(X, Y), d(Y, X)\},$$
$$d(X, Y) = \max_{x \in X} d(x, Y) \tag{2}$$

Another metric used is the Hausdorff metric, which turns the set of all non-empty compact subsets of a metric space into a metric space:

$$f = max\{d(X, Y), d(Y, X)\},$$
$$d(X, Y) = \frac{1}{N_x} \sum_{x \in X} d(x, Y).' \tag{3}$$

In [16–20], the authors present recommended requirements for the approach to solving the problem of graphic data reconstruction.

The approach proposed by the authors consists in calculating the forward and backward distance between cloud points, and, in contrast to the usual Hausdorff metric, not the maximum, but the average value of the distance is calculated. To improve the accuracy of calculating the metric, it is proposed to use the averaging method - calculating the weighted average of the normal vectors formed by pairs of adjacent triangles. algorithm in Fig. 1 describes how the metric works.

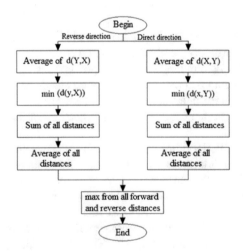

Fig. 1. Algorithm for determining the modified Hausdorff metrics

3   Experimental Part

Camera Calibrator App was loaded with four photos of the checkerboard from different angles. With the help of the program, the main parameters of the camera were determined (in our case, the camera of the phone). The camera data is saved in a.mat file, which contains data about the internal, external and distortion of the camera lens. Internal parameters: IntrinsicMatrix - projection matrix (contains a matrix with the coordinates of the optical center), PrincipalPoint - the center itself, Focal length - focal length, Skew - tilt of the camera axes, specified as a scalar (0 by default). Lens distortion: RadialDistortion - lens distortion (radial and tangential distortion). External parameters: RotationMatrices - rotation matrix, RotationVectors - rotation vectors and TranslationVectors – rotation (Fig. 2).

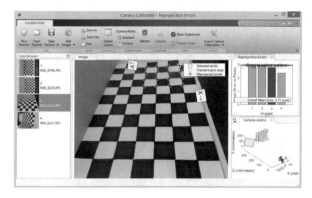

Fig. 2. Determination of points in a given plane of the checkerboard

The algorithm shown in Fig. 1 shows the algorithm of the main program loop, which performs pairwise transformation of 2D images into a three-dimensional data array in the form of a point cloud. The program is loaded with initial images (Fig. 3) and camera parameters obtained in advance using the Camera Calibrator App. Further, common points on a pair of images are found using the minimum eigenvalue algorithm, which returns characteristic points containing information found in b/w images. Then he connects these points visually using the image movement tracker. After that, outliers are eliminated using the MSAC algorithm (one of the variants of the RANSAC algorithm) - a stable method for estimating model parameters based on random samples. Only inliers are saved - points that are not outliers. Then the rotation matrix of the calibrated camera R is calculated, as well as the transfer vector t. The final step is to reconstruct the 3D locations of the matching pairs of undistorted image points (Fig. 4).

We tested the work of the usual and modified Hausdorff metrics. The results obtained were compared with the values obtained in the CloudCompare program. The decision to compare the values of the metrics with the program is based on the wide application and accuracy of this software when processing three-dimensional point clouds.

Fig. 3. Original image

Fig. 4. Resulting point cloud

Figure 5 shows graphs of the distribution of metric values for 10 tests. Analyzing these values, we can conclude that the values of the modified Hausdorff metric (yellow graph) turned out to be much closer to the CloudCompare values (red graph) than the values of the usual Hausdorff metric (blue graph).

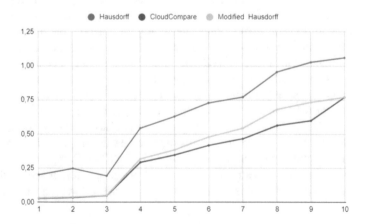

Fig. 5. Graph of distribution of metric values for 10 tests

Figure 6 shows the average running time of the algorithm with the modified Hausdorff metric for 10 tests.

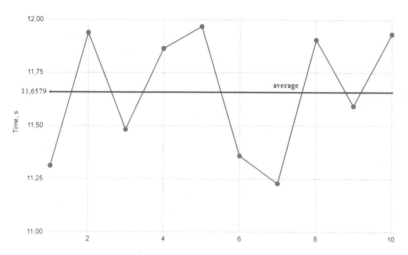

Fig. 6. Graph of the time distribution of the algorithm with a modified Hausdorff metric

4 Conclusion

Thus, the proposed modified iterative algorithm for nearest ICP points can be successfully applied in three-dimensional modeling and visualization of objects using Point Cloud construction technology. We tested the Camera Calibrator App, which allows you to get camera parameters and upload them to the MATLAB working environment. Two additional metrics were tested to prove the benefits of the modified Hausdorff metric.

In the future, it is planned to improve the algorithm in order to reduce the running time of the program. It is also planned to expand the capabilities of the algorithm for simultaneous processing of a large number of images.

In addition, it is planned to develop its own solutions (algorithms) to improve the synthesis of point clouds (especially in conditions of incomplete data) and improve existing approaches to evaluating the quality of synthesis and test them in the MATLAB environment.

References

1. Mezhenin, A.V., Izvozchikova, V.V., Burlov, D.I.: Virtual environment Modeling in cyber visualization and virtual presence technologies. Cybern. Program. **4**, 26–35 (2019)
2. Maron, H., Galun, M., Aigerman, N., Trope, M., Dym, N., Kim, V.G., Yumer, E., Lipman, Y.: Convolutional neural networks on surfaces via seamless toric covers. ACM Trans. Graph. **36**(4), 1–71 (2017)

3. Kormoczi, L., Kato Z.: Filling missing parts of a 3D mesh by fusion of incomplete 3D data. In: International Conference on Advanced Concepts for Intelligent Vision Systems. Springer International Publishing AG, pp. 711–722 (2017)
4. Arunava, S., Bhowmick, A.: Performance Analysis of Iterative Closest Point (ICP) Algorithm using Modified Hausdorff Distance (2017)
5. Chaughuri, S., Kalogerakis, E., Giguere, S., Funkhouser, T.: AttribIt: content creation with semantic attributes. In: 26th Annual ACM Symposium on User Interface Software and Technology (UIST), pp. 193–202 (2013)
6. Zou, C., Yumer, E., Yang, J., Ceylan, D., Hoiem, D.: 3D-PRNN: generating shape primitives with recurrent neural networks. In: IEEE International Conference on Computer Vision (ICCV), pp. 900–909 (2017)
7. Kormoczi, L., Kato, Z.: Filling missing parts of a 3D mesh by fusion of incomplete 3D data, pp. 711–722. Springer International Publishing AG, (2017)
8. Zhou, J., Fu, X., Zhou, S., Zhou, J., Ye, H., Nguyen, H.T.: Automated segmentation of soybean plants from 3D point cloud using machine learning, vol. 162, pp. 143–153. Published by Elsevier B.V. (2019)
9. Dupre, R., Argyriou, V., Greenhill, D., Tzimiropoulos, G.: A 3D scene analysis framework and descriptors for risk evaluation. In: International Conference on 3D Vision, pp. 19–22 (2015)
10. Hisahara, H., Hane, S., Takemura, H., Ogitsu, T., Mizoguchi, H.: 3D point cloud–based virtual environment for safe testing of robot control programs. In: 6th International Conference on Intelligent Systems, Modelling and Simulation, pp. 24–27 (2015)
11. Zhou, J., Fu, X., Zhou, S., Zhou, J., Ye, H., Nguyen, H.T.: Automated segmentation of soybean plants from 3D point cloud using machine learning. Published by Elsevier B.V. (2019)
12. Razakova, A.N., Salimgareev, A.M., Shamin, A.M., Luguev, T.S.: Methods for forming a three-dimensional representation of scene objects in the form of a point cloud. In the world of scientific discovery (2014)
13. Medvedev, V.I., Raikova, L.S.: Programs for processing data of laser scanning of the terrain. CAD and GIS of highways, no. 2(9) (2017)
14. Dupre, R., Argyriou, V., Greenhill, D., Tzimiropoulos, G.: A 3D scene analysis framework and descriptors for risk evaluation. In: International Conference on 3D Vision (2015)
15. Hisahara, H., Hane, S., Takemura, H., Ogitsu, T., Mizoguchi, H.: 3D point cloud–based virtual environment for safe testing of robot control programs. In: 6th International Conference on Intelligent Systems, Modelling and Simulation (2015)
16. Dupas, R., Grebennik, I., Lytvynenko, O., Baranov, O.: An heuristic approach to solving the one-to-one pickup and delivery problem with three-dimensional loading constraints. IJITCS 9(10), 1–12 (2017)
17. Kubiak, I.: The unwanted emission signals in the context of the reconstruct possibility of data graphics. IJIGSP 6(11), 1–9 (2014)
18. Mohan, R., Gopalan, N.P.: Dynamic load balancing using graphics processors. IJISA 6(5), 70–75 (2014)
19. Sharma, K.: GPU optimized stereo image matching technique for computer vision applications. IJMECS 7(5), 37–42 (2015)
20. Afif, M., Said, Y., Atri, M.: Efficient 2D convolution filters implementations on graphics processing unit using NVIDIA CUDA. Int. J. Image Graph. Signal Process. (IJIGSP), 10(8), 1–8 (2018)

Calculation of the Current Distribution Function Over a Radiating Structure with a Chiral Substrate Using Hypersingular Integral Equations

Veronika S. Beliaeva[1], Dmitriy S. Klyuev[2], Anatoly M. Neshcheret[2], Oleg V. Osipov[2], and Alexander A. Potapov[3,4(✉)]

[1] Lomonosov Moscow State University,
1-63, Leninskie Gory Street, 119991 Moscow, Russian Federation
shfordogs@gmail.com
[2] Povolzhskiy State University of Telecommunications and Informatics,
23, L. Tolstoy Street, 443010 Samara, Russian Federation
klyuevd@yandex.ru, neshchereta@gmail.com
[3] V.A. Kotelnikov Institute of Radio Engineering and Electronics,
11-7, Mokhovaya Street, 125009 Moscow, Russian Federation
potapov@cplire.ru
[4] Joint-Lab of JNU-IREE RAS, Jinan University, Guangzhou, China

Abstract. This article is devoted to the development of a method for electro-dynamic analysis of strip radiating structures based on chiral metamaterials, which provides high accuracy in calculating their characteristics with small computational resources. A mathematical model of investigated strip radiating structure based on chiral metamaterials is presented. The calculation of the current density distribution function over the emitter of such a structure are shown. The calculations were carried out using of above method and the Feko electrodynamic modeling software package in order to verify the obtained results. Good qualitative agreement of the obtained results is shown. This method can be used in high-efficiency computer-aided design tools for the development of new-generation antennas used in modern communication and telecommunications systems, including medical monitoring systems.

Keywords: Antenna · Chiral metamaterial · Hypersingularity · Method integral equations · Current density

1 Introduction

In connection with the improvement of radio-electronic technology, the search for new approaches to the creation of new-generation UHF, microwave and EHF antennas, which are an integral part of them and have improved electrical and mass-dimensional characteristics, has also been significantly intensified recently. An analysis of the literature shows that one of the promising approaches to creating such antennas is the using of chiral metamaterials in their design [1, 2].

© The Author(s), under exclusive license to Springer Nature Switzerland AG 2021
Z. Hu et al. (Eds.): AIMEE 2020, AISC 1315, pp. 101–113, 2021.
https://doi.org/10.1007/978-3-030-67133-4_10

However, due to the rather complex design of such antennas based on metamaterials, there are a number of difficulties associated with their correct electrodynamic analysis and synthesis. Currently, most studies of antennas based on metamaterials are carried out using various software systems for electrodynamic modeling (Feko, HFSS, etc.), which, generally speaking, do not always give the correct result and, in addition, are very demanding on computing resources.

Another approach to the analysis and synthesis of such antennas is associated with the using of integral equation methods, which in most cases using the Fredholm integral equation of the first kind, the solution of which belongs to the class of incorrect mathematical problems in the Hadamard sense, which leads to unstable solutions. This approach is presented in [3–6].

In this regard, there is an urgent scientific and technical problem of developing correct methods for analyzing and synthesizing antennas based on chiral metamaterials that provide a sufficiently high accuracy of calculations and do not require very large computing resources.

In [7–9], a method for analyzing data from narrow-emitter antennas based on the using of singular integral equations is presented. Numerical solution of such equations is a correct mathematical problem in comparison with the above methods based on the using of the first kind Fredholm equation. In addition, in contrast to the approach [10], which also uses a similar approach, but without highlighting a mathematical feature (singularity), the method [7–9] assumes a numerical solution only for a small range of values of the desired function, and the rest of it is calculated analytically, which can significantly increase computational efficiency. However, this method assumes a quasi-static distribution of the transverse variation of the longitudinal component of the current distribution density, which imposes a limit on the width of the emitter.

The purpose of this work is development of method for analyzing radiating structures based on the apparatus of hypersingular and singular equations, which allow to correctly calculate their characteristics and at the same time allow to analyze structures with large emitter sizes.

The results of the study can be used in the development of highly efficient computer-aided design tools that allow the development of a new generation of antennas based on chiral metamaterials.

The article consists of an introduction that justifies the relevance of the work, a review of the literature, which contains works aimed at the research and using of antennas based on metamaterials in communication and telecommunications systems, a methodology section that provides the method of analyzing antenna data based on chiral metamaterials, as well as a section that presents the results of calculations and their discussion.

2   Literature Review

The analysis of modern literature shows that traditional approaches to the development of strip and microstrip antennas have almost reached their limit regarding the further increase in their electrical and mass-dimensional characteristics, and therefore, new solutions and approaches to improving their characteristics are currently being sought.

One of these approaches is the using of artificial composite materials in the antenna design, which are called metamaterials [11–16].

Research results show that the using of metamaterials in antenna-feeder devices can significantly improve their characteristics [17], in particular, compensate for the reactivity of electrically small antennas [18], significantly increase directional properties [19], improve coordination, reduce the mutual influence between emitters in the antenna arrays of MIMO systems [20]. In [21], D. M. Pozar noted that the using of more complex metamaterials (bianisotropic and biisotropic media) makes it possible to improve the characteristics of antennas even better. Such metamaterials, for example chiral media, consist of a dielectric container with conducting elements of a mirror-asymmetric shape (helices, open rings, S-elements, etc.) are uniformly located. One of the important features of such media is the possibility of radiation redistribution (incident radiation transforms into azimuthal) [22], which can be used in creating highly efficient antennas for mobile radio networks' subscriber stations.

3 Problem Formulation

Consider a radiating structure with a chiral substrate on which an infinitely thin and perfectly conducting emitter is located, The emitter length is $2l$ and width is $2a$. The substrate is a chiral (metamaterial) layer with macroscopic parameters ε_1, μ_1, and χ, the thickness equal to d. A dielectric half-space with macroscopic parameters ε_2, μ_2 is located above the chiral layer. The parameters ε_1, ε_2, μ_1, μ_2 is the relative dielectric permittivity and permeability of the chiral and dielectric media, respectively, and χ is the chiral parameter [23–25]. The geometry of this radiating structure is shown in Fig. 1.

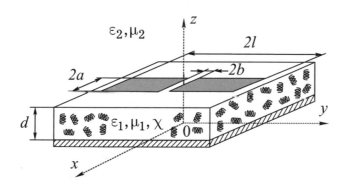

Fig. 1. Strip radiating structure with chiral substrate

To excite strip radiating structure, a harmonic electromagnetic field (EMF) source applied to the gap of the emitter is used. The width of gap is $2b$. As a result of excitation, the emitter generates currents distributed so that the resulting EMF satisfies the Maxwell equations, boundary conditions, and the Sommerfeld condition. In addition, the EMF source is connected in such a way that the surface current density function $\eta(x, y)$, where $x \in [-a, a]$, $y \in [-l, l]$, is continuous both on the surface of the vibrator and in the gap area.

Note also that in order to neglect the transverse component of the current density $\vec{\eta} = (\eta_x = 0, \eta_y)$ we will assume that the emitter will be quite narrow $(2a < l, \lambda)$. It is also assumed that the tangential component of the electric field has only one component $\vec{E}_\tau^{ext} = \left\{ E_x^{ext} = 0, E_y^{ext} \right\}$.

4   Methodology

The boundary conditions for this radiating structure will be written as follows:

$$\eta_y(x, -l) = \eta_y(x, +l) = 0,$$
$$\vec{E}_\tau(x, y) = 0 \text{ for } x \in [-a, a], \ y \in [-l, -b] \cup [b, l] \qquad (1)$$
$$\vec{E}_\tau(x, y) = -\vec{E}_\tau^{CT} \text{ for } x \in [-a, a], \ y \in [-b, b]$$

In [7, 8] it was shown that for narrow emitters, where the transverse component of the current density is equal to zero $\vec{\eta} = (\eta_x = 0, \eta_y)$, the connection between the Fourier image of the tangential component of the electric field strength and the Fourier image of the longitudinal component of the surface current density is carried out through the element of the surface impedance matrix $Z_{11}(\beta, h)$. In this regard, the integral equation for an unknown component of the surface current density will have the following form:

$$E_y(x, y, z = d) = \int_{-a}^{a} \int_{-l}^{l} \eta_y(x', y') Z_{11}^{\Sigma}(x', y'; x, y) dx' dy', \qquad (2)$$

where

$$Z_{11}^{\Sigma}(x', y', x, y) = \frac{1}{4\pi^2} \int_{-\infty}^{\infty} \int_{-\infty}^{\infty} Z_{11}(\beta, h) e^{-i\beta(x-x')} e^{-ih(y-y')} d\beta dh.$$

Taking into account the boundary condition, we take the integral in parts by y' from expression (2). The result is the following equation for a new unknown function $\eta_y'(x', y') = d\eta_y(x', y')/dy'$. In addition, integrals over infinite limits in the expression (2) are divergent, so the procedure for subtracting and adding summands with an asymptotic co-factor $Z_{11}^{\infty}(\beta, h)$, which is an element $Z_{11}(\beta, h)$ for $|\beta| \to \infty$ and

$|h| \to \infty$, was performed. The analytical expression of the asymptotic factor is given below:

$$Z_{11}^{\infty}(\beta, h) = i\omega\mu_0 \left(\frac{C1_{\varepsilon,\mu,\chi}}{k^2} \frac{h^2}{\sqrt{\beta^2 + h^2}} - C2_{\varepsilon,\mu,\chi} \frac{1}{\sqrt{\beta^2 + h^2}} \right) \qquad (3)$$

Given the above, expression (2) will have the form:

$$\begin{aligned}
E_y(x,y) &= -\frac{1}{4\pi^2} \int\limits_{-a}^{a}\int\limits_{-l}^{l} \eta_y'(x',y') \int\limits_{-\infty}^{\infty}\int\limits_{-\infty}^{\infty} \frac{1}{ih}[Z_{11}(\beta,h) - Z_{11}^{\infty}(\beta,h)]e^{-i\beta(x-x')}e^{-ih(y-y')}d\beta dh dx' dy' \\
&+ \frac{i\omega\mu_0}{4\pi^2 i}\frac{C1_{\varepsilon,\mu,\chi}}{k^2} \int\limits_{-a}^{a}\int\limits_{-l}^{l} \eta_y'(x',y') \int\limits_{-\infty}^{\infty}\int\limits_{-\infty}^{\infty} \frac{h}{\sqrt{\beta^2 + h^2}} e^{-i\beta(x-x')}e^{-ih(y-y')}d\beta dh dx' dy' \\
&- \frac{i\omega\mu_0}{4\pi^2 i} C2_{\varepsilon,\mu,\chi} \int\limits_{-a}^{a}\int\limits_{-l}^{l} \eta_y'(x',y') \int\limits_{-\infty}^{\infty}\int\limits_{-\infty}^{\infty} \frac{1}{h}\frac{1}{\sqrt{\beta^2 + h^2}} e^{-i\beta(x-x')}e^{-ih(y-y')}d\beta dh dx' dy'
\end{aligned} \qquad (4)$$

where

$$C1_{\varepsilon,\mu,\chi} = \frac{(\mu_1 + \mu_1)}{(\varepsilon_2 + \varepsilon_1)(\mu_1 + \mu_2) - \chi^2}, \quad C2_{\varepsilon,\mu,\chi} = \frac{(\varepsilon_2 + \varepsilon_1)\mu_1\mu_2}{(\varepsilon_2 + \varepsilon_1)(\mu_1 + \mu_2) - \chi^2}, \quad k = \frac{2\pi}{\lambda}.$$

The above expression (3) contains a number of integrals that can be calculated analytically [26], in particular:

$$\int\limits_{-\infty}^{\infty} \frac{h}{\sqrt{\beta^2 + h^2}} e^{ih(y'-y)}dh = 2i\sqrt{2\pi}\,\mathrm{sgn}(y' - y)K_1(|\beta||y' - y|),$$

where $K_1(|\beta||y' - y|)$ is 1-st order Macdonald function;

$$2i\sqrt{2\pi}\,\mathrm{sgn}(y' - y) \int\limits_{-\infty}^{\infty} |\beta|K_1(|\beta||y' - y|)e^{i\beta(x'-x)}d\beta = 2i\sqrt{2\pi}\pi \frac{(y' - y)}{\left[(x' - x)^2 + (y' - y)^2\right]^{\frac{3}{2}}};$$

$$\int\limits_{-\infty}^{\infty} \frac{1}{\sqrt{\beta^2 + h^2}} e^{i\beta(x'-x)}d\beta = 2K_0(|h||x' - x|),$$

where $K_0(|h||x' - x|)$ is 0 order Macdonald function;

$$2 \int\limits_{-\infty}^{\infty} \frac{K_0(|h||x' - x|)}{h} e^{ih(y'-y)}dh = 2i\pi\,\mathrm{arcsh}\left(\frac{y' - y}{|x' - x|}\right)$$

In addition, we decompose exponential functions in the first term of expression (4) by Bessel functions and Chebyshev polynomials of the 1-st kind:

$$e^{i\beta x'} = 2\sum_{p=0}^{\infty} \frac{i^p}{1+\delta_{p,0}} J_p(\beta a) T_p(x'/a) \quad and \quad e^{ihy'} = 2\sum_{k=0}^{\infty} \frac{i^k}{1+\delta_{k,0}} J_k(hl) T_k(y'/l),$$

where

$$\delta_{p,0} = \delta_{k,0} = \begin{cases} 1, & p,k = 0 \\ 0, & p,k \neq 0 \end{cases}$$

is the symbol of Kronecker.

We present the derivative of the current distribution function as a series that is a product of Chebyshev polynomials of the 1-st kind, taking into account the boundary conditions (1):

$$\eta_y'(x',y') = \sum_{m=0}^{N_m^x} \sum_{n=0}^{N_n^y} B_m^x B_n^y \frac{T_m(x'/a)}{\sqrt{1-(x'/a)^2}} \frac{T_n(y'/l)}{\sqrt{1-(y'/l)^2}}. \tag{5}$$

After substituting the above computed integrals and decompositions in (4), and limiting the infinite integrals to β_{max} and h_{max}, we obtain the following hypersingular integral representation (HSIR) of a field with a third-degree singularity:

$$E_y(x,y) = -\sum_{m=0}^{N_m^x} \sum_{n=0}^{N_n^y} B_m^x B_n^y i^{m+n-1}$$

$$\times \int_0^{h_{max}} \left[\int_0^{\beta_{max}} \left[Z_{11}(\beta,h) - Z_{11}^{\infty}(\beta,h) \right] J_m(\beta) e^{-i\beta x} d\beta \right] \frac{J_n(h)}{h} e^{-ihy} dh$$

$$+i\omega\mu_0 \frac{C1_{\varepsilon,\mu,\chi}}{k^2} \frac{\sqrt{2\pi}}{2\pi} \sum_{m=0}^{N_m^x} \sum_{n=0}^{N_n^y} B_m^x B_n^y$$

$$\times \int_{-a}^{a} \int_{-l}^{l} \frac{T_m\left(\frac{x'}{a}\right)}{\sqrt{1-\left(\frac{x'}{a}\right)^2}} \frac{T_n\left(\frac{y'}{l}\right)}{\sqrt{1-\left(\frac{y'}{l}\right)^2}} \frac{(y'-y)}{\left[(x'-x)^2+(y'-y)^2\right]^{\frac{3}{2}}} dx'dy'$$

$$-C2_{\varepsilon,\mu,\chi} \frac{i\omega\mu_0}{2\pi} \sum_{m=0}^{N_m^x} \sum_{n=0}^{N_n^y} B_m^x B_n^y \int_{-a}^{a} \int_{-l}^{l} \frac{T_m\left(\frac{x'}{a}\right)}{\sqrt{1-\left(\frac{x'}{a}\right)^2}} \frac{T_n\left(\frac{y'}{l}\right)}{\sqrt{1-\left(\frac{y'}{l}\right)^2}} \operatorname{arcsh}\left(\frac{y'-y}{|x'-x|}\right) dx'dy'. \tag{6}$$

To calculate the hypersingular integral with a third-degree singularity in HSIR (6), the discrete vortex method was used in the same way [27–29]. After applying this method and taking into account the boundary conditions (1), we obtain a hypersingular integral equation:

$$
-E_y^{ext}(x,y) = -\sum_{m=0}^{N_m^x}\sum_{n=0}^{N_n^y} B_m^x B_n^y i^{m+n-1}
$$

$$
\times \int_0^{h_{max}}\left[\int_0^{\beta_{max}}\left[Z_{11}(\beta,h)-Z_{11}^\infty(\beta,h)\right]J_m(\beta)e^{-i\beta x}d\beta\right]\frac{J_n(h)}{h}e^{-ihy}dh
$$

$$
+i\omega\mu_0\frac{C1_{\varepsilon,\mu,\chi}}{k^2}\frac{\sqrt{2\pi}}{2\pi}\sum_{m=0}^{N_m^x}\sum_{n=0}^{N_n^y} B_m^x B_n^y \sum_{k=0}^{N_k^{x'}}\sum_{p=0}^{N_p^{y'}}\frac{T_m(t_{0k}/a)}{\sqrt{1-(t_{0k}/a)^2}}\frac{T_n(\xi_{0p}/l)}{\sqrt{1-(\xi_{0p}/l)^2}}
$$

$$
\times \ln\left(x'-x+\sqrt{(x'-x)^2+(y'-y)^2}\right)\Big|_{t_{0k}}^{t_{0k+1}}\Big|_{\xi_{0p}}^{\xi_{0p+1}} \tag{7}
$$

$$
-i\omega\mu_0 C2_{\varepsilon,\mu,\chi}\frac{1}{2\pi}\sum_{m=0}^{N_m^x}\sum_{n=0}^{N_n^y}B_m^x B_n^y
$$

$$
\times \int_{-a}^{a}\int_{-l}^{l}\frac{T_m\left(\dfrac{x'}{a}\right)}{\sqrt{1-\left(\dfrac{x'}{a}\right)^2}}\frac{T_n\left(\dfrac{y'}{l}\right)}{\sqrt{1-\left(\dfrac{y'}{l}\right)^2}}\,\mathrm{arcsh}\left(\frac{y'-y}{|x'-x|}\right)dx'dy'.
$$

To calculate the current density distribution function, the collocation method was used, where Gaussian nodes (zeros of Legendre polynomials) were used as collocation points. This approach allows for faster convergence.

We introduce the following notation:

$$
A_j^x = -E_y^{ext}(x_j),
$$
$$
A_q^y = -E_y^{ext}(y_q),
$$

$$C_{j,q} = -\sum_{m=0}^{N_m^x} \sum_{n=0}^{N_n^y} i^{m+n-1}$$

$$\times \int_0^{h_{max}} \left[\int_0^{\beta_{max}} \left[Z_{11}(\beta,h) - Z_{11}^\infty(\beta,h) \right] J_m(\beta) e^{-i\beta x_j} d\beta \right] \frac{J_n(h)}{h} e^{-ihy_q} dh$$

$$+ i\omega\mu_0 \frac{C1_{\varepsilon,\mu,\chi}}{k^2} \frac{\sqrt{2\pi}}{2\pi} \sum_{m=0}^{N_m^x} \sum_{n=0}^{N_n^y} \sum_{k=0}^{N_k^{x'}} \sum_{p=0}^{N_p} \frac{T_m(t_{0k}/a)}{\sqrt{1-(t_{0k}/a)^2}} \frac{T_n(\xi_{0p}/l)}{\sqrt{1-(\xi_{0p}/l)^2}}$$

$$\times \ln\left(x'-x_j + \sqrt{(x'-x_j)^2 + (y'-y_q)^2} \right)\Bigg|_{t_{0k}}^{t_{0k+1}} \Bigg|_{\xi_{0p}}^{\xi_{0p+1}}$$

$$-i\omega\mu_0 C2_{\varepsilon,\mu,\chi} \frac{1}{2\pi} \sum_{m=0}^{N_m^x} \sum_{n=0}^{N_n^y} \int_{-a}^{a} \int_{-l}^{l} \frac{T_m\left(\frac{x'}{a}\right)}{\sqrt{1-\left(\frac{x'}{a}\right)^2}} \frac{T_n\left(\frac{y'}{l}\right)}{\sqrt{1-\left(\frac{y'}{l}\right)^2}} \operatorname{arcsh}\left(\frac{y'-y_q}{|x'-x_j|}\right) dx'dy',$$

where $x_j = a\zeta_j$, $j = \overline{0\ldots N_x}$, $y_q = l\zeta_q$, $q = \overline{0\ldots N_y}$, ζ_j, ζ_q are Gaussian nodes.

Unknown coefficients B^x and B^y are determined from the solution of the following system of linear algebraic equations (SLAE):

$$[C^{x,y}][B^{x,y}] = [A^{x,y}] \tag{8}$$

Some direct and iterative methods were used to solve this problem, in particular, direct methods of LU (LUP) decompositions.

Due to the fact that the expression (5) is a derivative of the current density distribution function, to determine the function itself, it is necessary to calculate the integral of this expression for y:

$$\eta_y(x',y') = \sum_{j=0}^{N_j^x} \sum_{q=1}^{N_q^y} B_j^x \frac{T_j(x'/a)}{\sqrt{1-(x'/a)^2}} \frac{B_{q-1}^y}{q} U_{q-1}(y'/l)\sqrt{1-(y'/l)^2}, \tag{9}$$

where $U_{q-1}(y')\sqrt{1-(y')^2}$ are Chebyshev polynomials of the 2-nd kind.

5   Results and Discussion

Figure 2 presents the results of calculating the real and imaginary parts (a) of the current density distribution function and its module (b), along the center of the radiator $(x = 0)$, length $l = 0.25\lambda$ and width $a = 0.025\lambda$, located on the dielectric substrate

$(\chi = 0)$, using the HSIE method, the method of singular integral equations (SIE) [7–9] and the Feko software package.

It should be noted that the singular integral equation (SIE) obtained using SIE was solved in the same way as in HSIE by the collocation method. Moreover, the same number of collocation points was selected in both methods. Figure 2 presents a relatively small discrepancy between the imaginary part of the current density distribution function and the results obtained in Feko. In [7–9], it is shown that in the case of solving the SIE by partial operator inversion, the results of calculating the current distribution density function coincide. In this regard, if the number of collocation points increases further, these functions will match. However, the HSIE method becomes much more demanding on computing resources. One approach to solving this problem is to use the method of moments [30] to solve Eq. (7).

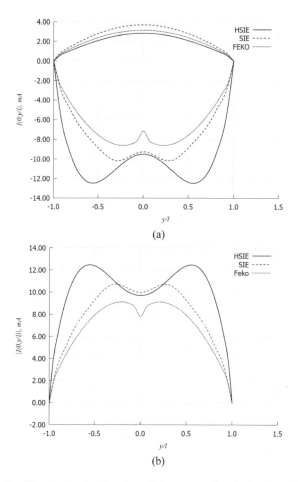

Fig. 2. The distribution function of the current density in the emitter

The results of calculating the current density distribution function with the chirality parameter equal to zero coincide with the results obtained for a microstrip antenna with a dielectric substrate [31], which confirms the adequacy of the developed method.

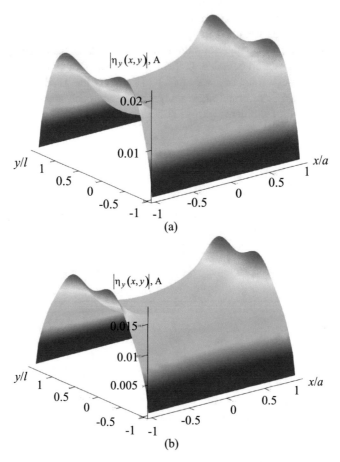

Fig. 3. Current density distribution functions over the emitter of a structure with an isotropic chiral substrate

Figure 3a and 3b shows graphs of current density distribution functions for the emitter located on an isotropic chiral substrate $(\chi = 0.1)$ obtained by using HSIE and SIE methods, respectively.

In the case of SIE, since a narrow emitter was used $(2a \ll l, \lambda)$, the quasi-static approximation [7–9] was used to describe the current density distribution function, which assumes a uniform distribution of the transverse variation of the longitudinal component of the current density:

$$\eta(x, y) = \frac{f(y)}{\sqrt{1 - (x/a)^2}} \tag{10}$$

where $f(y')$ is function describing the longitudinal distribution of current density.

As you can see, these functions coincide quite accurately, which in turn indicates the validity of using the quasi-static approximation for narrow emitters located on substrates of chiral metamaterials. In this case, there is a greater gain in computational efficiency compared to HSIE.

6 Conclusions

Thus, the article presents a method for electrodynamic analysis of radiating structures based on chiral metamaterials, which has high computational efficiency and allows correct calculation of their characteristics. The results of calculating the current distribution function over the radiating structure were verified by comparing them with the results obtained in the Feko electrodynamic modeling software package and with the results of other authors. Based on the calculation of the current distribution functions over the radiating structure with a chiral metamaterial substrate, it was concluded that in the case of narrow emitters, it is advisable to using by the method of singular SIE equations and in the case of wider ones to using by the HSIE. The obtained results can be used in computer-aided design tools for the development of new-generation antennas used in modern communication and telecommunications systems, including medical monitoring systems. Further research is related to the development of mathematical models of other types of antennas and antenna arrays based on chiral metamaterials, which make it possible to correctly calculate their characteristics.

Acknowledgments. One of the authors (A.A. Potapov) thanks the Russian Foundation for Basic Research (project no. 18-08-01356-a) for the support, as well as the project "Leading Talents", no. 00201502 (2016–2020) at JiNan University (Guangzhou, China).

References

1. Engheta, N.: The theory of chirostrip antennas. Proceedings of the 1988 URSI International Radio Science Symposium Syracuse, New York, p. 213 (1988)
2. Engheta, N., Pelet, P.: Modes in chirowaveguides. Opt. Lett. **14**(11), 593–595 (1989)
3. Toscano, A., Vegni, L.: Evaluation of the resonant frequencies and bandwidth in microstrip antennas with a chiral grounded slab. Int. J. Electron. **81**(6), 671–676 (1996)
4. Zebiri, C., Benabdelaziz, F., Lashab, M.: Bianisotropic superstrate effect on rectangular microstrip patch antenna parameters. In: META, January 2012 (2012)

5. Zebiri, C., Lashab, M., Benabdelaziz, F.: Asymmetrical effects of bi-anisotropic substrate-superstrate sandwich structure on patch resonator. Progress Electromag. Res. B **49**, 319–337 (2013)
6. Zebiri, C., Daoudi, S., Benabdelaziz, F., Lashab, M., Sayad, D., Ali, N.T., Abd-Alhameed, R.A.: Gyro-chirality effect of bianisotropic substrate on the operational of rectangular microstrip patch antenna. Int. J. Appl. Electromagn. Mech. **51**, 249–260 (2016)
7. Klyuev, D.S., Neshcheret, A.M., Osipov, O.V., Potapov, A.A., Sokolova, Yu.V.: The method of singular integral equations in the theory of microstrip antennas based on chiral metamaterials. In: 12th Chaotic Modeling and Simulation International Conference, Springer Proceedings in Complexity, pp. 267–294 (2020). https://doi.org/10.1007/978-3-030-39515-5_22
8. Abramov, V.Y., Klyuev, D.S., Neshcheret, A.M., Osipov, O.V., Potapov, A.A.: Input impedance of a microstrip antenna with a chiral substrate based on left-handed spirals. J. Eng. JOE **19**, 6218–6221 (2019)
9. Buzov, A.L., Buzova, M.A., Klyuev, D.S., Mishin, D.V., Neshcheret, A.M.: Calculating the input impedance of a microstrip antenna with a substrate of a chiral metamaterial. J. Commun. Tech. Electron. **63**(11), 1259–1264 (2018)
10. Ni, J.: Analysis of Shielded and Open Microstrip Lines of Double Negative Metamaterials Using Spectral Domain Approach (SDA), p. 65. Iowa State University, Ames (2008)
11. Dwivedi, S., Mishra, V., Kosta, Y.P.: Directivity enhancement of miniaturized microstrip patch antenna using metamaterial cover. Int. J. Appl. Electromagn. Mech. **47**(2), 399–409 (2015)
12. Kenari, M.A.: Printed planar patch antennas based on metamaterial. International Journal of Electronics Letters **2**(1), 37–42 (2014)
13. Deepak, D., Kaur, J.: Dual band high directivity microstrip patch antenna rotatedstepped-impedance array loaded with CSRRs for WLAN applications. Int. J. Wireless Microwave Technol. **4**, 1–11 (2016). http://www.mecs-press.net. https://doi.org/10.5815/ijwmt.2016.04.01. Accessed July 2016 in MECS
14. Kaur, P., Aggarwal, S.K., De, A.: Design and investigation of circularly polarized RMPA with chiral metamaterial cover. Int. J. Wireless Microwave Technol. **3**, 61–70 (2016). http://www.mecs-press.net. https://doi.org/10.5815/ijwmt.2016.03.07. Accessed May 2016 in MECS
15. Shareef, A.N., Shaalan, A.B.: Fractal peano antenna covered by two layers of modified ring resonator. Int. J. Wireless Microwave Technol. **5**(2), 1–11 (2015). http://www.mecs-press.net. https://doi.org/10.5815/ijwmt.2015.02.01. Accessed Apr 2015 in MECS
16. Saleh, A.A., Abdullah, A.S.: A novel design of patch antenna loaded with complementary split-ring resonator and L-shape slot for (WiMAX/WLAN) applications. Int. J. Wireless Microwave Technol. **4**(3), 16–25 (2014). http://www.mecs-press.net. https://doi.org/10.5815/ijwmt.2014.03.02. Accessed Oct 2014 in MECS
17. Caloz, C., Itoh, T., Rennings, A.: CRLH metamaterial leaky-wave and resonant antennas. IEEE Antennas Propag. Mag. **50**(5), 25–39 (2008)
18. Erentok, A., Ziolkowski, R.W.: Metamaterial-inspired efficient electrically small antennas. IEEE Trans. Antennas Propag. **56**(3), 691–707 (2008)
19. Mao, S.-G., Chen, C.-M., Chang, D.-C.: Modeling of slow-wave EBG structure for printed-bowtie antenna array. IEEE Antennas Wireless Propag. Lett. **1**, 124–127 (2002)
20. Alibakhshikenari, M., Khalily, M., Virdee, B.S., See, C.H., Abd-Alhameed, R., Limiti, E.: Mutual coupling suppression between two closely placed microstrip patches using EM-bandgap metamaterial fractal loading. IEEE Access (2019). https://doi.org/10.1109/ACCESS.2019.2899326

21. Pozar, D.M.: Radiation and scattering from a microstrip patch on a uniaxial substrate. IEEE Trans. Antennas Propagat. **35**(6), 613–621 (1987)
22. Kolmakova, N., Prikolotin, S., Kirilenko, A., Perov, A.: Simple example of polarization plane rotation by the fringing fields interaction. In: IEEE European Microwave Conference (EuMC), pp. 936–938 (2013)
23. Caloz, C., Sihvola, A.: Electromagnetic chirality, part 1: the macroscopic perspective. IEEE Antennas Propagat. Mag. **62**(1), 58–71 (2020)
24. Caloz, C., Sihvola, A.: Electromagnetic chirality, part 2: the macroscopic perspective. IEEE Antennas Propagat. Mag. **62**(2), 58–71 (2020). https://doi.org/10.1109/MAP.2020.2969265
25. Capolino, F.: Theory and Phenomena of Metamaterials, p. 992. Taylor & Francis/CRC Press, Boca Raton (2009)
26. Prudnikov, A.P., Brychkov, Y.A., Marichev, O.I.: Integrals and series of elementary functions. Math. Comput. **40**(161), 413–414 (1983)
27. Lifanov, I.K.: Singular Integral Equations and Discrete Vortices, p. 475. VSP, Utrecht (1996)
28. Setukha, A.V., Bezobrazova, E.N.: The method of hypersingular integral equations in the problem of electromagnetic wave diffraction by a dielectric body with a partial perfectly conducting coating. Russ. J. Numer. Anal. Math. Model. **32**(6), 371–380 (2017)
29. Saffman, P.G.: Vortex Dynamics, p. 311. Cambridge University Press, Cambridge (1992)
30. Gibson Walton, C.: The Method of Moments in Electromagnetics, p. 448. CRC Press/Taylor & Francis Group, Boca Raton (2015)
31. Klyuev, D.S., Sokolova, Y.V.: A hypersingular integral equation for current density on the surface of a microstrip vibrator. Tech. Phys. Lett. **40**(2), 112–115 (2014). https://doi.org/10.1134/S1063785014020102

Influence of Image Pre-processing Algorithms on Segmentation Results by Method of Persistence Homology

Sergey Eremeev and Semyon Romanov$^{(\boxtimes)}$

Vladimir State University, Vladimir, Russia
sv-eremeev@yandex.ru, cwwc@bk.ru

Abstract. In this paper, the problem of image segmentation for detecting defects is considered. It is proposed to use the method of persistent homology for image segmentation using topological features. The influence of various image preprocessing methods on the result of segmentation by persistent homology is investigated. A mathematical model of segmentation is demonstrated. Examples of the algorithm for detecting defects in wood images are shown. The results of segmentation and comparison tables of characteristics depending on preprocessing methods are presented. The advantage of using the persistent homology method with Gauss filtering for detecting wood defects is demonstrated.

Keywords: Topological analysis · Defect detection · Persistent homology · Segmentation

1 Introduction

Numerous studies of the application of persistent homology in various fields have shown their effectiveness [1–6]. Moreover, in one of our previous works we used the method to solve the problem of metal defect detection in an image [7]. The method of persistent homology allows us to get the history of formation of image segments. We can identify different types of material defects in one processing cycle. In [7] we describe the advantage of the persistence homology method over other methods of segmentation and image processing. We expect to further improve the algorithm by combining it with other image pre-processing methods.

The aim of the work is to modernize the method of persistent homology by applying various methods of image pre-processing to improve the quality of segmentation and defect detection.

The algorithm will be designed to increase product quality in enterprises with continuous production in the metallurgy and woodworking industries.

© The Author(s), under exclusive license to Springer Nature Switzerland AG 2021
Z. Hu et al. (Eds.): AIMEE 2020, AISC 1315, pp. 114–123, 2021.
https://doi.org/10.1007/978-3-030-67133-4_11

2   Related Work

Currently, there are approaches for solving segmentation problems based on derived functions, binaryization, Sobel and Previtt filters or using RGB component. But all these methods have different parameters such as threshold values or number of clusters [8–11]. Therefore, the use of these methods is difficult under conditions of constant environmental variability such as changes in lighting and material type [12–14]. In addition, the use of parameters may require additional running of the algorithm to detect another type of defect. It increases the number of resources and sometimes makes the system unusable in real time due to high time delays.

There are many different segmentation methods aimed at selecting zones of interest. For example, the authors in [9] suggest using the k-means method. The algorithm is a version of the EM algorithm used to separate a mixture of Gaussian. It divides the set of elements of a vector space into a known number of clusters k. However, we do not know the number of clusters that the image will contain, because the image may include a combination of several defects. Application of the method at a sufficiently high number of clusters may have an interesting effect in its further processing by the method of persistent homology.

Work [9] analyzes the application of different filters to track an object, the analysis also includes the application of the filter MeanShift. MeanShift is a procedure for determining the location of the maximum probability density set by a discrete sample. In [10] the entertaining effect of Gauss filter application is considered. At certain dispersion values, pre-processing significantly accelerated the image recovery process, which may be useful in the segmentation problem. The method of persistence homology is based on the distance between neighboring pixels, and the Gaussian filter leads to alignment of neighboring values.

3   Methodology

Persistent homology processes a group of pixels combined in color characteristics and called holes. To combine pixels into groups, the image is represented as $P = \{p_1, p_2, p_3, \ldots, p_n\}$, where n is the number of pixels in the image. Each element of the set contains information about the pixel position, its brightness and information about its belonging to a certain hole. Let $c = (r, g, b)$ be a triple of pixel color channels. A list of relations of all pixels is formed and denoted as $R = \{r_1, r_2, r_3, \ldots, r_m\}$, where m is the total number of connections. Each element of the set contains image points between which a connection is established. Let l be a distance between points which is calculated using the expression (1) and depends on the brightness value of the color channels of the selected points:

$$l = avg(|c_1 - c_2| + |c_1 - c_2| - avg(|c_1 - c_2|)) \tag{1}$$

The function avg means calculating the average value.

This approach allows us to select areas with changing color intensity such as object boundaries. Relations are established only between neighboring points located on one of the axes. Relations between diagonal neighbors are not established. Thus, we can say that each point, depending on its location, will have from two to four connections.

The list of relations should be sorted by increasing lengths of distances between points. The visual representation of all relations is shown in Fig. 1a. The drawing is an enlarged image, where each square is a pixel. White lines represent the relations between the lines, and a bold white line indicates a relation to the lowest length value.

The algorithm processes elements from the list of relations by merging pixels into groups, i.e. holes. The hole h has the following structure:

$$h = (n, Ls, Ld) \tag{2}$$

where

n is hole number,

Ls is iteration of the algorithm (distance) at which the hole appeared,

Ld is iteration at which the hole ceased to exist.

The new color value for the hole is calculated, which is also set for all pixels belonging to the hole. The color value of the resulting hole is calculated using the expression: $c = (c_1 + c_2)$. Figure 1 shows the process of hole formation.

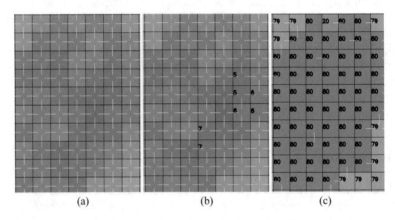

Fig. 1. Hole formation: pattern of connections (a); formation of new holes (b); result of transformation (c)

In the case of processing the relation that connects two holes the length of the distance between these objects is recalculated by expression (1). Then the connection is moved through the list of relations according to its new length. Two holes are combined if the length of the relation between the holes cannot be moved to a new position in the list according to the rules of sorting. A new hole with a number is formed, and the color of the hole will be redefined by expression:

$$c = c_1 * d_1 + c_2 * d_2, \tag{2}$$

where

d_1 is the ratio of the total number of points of two holes to the number of points of the first hole,

d_2 is the ratio of the total number of points of two holes to the number of points of the second hole.

To improve efficiency before segmenting an image, various pre-processing methods such as Mean-shift, Gauss Filtering, kMeans and removal of range colors will be applied.

4 Experiments

The developed algorithm will allow us to segment images with defects preserving the entire history of transformations. Thus, we can say that the result is segmentation containing all variations when the brightness threshold value changes. The software provides visualization with selection of the current iteration which is equivalent to the brightness threshold value. It is also possible to select the hole area in the current iteration (Fig. 2).

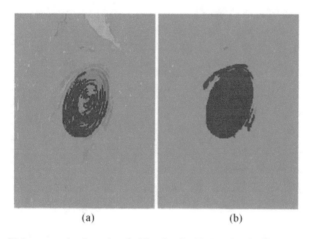

(a)   (b)

Fig. 2. Hole area selection: threshold value is 10 (a); threshold value is 26 (b)

Experiments have shown that the ratio of hole sizes and their color values can be a feature of defect detection. The selected parameters will differ significantly from the defect type and material.

For experiments we used images of wood with and without defects (Fig. 3).

To find the threshold at which a defect appears, we use the numerical characteristic. It shows the number of holes of a certain area in the image for different thresholds (see Table 1). In the table, we use the following features. These are the threshold and the size of the hole. The threshold displays the value of the distance between pixels. For example, all pixels with a distance less than 22 will be combined into holes. The size of the hole is the ratio of the area of the hole to the total area of the image. Thus, we have a table showing the number of holes of some area for different thresholds.

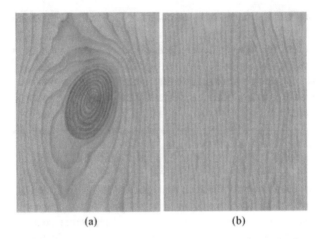

Fig. 3. Image of wood with a defect (a) and without a defect (b)

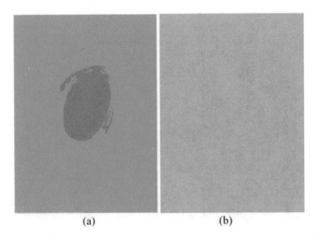

Fig. 4. Segmented images of wood with defect (a) and without defect (b)

Figure 4 shows segmented variants of images from Fig. 3 with a threshold value of brightness equal to 26.

Table 1 clearly shows that the total number of holes decreases with increasing threshold value, but the area of remaining holes increases. Segmentation of the image by methods of persistent homology without preprocessing showed that there is a sufficient number of small noises. Therefore, it is necessary to study the effect of pre-processing on the algorithm. Table 2 shows the results of segmentation after prepro-cessing of the source image by the Gauss filter with a convolution kernel equal to 3. Table 3 shows the results of segmentation after processing the source image with the Mean-shift algorithm, where the spatial mask radius is 20, the color mask radius is 25, and the maximum level of the segmentation pyramid is 3. Table 4 shows the results of segmentation with preliminary removal of colors in the range: red from 139 to 256, green from 144 to 215, blue from 62 to 173. Table 5 shows the results of segmenting after preliminary processing of the source image by kMeans algorithm with the number of segments equal to 3.

Table 1. Number of holes in the area ratio ranges for the specified threshold value without preliminary processing

Threshold	Number of holes at the percentage of hole area to total area					
	0–0,05	0,05–0,5	0,5–1	1–9	9–20	50–100
3	698	28	0	1	0	0
8	597	28	2	2	0	1
15	159	10	0	3	0	1
22	38	3	0	0	1	1
26	16	1	0	0	1	1
40	0	0	0	0	0	1

Table 2. Number of holes in the area ratio ranges for the specified threshold value with Gaussian pre-filtration

Threshold	Number of holes at the percentage of hole area to total area					
	0–0,05	0,05–0,5	0,5–1	1–9	9–20	50–100
3	487	24	1	1	0	1
8	131	22	1	2	0	1
15	18	7	2	1	0	1
22	4	3	0	0	1	1
31	0	0	0	0	1	1
40	0	0	0	0	0	1

Table 3. Number of holes in the area ratio ranges for the specified threshold value with Mean-shift preprocessing

Threshold	Number of holes at the percentage of hole area to total area					
	0–0,05	0,05–0,5	0,5–1	1–9	9–20	50–100
3	47	14	0	1	0	1
8	23	5	0	0	1	1
15	5	2	0	0	1	1
22	3	1	0	0	1	1
31	3	0	0	0	1	1
40	0	0	0	0	0	1

Table 4. Number of holes in the area ratio ranges for the specified threshold value with preliminary removal of the color range

Threshold	Number of holes at the percentage of hole area to total area					
	0–0,05	0,05–0,5	0,5–1	1–9	9–20	50–100
3	551	32	0	0	0	1
8	467	38	0	1	0	1
15	203	14	1	1	0	1
22	158	12	0	0	1	1
31	153	11	0	0	1	1
49	46	0	0	0	1	1
50	0	0	0	0	0	1

Table 5. Number of holes in the area ratio ranges for the specified threshold value with kMeans pre-processing

Threshold	Number of holes at the percentage of hole area to total area					
	0–0,05	0,05–0,5	0,5–1	1–9	9–20	50–100
3	480	62	4	3	2	0
8	480	62	4	2	3	0
15	118	7	0	1	0	1
22	116	7	0	0	1	1
31	60	6	0	0	1	1
40	0	0	0	0	0	1

5   Discussion

Tables 2, 3, 4, and 5 clearly show that in most cases the defect is identified at 31 iterations of the algorithm. At this iteration, the amount of noise is close to zero and 2 main holes are clearly visible. They are a hole representing the main material, i.e. wood without defects, which occupies about 80% of the total area, and a hole with defect, which occupies 9–20% of the total area. Processing with color range removal has no smoothing effects. The iteration on which the defect becomes fully visible is 49. There is also a lot of noise for this method (Table 4), which indicates that many details of the image are preserved. Figure 5 shows images of materials at preprocessing and at key iteration of segmentation.

Although the Gaussian filter makes minimal visible changes to the original image in Fig. 5a, it gives the best visible result in Fig. 5b. This fact is also confirmed by the data in Table 2, which shows that at 31 iterations of the algorithm there are only 2 holes, one of which indicates the area of the defect and the other shows the area of the material. The obtained effect is explained by a general decrease in the distance between neighboring pixels. Thus, in the early stages, little distinguishable material details can be combined with a common background. However, this approach may have a negative effect on detecting defects whose color is close to the color of the base material.

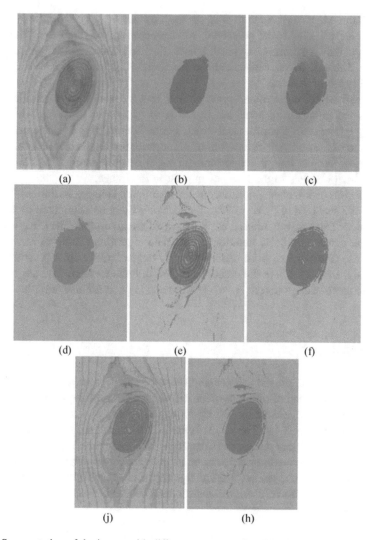

Fig. 5. Segmentation of the image with different preprocessing: blurring of Gaussian (a), where the convolution dimension is 3, iteration of the main algorithm is 0; blurring of Gaussian, where the convolution dimension is 3, iteration of the main algorithm is 31 (b); Mean-Shift, where the spatial mask radius is 20, color mask radius is 25, maximum level of the segmentation pyramid is 3, iteration of the main algorithm is 0 (c); Mean-Shift, where the spatial mask radius is 20, the color mask radius is 25, the maximum level of the segmentation pyramid is 3, iteration of the main algorithm is 31 (d); removal of background color in the range red from 139 to 256, green from 144 to 215, blue from 62 to 173, iteration of the main algorithm is 0 (e); removal of background color in the range red from 139 to 256, green from 144 to 215, blue from 62 to 173, iteration of the main algorithm is 49 (f); k Means, where the number of clusters is 3, iteration of the main algorithm is 0 (j); kMeans, where the number of clusters is 3, iteration of the main algorithm is 31 (h)

Using the Mean-shift method distorts the original image quite strongly, but the areas of interest in Fig. 5c are preserved. After processing with the basic algorithm, we get an image with the presence of small noises (Fig. 5d), which is also confirmed by the data from Table 3. These noises can be discarded, but can also be used to detect additional types of defects.

The method of background removal requires manual installation, but largely eliminates the influence of areas not in the zone of interest (Fig. 5e). The result of processing by the main algorithm is a clearly defined defect (Fig. 5f). The main noise is the structure of the defect. The iteration of the algorithm with the selected defect area is 49 (Table 4). Despite this, the algorithm seems to be quite interesting, since it introduces minimal changes to the areas of interest in the image.

Application of k-means algorithm makes significant changes in the original image (Fig. 5j). We see a lot of noise in Fig. 5h. This effect can be eliminated by increasing the number of clusters, but this is a weak feature of the algorithm. The k-means method denotes the defect area at 31 iteration of the main algorithm (Table 5).

Thus, the Gauss filter can be used to identify defects whose color parameters differ significantly. In cases where the defect has small differences from the base material, it is better to use a background removal algorithm. It is good for analyzing low-visibility parts.

6   Conclusions

The algorithm for building a segmentation history has been developed in the article. The results of influence of different preprocessing methods on its work are given. The algorithm creates a segmentation history for each iteration by establishing topological relations between segments. The result is the construction and analysis of a segmentation history to sift out noise and identify stable segments that are defects. The topological segmentation algorithm can be used to analyze different types of materials for defect detection. It does not require significant changes for use in other areas of production, and can also be used for medical tasks. For example, it can detect inflamed areas or cancers.

References

1. Kurlin, V.: A fast persistence-based segmentation of noisy 2D clouds with provable guarantees. Pattern Recogn. Lett. **83**, 3–12 (2016)
2. Eremeev, S.: Analysis of in changes in topological relations between spatial objects at different times. In: Hu, Z., Petoukhov, S., He, M. (eds) Advances Artificial Systems for Medicine and Education III. AIMEE2019. Advances in Intelligent Systems and Computing, vol. 1126. Springer, Cham, pp. 1–11 (2020)
3. Eremeev, S., Romanov, S.: An algorithm for constructing a topological skeleton for semi-structured spatial data based on persistent homology. In: van der Aalst, W.M.P., Batagelj, V., Ignatov, D.I., Khachay, M., Kuskova, V., Kutuzov, A., Kuznetsov, S.O., Lomazova, I.A., Loukachevitch, N., Napoli, A., Pardalos, P.M., Pelillo, M., Savchenko, A.V., Tutubalina, E. (eds.) AIST 2019. CCIS, vol. 1086, pp. 16–26. Springer, Cham (2020)

4. Eremeev, S., Seltsova, E.: Algorithms for topological analysis of spatial data. In: Hu, Z., Petoukhov, S.V., He, M. (eds.) AIMEE2018 2018. AISC, vol. 902, pp. 81–92. Springer, Cham (2020)

5. Eremeev, S.V., Andrianov, D.E., Titov, V.S.: An algorithm for matching spatial objects of different-scale maps based on topological data analysis. Comput. Opt. **43**(6), 1021–1029 (2019)

6. Eremeev, S., Kuptsov, K., Romanov, S.: An approach to establishing the correspondence of spatial objects on heterogeneous maps based on methods of computational topology. In: van der Aalst, W.M.P., Ignatov, D.I., Khachay, M., Kuznetsov, S.O., Lempitsky, V., Lomazova, I.A., Loukachevitch, N., Napoli, A., Panchenko, A., Pardalos, P.M., Savchenko, A.V., Wasserman, S. (eds.) AIST 2017. LNCS, vol. 10716, pp. 172–182. Springer, Cham (2018)

7. Eremeev, S.V., Romanov, S.A.: Algorithm of image segmentation based on persistent homology for solving problems of searching for defects. Bull. South-West State Univ. **24**(1), 144–158 (2020)

8. Kama, R., Chinegaram, K., Tummala, R.B., Ganta, R.R.: Segmentation of soft tissues and tumors from biomedical images using optimized K-means clustering via level set formulation. Int. J. Intell. Syst. Appl. **11**, 18–28 (2019)

9. Jatoth, R., Gopisetty, S., Hussain, M.: Performance analysis of alpha beta filter, kalman filter and meanshift for object tracking in video sequences. Int. J. Image Graph. Signal Process. **7**, 24–30 (2015)

10. Goel, V., Raj, K.: Removal of image blurring and mix noises using gaussian mixture and variation models. Int. J. Image Graph. Signal Process. **10**, 47–55 (2018)

11. Hossain, A.B.M.A., Bhakta, R.: Lung tumor segmentation and staging from ct images using fast and robust fuzzy C-Means clustering. Int. J. Image Graph. Signal Process. **12**, 38–45 (2020)

12. Feng, Y., Zhao, H., Li, X., Zhang, X., Li, H.: A multi-scale 3D Otsu thresholding algorithm for medical image segmentation. Digit. Signal Proc. **60**, 186–199 (2017)

13. Berthon, B., Evans, M., Marshall, C., Palaniappan, N., Cole, N., Jayaprakasam, V., Rackley, T., Spezi, E.: Head and neck target delineation using a novel PET automatic segmentation algorithm. Radiother. Oncol. **122**, 242–247 (2017)

14. Shen, J., Hao, X., Liang, Z., Liu, Y., Wang, W., Shao, L.: Real-time superpixel segmentation by DBSCAN clustering algorithm. IEEE Trans. Image Process. **25**, 5933–5942 (2016)

Influence of Delays on Self-oscillations in System with Limited Power-Supply

Alishir A. Alifov[✉]

Mechanical Engineering Research Institute of Russian Academy of Sciences,
Moscow 101990, Russia
a.alifov@yandex.ru

Abstract. Self-oscillations under delays in elasticity and damping in a system with an energy source of limited power are considered. The dynamics of the system is described by nonlinear differential equations. The method of direct linearization of non-linearity is used for their solution. It has a number of advantages over the known methods of analysis of nonlinear systems: labor and time costs are reduced by several orders of magnitude; regardless of the specific type and degree of nonlinearity, you can get the final calculated ratios; there are no labor-intensive and complex approximations of various orders inherent in known methods. On the basis of this method, the equations of non-stationary and stationary movements are obtained, and the conditions for stability of stationary oscillations are derived. An analysis was performed to obtain information about the influence of delays on the parameters of stationary oscillations. The influence of delays on the amplitude of oscillations and the load on the energy source is shown. It turned out that depending on the delay value, the amplitude and load curves shift in the speed range of the energy source.

Keywords: Self-oscillations · Method · Non-linearity · Direct linearization · Limited power-supply · Stability · Elasticity · Damping · Delay

1 Introduction

The functioning of a variety of technical systems for various purposes is often accompanied by the occurrence of frictional self-oscillations. Among these systems, we can note, for example, textile equipment, guides of metal-cutting machines, brakes, and a number of other objects [1, 2]. Self-oscillation often has a negative impact on the functioning of a technical object. Therefore, their study, in addition to scientific interest, is of important practical interest.

Imperfection of elastic properties of materials and internal friction in them cause lag in mechanical systems [3–6]. The presence of a delay can be useful or harmful, lead to fluctuations in the object (in tracking systems, belt conveyors, regulators, rolling mills, etc.). The delay is widespread in transportation, electronics and radio engineering, non-ferrous metallurgy, paper and glass production, automatic control systems with logic devices, etc. Lag has a great impact on the stability and regulation of the system. Therefore, its influence, in contrast to low-frequency systems, must be taken into

Z. Hu et al. (Eds.): AIMEE 2020, AISC 1315, pp. 124–132, 2021.
https://doi.org/10.1007/978-3-030-67133-4_12

account in high-frequency systems. Approximate methods are used to study systems with delay, in particular, the well-founded asymptotic averaging method [7].

As you know, energy issues (economy, consumption, and environmental impact) have come to the fore all over the world. The connection of environmental problems with the level of energy consumption, accuracy of system calculation models, metrology, and accuracy of parts processing is shown in [8]. In this regard, the theory of interaction between an energy source and an oscillatory system, consistently created by V.O. Kononenko and published in her well-known monograph [9], becomes relevant. As a result of V.O. Kononenko's work, a new direction in the theory of vibrations has appeared, and its development has been reflected in many works of a wide range of researchers around the world, including in the book [10]. The application of this theory can make a certain contribution to the solution of the noted problems related to the use of energy.

As noted in [11–15], one of the main problems of nonlinear dynamics of systems is the large labor costs in the analysis of coupled oscillator networks (it plays an important role in biology, chemistry, physics, electronics, neural networks, etc.). For analysis, a number of methods are used, described in many works, including, for example, [7, 16–23]. The method of direct linearization (MDL) described in [24–29] and other works of the author differs from these methods in principle. The properties of MDL (ease of use, lack of labor-intensive and complex approximations of various orders, the ability to obtain final calculation ratios regardless of the specific type and degree of nonlinearity, several orders of magnitude less than when using known methods of nonlinear mechanics, labor and time costs) determine its advantage in comparison with known methods of calculating nonlinear systems. The purpose of this work is to consider on the basis of MDL self-oscillations with delays in elasticity and damping in the case of an imperfect energy source.

2 Model, Equations of Motion

Frictional self-oscillations are described fairly well by the model shown in Fig. 1. Body *1* with mass m lies on the belt, which is driven by an engine having a torque characteristic $M(\dot{\phi})$, where $\dot{\phi}$ is the speed of rotation of the motor rotor. The friction force $T(U)$ arising between the body *1* and the tape depends on the relative speed $U = V - \dot{x}$, $V = r_0\dot{\phi}$. Under certain conditions, it can cause self-oscillations of body *1*. The elastic and damping forces depend on the delay, i.e. respectively, $c_1 x_\tau$ and $k_0 \dot{x}_\Delta$. Here $x_\tau = x(t - \tau)$, $\dot{x}_\Delta = \dot{x}(t - \Delta)$, where τ and Δ are constant time delay factors.

Nonlinear differential equations describing the system's motion have the form

$$m\ddot{x} + k_0\dot{x}_\Delta + c_0 x = T(U) - c_1 x_\tau,$$
$$I\ddot{\phi} = M(\dot{\phi}) - r_0 T(U) \tag{1}$$

where the values m, k_0, c_0, λ, v, I, r_0 are constant, c_0 is the spring stiffness coefficient 2, k_0 is the drag coefficient of the damper 3.

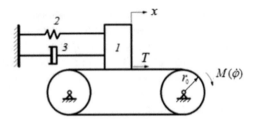

Fig. 1. System model

Let's represent the characteristic of the force $T(U)$ as a function

$$T(U) = T_0(\text{sgn } U - \alpha_1 U + \alpha_3 U^3) \tag{2}$$

Here T_0 is the normal reaction force, α_1 and α_3 are positive constants, $\text{sgn } U = 1$ at $U > 0$ and $\text{sgn } U = -1$ at $U < 0$. Form (2) is widely used in practice. It was also observed when considering the problem of measuring friction forces in space conditions [30].

3 Solving Equations by Direct Linearization

Using the direct linearization method [24], we replace the $f(\dot{x}) = -\alpha_1 U + \alpha_3 U^3$ part in function (2) with the function

$$f_*(\dot{x}) = B_f + k_f \, \dot{x} \tag{3}$$

where $B_f = -\alpha_1 V + \alpha_3 V^3 + 3\alpha_3 N_2 V v^2$, $k_f = \alpha_1 - 3\alpha_3 V^2 - \alpha_3 \bar{N}_3 v^2$.

Here is $v = \max |\dot{x}|$, $N_2 = (2r+1)/(2r+3)$, $\bar{N}_3 = (2r+3)/(2r+5)$. The symbol r is a *linearization accuracy parameter* and the choice of the interval of its value r is not limited. As shown in [24], it can be selected in the interval $(0, 2)$.

Equations (1) based on (3) take the form

$$m\ddot{x} + k_0\dot{x}_\Delta + c_0 x = T_0 \left(\text{sgn } U + B_f + k_f\dot{x}\right) - c_1 x_\tau \\ I\ddot{\varphi} = M(\dot{\varphi}) - r_0 T_0 \left(\text{sgn } U + B_f + k_f\dot{x}\right) \tag{4}$$

Let's consider solutions (4), for which we use the method of replacing variables with averaging [24] using the procedure for calculating oscillatory systems with non-ideal energy sources [29]. Since the character of solutions for x and $\dot{x}$ under $U > 0$ and $U < 0$ are fundamentally different [10], we will consider the solution for $u \geq v$ and $u < v$ separately.

The desired solutions, taking into account $\dot{x}_\Delta = -ap \sin (\psi - p\Delta)$, $x_\tau = a \cos (\psi - p\tau)$, have the form

$$x = a \cos \psi, \quad \dot{x} = -ap \sin \psi, \quad \psi = pt + \xi, \quad \dot{\varphi} = \Omega$$

where $\upsilon = ap$ comes from.

The solutions for determining the amplitude a, the phase ξ of the oscillations, and the velocity of the energy source Ω at $u \geq \upsilon$ are as follows

$$\frac{da}{dt} = -\frac{aA}{2pm}, \quad \frac{d\xi}{dt} = \frac{D}{2pm}$$

$$\frac{du}{dt} = \frac{r_0}{I} \left[M(\frac{u}{r_0}) - r_0 T_0 (1 + B_f) \right]$$

$$(5)$$

where $\quad A = p(k_0 \cos p\Delta - T_0 k_f) - c_1 \sin p\tau$, $\quad D = m(\omega_0^2 - p^2) + c_1 \cos p\tau$, $u = r_0 \Omega$.

From (5) for $\dot{a} = 0$, $\dot{\xi} = 0$ follow the equations for determining the characteristics of stationary self-oscillations. Their amplitude is determined by the ratio

$$a^2 p^2 = h - \mu \quad (6)$$

where $h = 3(u_0^2 - u^2)/\bar{N}_3$, $u_0^2 = \alpha_1/3\alpha_3$, $\mu = (k_0 \cos p\Delta - p^{-1} c_1 \sin p\tau)/\alpha_3 T_0 \bar{N}_3$.

It follows from (6) that the amplitude:

- increases for delays with which $\cos p\Delta < 0$ ($p\Delta = \pi/2 \div 3\pi/2$) and decreases if $\cos p\Delta > 0$ ($p\Delta = 0 \div \pi/2$; $3\pi/2 \div 2\pi$);
- increases with delays, with which $\sin p\tau > 0$ ($p\tau = 0 \div \pi$; $3\pi/2 \div 2\pi$) and decreases if $\sin p\tau < 0$ ($p\tau = \pi/ \div 2\pi$);
- increases strongly with delays with which $\cos p\Delta < 0$, $\sin p\tau > 0$ and decreases strongly if $\cos p\Delta > 0$, $\sin p\tau < 0$;
- the same as in the absence of delays, that is, $a^2 p^2 = h$, if $k_0 \cos p\Delta = -p^{-1} c_1 \sin p\tau$.

There is a certain speed range u in which the $u \geq ap$ condition is violated. The boundary velocity u_b is defined by the expression

$$u_b^2 = 3 \tilde{u}_0^2/(3 + \bar{N}_3) \quad (7)$$

The amplitude is zero at $h = \mu$, that is, the speed

$$\tilde{u}_0^2 = u_0^2 - \frac{1}{3} \bar{N}_3 \mu \quad (8)$$

Note that, with linearization accuracy $r = 1.5$, we have $\bar{N}_3 = 3/4$. This gives the following relations, which completely coincide with those obtained by the averaging method and other known methods of nonlinear mechanics:

$$h = 4(u_0^2 - u^2), \quad \mu = 4(k_0 \cos p\Delta - p^{-1}c_1 \sin p\tau)/3\alpha_3 T_0$$
$$\tilde{u}_0^2 = u_0^2 - 0.25\mu, \quad u_b = 0.894427191\tilde{u}_0$$

The second Eq. (5) delivers an expression for frequency

$$p^2 = \omega_0^2 + \omega_1^2 \cos p\tau$$

where $\omega_1^2 = c_1/m$.

Given $p^2 - \omega_0^2 \approx 2\omega_0(p - \omega_0)$, can approximately take

$$p \approx \omega_0 + \frac{\omega_1^2}{2\omega_0} \cos p\tau$$

The amplitude of stationary self-oscillations is determined at $u < ap$ by the $ap \approx u$ dependence.

The graph of the dependence of the amplitude on the speed over the entire range of its change is shown in Fig. 2 with $\tau = 0$ lag and different Δ lag values. The boundary velocity u_b and the velocity $\tilde{u}_0$, at which the amplitude is zero, are shown only for curve 1 ($\Delta = 0$). Similar velocities hold for curves 2 and 3.

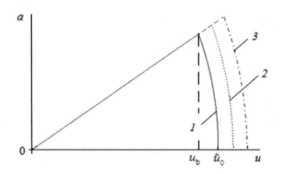

Fig. 2. The amplitude curves: $\tau = 0$ curve $1 - \Delta = 0$, curve $2 - \Delta = \pi/2$, curve $3 - \Delta = \pi$

To check the results of the direct linearization method, the first Eq. (1) was numerically integrated. Figure 3 shows one of the oscillograms obtained by integration with the parameters $\omega_0 = 1\,c^{-1}$, $m = 1\,\text{kgf} \cdot c^2 \cdot \text{cm}^{-1}$, $R = 0.5\,\text{Kgf}$, $k_0 = 0.02\,\text{kgf} \cdot c \cdot \text{cm}^{-1}$, $\alpha_1 = 0.84\,c \cdot \text{cm}^1$, $\alpha_3 = 0.18\,c^3 \cdot \text{cm}^{-3}$, $V = 1.2\,\text{cm} \cdot c^{-1}$, $\Delta = 0$, $\tau = 0$. The magnitude of the amplitude is ~ 0.43, which corresponds well to the value ~ 0.41 obtained from Eq. (6) and also occurs in the case of applying the asymptotic averaging method [7] to solve (1).

Note that the case with a delay in elasticity τ and no delay in damping Δ was studied in [10]. In this case, the presence of a delay in elasticity at $\tau = \pi/2$ increases the amplitude of self-oscillations (the range of speeds u at which self-oscillations occur expands), at $\tau = 3\pi/2$ decreases (the range of speeds u narrows), and at $\tau = \pi$ coincides

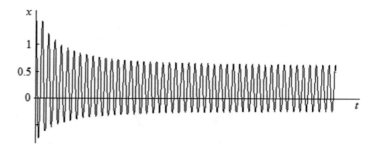

Fig. 3. Oscillogram

with it in comparison with its absence. That is, for $\tau = \pi/2$, the delay plays the role of additional excitation, and for $\tau = 3\pi/2$, it plays the role of additional resistance. As can be seen from Fig. 2, a change in the velocity range u, at which self-oscillations occur, also occurs with delayed damping. However, it differs from the case with delayed elasticity. The speed range u is the widest at $\Delta = \pi$. The amplitude curves "walk" between the extreme curves 1 and 3 at different values of the delay from the $\Delta = 0 \div 2\pi$ interval.

The third Eq. (5) gives an expression for $\dot{u} = 0$ to calculate the load $S(u)$ on the energy source. In the case of $u \geq ap$ it has the form

$$S(u) = r_0\, T_0\, (1 + B_f)$$

Equation

$$M(u/r_0) - S(u) = 0 \tag{9}$$

allows you to calculate the stationary value of the speed u.

Equation (9) also holds for $u < ap$, but with $ap \approx u$ speeds taken into account. For Fig. 4 shows a graph of the load on the power source over the entire speed range with different delay values. The forms and locations of the characteristics of M_1 and M_2 energy source $M(u/r_0)$ are shown conditionally. The stationary speed value can be determined, except (9), also by the intersection point of the curves $M(u/r_0)$ and $S(u)$ as shown for the speed u_1 under the characteristic M_1. In the case of the M_2 characteristic, depending on the initial conditions, it is possible to implement self-oscillations at speeds corresponding to points A, B, C.

The implementation of the form of the amplitude curve depends on the steepness of the characteristics of the energy source and, depending on it, transitions from one dynamic state to another are possible with a slow change in the parameters in the system, both in the absence and in the presence of delay. Transitions will occur only with gentle characteristics of the energy source, but with sufficiently steep ones they will not. If you move the characteristic $M(u/r_0)$ from left to right, the transition will occur from point D (corresponds to the boundary speed u_b) to point E, where the speed u_E, and from right to left – from point K (corresponds to the speed $\tilde{u}_0$) to point R, where the speed u_R. Thus, with increasing speed, the portion of the amplitude curve

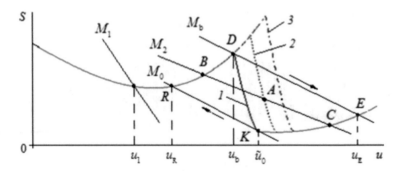

Fig. 4. Load curves for the power source: $\tau = 0$; curve $1 - \Delta = 0$, curve $2 - \Delta = \pi/2$, curve $3 - \Delta = \pi$

corresponding to the portion of the load between points D and K will not be realized. Only its linear portion in the $0 \div u_b$ range in Fig. 2 will be realized. This linear section will be shorter when transition from point K to point R.

4   Stability of Stationary Oscillations

To determine the stability of stationary movements, we create equations in variations for (5) and use the Routh-Hurwitz criteria.

The stability conditions are

$$D_1 > 0, \quad D_2 > 0 \tag{10}$$

where $D_1 = -(b_{11} + b_{22})$, $D_2 = b_{11}b_{22} - b_{12}b_{21}$.

In the case of $u \geq a\,p$ speeds we have

$$b_{11} = \tfrac{r_0}{J}[Q + r_0T_0(\alpha_1 - 3\alpha_3u^2 - 3\alpha_3N_2\,a^2p^2)], \quad b_{12} = -\tfrac{6r_0^2}{J}T_0N_2\alpha_3uap^2$$
$$b_{21} = -\tfrac{3}{m}T_0\alpha_3u, \quad b_{22} = -\tfrac{1}{m}\,T_0\bar{N}_3\alpha_3a^2p^2$$

where $Q = \tfrac{d}{du}M(\tfrac{u}{r})$.

In the case of $u < a\,p$, the first criterion (10) is almost always satisfied due to the dependence of $ap \approx u$. The determining factor is the second criterion that can be represented in the form

$$\frac{d}{du}\Big[M(\tfrac{u}{r}) - S_-(u)\Big] < 0 \tag{11}$$

where $S_-(u) \approx r_0T_0\,[(1 - \alpha_1u + \alpha_3(1 + 3N_2)u^3)]$ taking into account $ap \approx u$.

Criterion (11) is fulfilled for a very wide range of characteristics of the energy source, which can be seen from the difference in the graphs of the source $M(u/r)$ and the load $S_-(u)$ in Fig. 4. Therefore, the velocity region u, where $ap \approx u$, is the best for the realization of self-oscillations, and the worst for elimination.

5 Conclusion

Self-oscillations with a non-ideal energy source in the case of delays in elasticity and damping are considered. The solution of the nonlinear system of equations is based on the method of direct linearization of nonlinearity. This method significantly reduces the cost of labor and time compared to the known methods of nonlinear mechanics. Reducing these costs and ease of application of the method increases the efficiency of practical calculations, which is important when conducting design and technological calculations of real objects. The results obtained show the influence of delayed elastic and damping forces on the parameters (amplitude, frequency) of self-oscillations and the load on the energy source. Depending on the value of the lag, the amplitude and load curves shift in the range of speeds of the energy source.

References

1. Kudinov, V.A.: Dynamics of Machine Tools. Mashinostroenie, Moscow (1967). (in Russian)
2. Korityssky, Ya. I.: Torsional Self-Oscillation of Exhaust Devices of Spinning Machines at Boundary Friction in Sliding Supports/in SB. Nonlinear Vibrations and Transients in Machines. Nauka, Moscow (1972). (in Russian)
3. Encyclopedia of mechanical engineering. https://mash-xxl.info/info/174754/
4. Rubanik, V.P.: Oscillations of Quasilinear Systems with Time Lag. Nauka, Moscow (1969). (in Russian)
5. Zhirnov, B.M.: On self-oscillations of a mechanical system with two degrees of freedom in the presence of delay. J. Appl. Mech. 9(10), 83–87 (1973)
6. Astashev, V.K., Hertz, M.E.: Self-oscillation of a visco-elastic rod with limiters under the action of a lagging force. Mashinovedeniye 5, 3–11 (1973). (in Russian)
7. Bogolyubov, N.N., Mitropolsky, Y.A.: Asymptotic Methods in the Theory of Nonlinear Oscillations. Nauka, Moscow (1974). (in Russian)
8. Alifov, A.A.: About application of methods of direct linearization for calculation of interaction of nonlinear oscillatory systems with energy sources. In: Proceedings of the Second International Symposium of Mechanism and Machine Science (ISMMS – 2017), Baku, Azerbaijan, 11–14 September 2017, pp. 218–221
9. Kononenko, V.O.: Vibrating Systems with Limited Power-Supply. Iliffe, London (1969)
10. Alifov, A.A., Frolov, K.V.: Interaction of Nonlinear Oscillatory Systems with Energy Sources. Hemisphere Publishing Corporation, Taylor & Francis Group, New York, Washington, Philadelphia, London, 327 p. (1990)
11. Bhansali, P., Roychowdhury, J.: Injection locking analysis and simulation of weakly coupled oscillator networks, In: Li, P., et al. (eds.) Simulation and Verification of Electronic and Biological Systems, pp. 71–93. Springer, Cham (2011)
12. Acebrón, J.A., et al.: The Kuramoto model: a simple paradigm for synchronization phenomena. Rev. Modern Phys. 77(1), 137–185 (2005)
13. Ashwin, P., Coombes, S., Nicks, R.: Mathematical frameworks for oscillatory network dynamics in neuroscience. J. Math. Neurosci. 6(1), 1–92 (2016)
14. Gourary, M.M., Rusakov, S.G.: Analysis of oscillator ensemble with dynamic couplings. In: AIMEE 2018. The Second International Conference of Artificial Intelligence, Medical Engineering, Education, pp. 150–160 (2018)

15. Ziabari, M.T., Sahab, A.R., Fakha-ri, S.N.S: Synchronization new 3D chaotic system using brain emotional learning based intelligent controller. Int. J. Inf. Tech. Comput. Sci. (IJITCS) 7(2), 80–87 (2015). https://doi.org/10.5815/ijitcs.2015.02.10
16. Moiseev, N.N.: Asymptotic Methods of Nonlinear Mechanics. Nauka, Moscow (1981). (in Russian)
17. Butenin, N.V., Neymark, Y.I., Fufaev, N.A.: Introduction to the Theory of Nonlinear Oscillations. Nauka, Moscow (1976). (in Russian)
18. He, J.H.: Some asymptotic methods for strongly nonlinear equations. Int. J. Modern Phys. B 20(10), 1141–1199 (2006)
19. Hayashi, Ch.: Nonlinear Oscillations in Physical Systems. Princeton University Press, New Jersey (2014)
20. Tondl, A.: On the interaction between self-exited and parametric vibrations. In: National Research Institute for Machine Design Bechovice. Series: Monographs and Memoranda, vol. 25, 127 p. (1978)
21. Karabutov, N: Structural identification of nonlinear dynamic systems. Int.J. Intell. Syst. Appl. 09, 1–11 (2015). Published Online August 2015 in MECS (http://www.mecs-press.org/). https://doi.org/10.5815/ijisa.2015.09.01
22. Wang, Q., Fu, F.L.: Numerical oscillations of runge-kutta methods for differential equations with piecewise constant arguments of alternately advanced and retarded type. Int. J. Intell. Syst. Appl. 4, 49–55 (2011). Published Online June 2011 in MECS (http://www.mecs-press.org/)
23. Chen, D.-X., Liu, G.-H.: Oscillatory behavior of a class of second-order nonlinear dynamic equations on time scales. J. Eng. Manufact. 6, 72–79 (2011). Published Online December 2011 in MECS (http://www.mecs-press.net). https://doi.org/10.5815/ijem.2011.06.11
24. Alifov, A.A.: Methods of Direct Linearization for Calculation of Nonlinear Systems. RCD, Moscow (2015). (in Russian). ISBN 978–5-93972-993-2
25. Alifov, A.A.: About direct linearization methods for nonlinearity. In: Hu, Z., Petoukhov, S., He, M. (eds.) AIMEE 2019. AISC, vol. 1126, pp. 105–114. Springer, Cham (2020). https://doi.org/10.1007/978-3-030-39162-1_10
26. Alifov, A.A.: On the calculation by the method of direct linearization of mixed oscillations in a system with limited power-supply. In: Hu, Z., Petoukhov, S., Dychka, I., He, M. (eds.) ICCSEEA 2019. AISC, vol. 938, pp. 23–31. Springer, Cham (2020). https://doi.org/10.1007/978-3-030-16621-2_3
27. Alifov, A.A.: Method of the direct linearization of mixed nonlinearities. J. Machine. Manufact. Reliab. 46(2), 128–131 (2017). https://doi.org/10.3103/S1052618817020029
28. Alifov, A.A.: Self-oscillations in delay and limited power of the energy source. Mech. Solids 54(4), 607–613 (2019). https://doi.org/10.3103/S0025654419040150
29. Alifov, A.A., Farzaliev, M.G., Jafarov, E.N.: Dynamics of a self-oscillatory system with an energy source. Russ. Eng. Res. 38(4), 260–262 (2018). https://doi.org/10.3103/S1068798X18040032
30. Bronovec, M.A., Zhuravljov, V.F.: On self-excited vibrations in friction force measurement systems. Izv. RAN, Mekh. Tverd. Tela 3, 3–11 (2012). (in Russian)

Cognitive Prediction Model of University Activity

A. Mikryukov$^{(\boxtimes)}$ and M. Mazurov

Plekhanov Russian University of Economics, Moscow, Russia
Mikrukov.aa@rea.ru, Mazurov37@mail.ru

Abstract. The relevance of the problem being solved is due to the need to develop scientifically grounded proposals to achieve the required values of the basic indicators of the university's activity in accordance with the international institutional rating QS to the required values necessary for the university to enter the TOP-500 universities by 2025. To solve this problem, an approach is proposed to the study of weakly structured systems, the class of which includes universities and their activities, based on scenario forecasting methods by building a cognitive model in order to determine the necessary increments of target indicators. The proposed approach makes it possible, under the given constraints, to find the most acceptable scenario for planning the increment of the basic indicators to the target values by identifying the latent factors affecting them and impulse influences (increments) on them, ensuring the guaranteed achievement of the set goal. The results obtained make it possible to subsequently justify the annual costs to ensure an unconditional increase in the values of latent factors with the aim of guaranteed obtaining the required values of the basic indicators by 2025. The novelty of the proposed approach lies in the use of correlations between latent factors, identified on the basis of factor analysis methods, with basic indicators in the construction of a cognitive model, as well as the application of an iterative approach to solving the problem, which makes it possible to update the set of initial data, as well as train the developed cognitive model taking into account the results of identifying latent factors and correcting their correlations. The results obtained make it possible to form the most preferable scenario plan for the necessary stepwise increase in the values of the basic indicators in the interval 2020–2025 subject to resource constraints.

Keywords: Cognitive model · Scenario forecasting · Baseline indicators · Institutional ranking

1 Introduction

The purpose of developing a cognitive model of scenario forecasting is to substantiate the list of necessary measures to ensure the increments of the values of the target indicators of the university's activities in accordance with the international institutional rating QS to the required values necessary for the university to enter the TOP-500 universities by 2025.

To achieve the stated goal of the study, the application of methods for solving poorly structured tasks was substantiated based on the development of a scenario

© The Author(s), under exclusive license to Springer Nature Switzerland AG 2021
Z. Hu et al. (Eds.): AIMEE 2020, AISC 1315, pp. 133–145, 2021.
https://doi.org/10.1007/978-3-030-67133-4_13

forecasting (planning) model using cognitive maps, which made it possible to determine the probable (possible) trends in the development of events by alternative options and the possible consequences of decisions made in order to choose the most preferable alternative. The proposed approach makes it possible, under the given constraints, to find the most acceptable scenario for planning the increment of the values of basic indicators and functional (university rating) to target values due to impulse influences on latent factors (assessing the feasibility of the necessary increments in the values of factors - reasons), ensuring guaranteed achievement of the goal.

In the course of the study, the following tasks were solved: a cognitive model of scenario forecasting of measures to achieve the required values of the target performance indicators of the university in the international institutional rating QS was developed, on the basis of the developed model, the calculation of the most preferable variant of the set of required increment of the value of the target factor (functional). The scenario of planning measures for guaranteed achievement of the target value, obtained on the basis of the cognitive model, made it possible to substantiate the value of the necessary increments in the values of the identified latent factors that affect the basic indicators. Based on the results obtained, in the future, it is possible to justify the necessary annual costs to ensure the guaranteed achievement of the required value of the rating functionality of the university by 2025.

The results obtained made it possible to formulate a scenario plan of the necessary stepwise increase in the values of indicators, considering the factors influencing them.

A feature of the developed approach is the identification of the most significant latent factors affecting the baseline indicators, as well as the assessment of the degree of their influence on the baseline indicators based on the methods of factor analysis. To ensure the adequacy of the cognitive model, an iterative approach was used, which makes it possible to build a learning cognitive model. The use of this approach provides for the adjustment of the cognitive model based on the results of assessing the activities of universities after the release of the annual report on the place of the university in the QS institutional ranking. The adjustment is carried out by clarifying the structure of the cognitive map based on the actualization of the sets of initial data, as well as the results of identification of latent factors. During the adjustment, the correlation dependences are assessed based on the factor model.

The use of an iterative approach allows for an increase in the reliability and accuracy of obtaining a solution.

Section 2 provides a literature review on the research topic, Sect. 3 presents the results of the development of a cognitive model, Sect. 4 presents the results of scenario forecasting of university performance indicators based on the developed cognitive model of scenario forecasting, Sect. 5 provides an interpretation of the results of a numerical experiment.

2 Literature Review

A fairly large number of works by domestic and foreign scientists are devoted to the problem of cognitive modeling [1–11]. An important task when using cognitive models is to ensure their reliability, adequacy, and accuracy. In papers [1–5], an approach based on classical cognitive models is proposed. In work [6], problem solving using fuzzy cognitive models is considered. The work [8] solves the problem of assessing the stability of cognitive models, as well as methods for their verification. In [9], the features of modeling nonlinear dynamic systems under uncertainty are considered.

The analysis of the sources showed that in the presented formulation, the problem of cognitive modeling of university development scenarios was not considered.

3   Methodology for Solving the Problem

The analysis of the problem posed showed that it belongs to the class of semi-structured problems, which is solved under the conditions of a limited amount of initial data and a number of uncertainties associated, among other things, with the absence of a linear relationship between the values of the functional and the place of the university in the QS ranking in the considered time interval.

To solve this class of problems, the methodology of cognitive modeling is used, designed for analysis and decision-making in poorly defined situations, when it is not possible to analytically describe and build formal mathematical models that take into account the specifics of the studied poorly structured system and, in particular, the method of scenario forecasting (planning) based on cognitive maps (cognition maps), which allows you to determine the probable (possible) trends in the development of events by alternative options and the possible consequences of decisions made in order to choose the most preferable alternative [12].

A cognitive model in the form of a cognitive map is a subjective model of a weakly structured dynamically developing situation.

As you know, a scenario is a certain relative, conditional assessment of the possible development of an object or situation under study, since it is always built within the framework of assumptions about future development conditions, which are often fundamentally unpredictable.

In contrast to classical forecasting methods, in which the main attention is paid to assessing the most probable variant of the system's development, scenario forecasting proceeds from the idea of uncertainty and ambiguity of the trajectory of this development. A feature of scenario forecasting is that it allows you to simultaneously consider several options for the development of a situation, analyze opportunities and risks [13].

To solve this problem, an approach to scenario forecasting based on a cognitive map is proposed, which includes the following stages [14]:

- scenario generation and risk assessment,
- scenario adjustment at discrete time intervals (in our case, annually), considering the achieved values of the place in the QS rating of the functional and indicators.

At the end of the next time interval, a new scenario is built to achieve the next set goal (adjusted considering current calculations).

At the first stage of the study, the identification and interpretation of the hidden (latent factors) that affect the baseline indicators and the assessment of their significance using the methods of factor analysis, which is a class of multivariate statistical analysis procedures aimed at identifying the latent variables (factors) responsible for the presence of linear statistical relationships (correlations) between the observed variables [15–18].

Factors are groups of certain variables that correlate with each other more than with the variables included in another factor. Thus, the meaningful meaning of the factors was revealed by examining the correlation matrix of the initial data.

At the next stage, a cognitive model was developed, based on which the problem of scenario forecasting of the university performance indicators was solved. The development of a cognitive model is based on the construction of a cognitive map in the form of a sign oriented graph (1), which is formed by structuring the knowledge of experts about the subject area on the basis of theoretical concepts, statistical data, as well as the use of various expert methods [19, 20].

$$G = \langle V, E \rangle, \tag{1}$$

where: G is a signed directed graph, $V = \{v_i\}$, $i = 1, 2,..., k$ is the set of vertices (concepts) of the cognitive map, $E = \{e_{ij}\}$ is the set of arcs connecting the vertices v_i and v_j.

The cognitive model can be represented as a functional graph:

$$\Phi = \langle G, X, F, \theta \rangle, \tag{2}$$

where: Φ is a functional graph, $X = \{x_i\}$ is a set of vertex parameters, $F = f\{v_i, v_j, e_{ij}\}$ is a function (or functional $f\{v_i, v_j, e_{ij}\}$, or coefficient f_{ij}) of connections between vertices, θ - the space of vertex parameters.

Taking into account the values of correlation dependences between the functional and basic indicators obtained on the basis of factor analysis in work [8], as well as estimates of the influence of factors on basic indicators and their mutual influence, a cognitive model (map) was built, reflecting the interconnection of factors, basic indicators and functional (Fig. 1).

Взаимосвязь базовых показателей и функционала определятся формулой (3) в соответствии с правилами международного институционального рейтинга QS [21].

$$R = \sum_{i=1}^{6} w_i x_i \tag{3}$$

г д е w_i– вес соответствующего показателя; x_i– его значение.

In general, the formalization of the task based on a cognitive map is as follows [20–22]. The model of a situation is given by the four $\langle F, X, X(0), W \rangle$, where F is a set of features (characteristics) of a situation, X is a set of states, $X(0) = (x_{11}^{0}, x_{12}^{0},..., x_{nm}^{0})$ is an initial state, W is a matrix of the adjacent functional organ, reflecting the relationship between the tops of the cognitive map, $X^{G} = (x_{11}^{g}, x_{12}^{g},..., x_{nm}^{g})$ is the desired state.

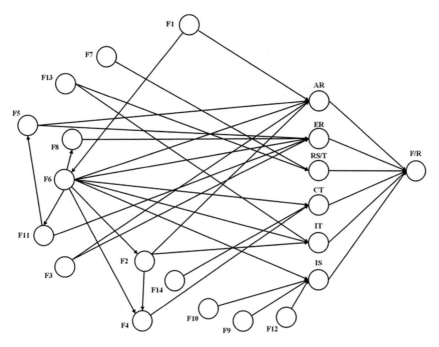

Fig. 1. Cognitive model of correlation of latent factors, indicators and functional. (Symbols: F - functional; R - university ranking, indicators: AR - academic reputation; ER - employer's reputation; RS/T - ratio of students to teachers; CT - teacher citation index; IT - number of international teachers; IS - number of international students; latent factors: F_1 - Factor "Scientific Schools and Dissertation Councils"; F_2 - Factor "Joint Scientific Projects"; F_3 - Presence of basic departments; F_4 - Number of publications in Scopus Database; F_5 - Demanded training directions; F_6 - Qualification level teachers; F_7 - Number of teachers; F_8 - Competence level of students; F_9 - teachers with language training; F_{10} - Dormitory places; F_{11} - Demand of graduates from employers; F_{12} - Areas for educational activities; F_{13} - teachers payment level; F_{14} - Incentive factors).

It is necessary to find the target grading vector $G = (p_{1j}, p_{2j}..., p_{nm})$, where $p_{11} = x_{11}^g - x_{11}^0$, $p_{12} = x_{12}^g - x_{12}^0$, etc. The target grading vector G shows in which direction and how much you need to change the values of features in the initial state X (0) to go to the target state X^G.

Let a set of control features $R \subset F$ be given, for each of which restrictions on their possible changes are defined, that is, control resources are defined in the form of a vector $P^R = (p_{11}^r, ..., p_{nm}^r)$. The task is to find a set of solutions $U = \{U_1, U_2, ..., U_v\}$ to transfer the situation from the current state to the target state. Evaluation and selection of alternatives is based on expert opinion or some criteria. For example, the best solution can be associated with finding either the shortest path or finding the path with the lowest or highest weight. In the problem under consideration, the best solution corresponds to the path with the minimum weight (the cost of the cost of incrementing the set of latent (partial) factors under the given constraints).

Let us denote latent factors affecting the indicators (Fig. 1), factors - causes, indicators - effects, and functional (target factor). The strength of the mutual influence of factors - causes and the strength of the influence of factors - causes on factors - consequences are determined on the basis of the method of expert assessment (the Saaty method of paired comparisons) [23].

The strength of influence for the factor-cause, which has the maximum strength of connection, is determined as the transfer ratio according to the formula (4)

$$w_{ij} = \frac{p_i^p}{p_j^r}, \tag{4}$$

where p_i^p is the increment of the factor-cause; p_j^r - increment of the effect factor; i - factor number. The strength of the influence of other factors will be determined from the relationship (2)

$$w_{il} = wij \frac{\lambda_i}{\lambda_j}, \tag{5}$$

where λ_i is the estimate of the factor in the matrix of paired comparisons Λ obtained by expert means (Table 1).

Table. 1. Matrix of the influence of private factors on basic indicators and the mutual influence of latent factors.

Latent factors/Base. Indicators	F_1	F_2	F_3	F_4	F_5	F_6	F_7	F_8	F_9	F_{10}	F_{11}	F_{12}	F_{13}	F_{14}
AP	0,8	0,7	0,4	0,7	0,6	0,6	0,3	0,6	0,3	–	0,4	0,2	0,2	–
ER	0,3	0,4	0,6	0,3	0,4	0,7	0,2	0,8	0,2	–	0,6	–	0,2	–
RS/T	–		–	–	–	0,2	–	–	–	–	–	–	–	–
CT	0,3	0,7	–	0,8	–	0,6	0,3	0,1	0,3	–	–	–	0,4	0,7
IT	0,3	0,4	–	–	0,2	0,5	0,2	0,1	0,3	–	0,2	–	0,6	–
IS	0,2	–	0,2	–	0,2	0,8	0,1	0,2	0,8	0,8	0,4	0,8	–	–
F_6	0,8	0,9												
F_{11}								0,8						
F_5				0,6			0,7							
F_4							0,9							
F_2							0,8							
F_8		0,6	0,7							0,6		0,8		

The second order impacts (indirect, synergetic, etc.) were considered using weighted additive convolution (6):

$$w_j = \sum_{i=1}^{n} \beta_i w_i \qquad (6)$$

where: n is the number of causal factors influencing the selected effect factor; β_i are weight coefficients ($0 \le \beta i \le 1$) satisfying condition (7)

$$\sum_{i=1}^{n} \beta_i = 1. \qquad (7)$$

4 Results of Scenario Forecasting of University Performance Indicators Based on the Cognitive Model

Scenario analysis of the system's behaviour was carried out on its cognitive model by means of pulsed modeling. The formula of the impulse process is as follow: [24, 25]:

$$x_i(n+1) = x_i(n) + \sum_{j=1}^{k-1} f_{ij} P_j(n) + Q_i(n), \qquad (8)$$

where $x_i(n)$ is the impulse value in the i-th vertex of the cognitive model at the previous moment (modeling tact) n, $x_i(n + 1)$ - at the moment of interest to the researcher $(n + 1)$; f_{ij} is the impulse conversion factor; $P_j(n)$ is the impulse value in the vertices adjacent to i-th vertex; $Q_i(n)$ is the vector of perturbations and control actions introduced in i-th vertex at the moment of n.

The set of implementations of impulse processes represents a "scenario of development" and indicates possible trends in the development of situations.

On a time, axis the increment of factor values in discrete moments of time can be represented as a linear dependence [14]:

$$x_i(t+1) = x_i(t) + \sum_{j \in I_i} a_{ij} \big(x_j(t) - x_j(t-1) \big), i = 1, \dots, N \qquad (9)$$

at known initial values of the factors $(x(0))_{i \in N}$ and their initial increments $(x_i(0) - x_i(1))_{i \in N}$.

Here $x_i(t + 1)$ and $x_i(t)$ are the values of the i-th factor at moments $t + 1$ and t respectively; $x_j(t) - x_j(t - 1) = \Delta x_j(t)$ - increments of the factor x_j at moments t; a_{ij} - weight of the influence of factor x_i on factor x_j; I_i- number of factors directly affecting factor x_i; N - number of factors (vertices of the graph).

Let us introduce impulses ("speeds") $p_j(t) = \Delta x_j(t)$, then the dynamics of the system can be represented by the following equation:

$$p_i(t+1) = x_i(t) + \sum_{j \in I_i} a_{ij} p_j(t) + p_j^0(t), \qquad (10)$$

where $p_j^0(t)$ is the external impulse introduced to the vertex i now t.

We denote by u_i the external entrance to the vertex i; in this case, we obtain

$$p_i^0(t+1) = u_i(t+1) - u(t). \tag{11}$$

Let us $x_i(t) = 0$, $t < 0$, $i = 1,...,N$, and external impulses are supplied starting from $t = 0$. Then, in accordance with (6) and (7), we obtain

$$x_i(t+1) = \sum_{i,j} a_{ij}x_j(t) + u_j^0(t), i = 1,...,N \tag{12}$$

or in matrix form

$$x(t+1) = (I+A)x(t) + Bu(t), \tag{13}$$

where $A = \|a_{ij}\|$ is an $N \times N$ adjacency matrix of the cognitive map graph, I is an $N \times N$ identity matrix; $u = (u_1,..., u_M)^T$ is the vector of controls (external control actions); B - (0,1) is an $N \times M$ matrix, the nonzero elements of which indicate the points of application of controls.

Thus, after describing the relationships between the factors by equations, setting the weights of their mutual influence and the values of the initial increments of factors, it is possible to analyze the dynamics of changes in the factors and the development of the system of indicators as a whole [26–30].

The values of the initial data that are not quantitative in nature (Qualification level teachers, the level of competencies of students, etc.) are determined by expert advice.

The initial data for solving the problem were obtained by normalizing the initial ("physical variables") according to the formula (14), take values in the interval [0,1] and are presented in Table 2.

$$f_i = \frac{f_i - f_{imin}}{f_{imax} - f_{imin}}, \tag{14}$$

где $f_{i\ min}$ и $f_{i\ max}$ – minimum and maximum value of a variable f_i.

Table 2. Source data table.

$f1$	f_2	f_3	f_4	f_5	f_6	f_7	f_8	f_9	f_{10}	f_{11}	f_{12}	f_{13}	f_{14}
0,59	0,19	0,57	0,44	0,54	0,52	0,86	0,77	0,32	0,38	0,81	0,44	0,56	0,32

Sequentially set "weak" increments in the values of the above factors at the level of 10%, made it possible to assess the sensitivity of the target to control actions in these areas of regulation, on the basis of which the most preferable alternative to scenario forecasting was chosen.

The preliminary values of the intensity of mutual influence between the measurable factors of the cognitive model were established based on correlation and factor analysis [18]. During simulation modeling of the controlled development of the system of indicators, the data presented in Table 2 were selected as assessments of the degree of influence of factors subject to control actions 1.

Further, the coefficients were refined in accordance with the logic of the transition of the system from one stationary state to another because of external impulse influences.

Verification of the cognitive model is performed based on the criterion of completeness and consistency of the influence of factors - causes on factors - consequences and violations of the transitivity of causal phenomena [31].

To check the adequacy and accuracy of the cognitive model, it was tested in the retrospective period 2014–2019. Based on available statistics on measurable factors in the model. The general correctness of the model at this stage was confirmed by the closeness of the growth rates of factors calculated on the model to the actual growth rates.

Scenario analysis of the development of the situation made it possible to choose the most preferable scenario, which, under the conditions of the given restrictions, ensures the achievement of the required planned value of the target indicator with minimal resource costs for the increment of latent factors.

5   Discussion. Interpretation of the Results of a Numerical Experiment

The required intensity of influence on the control factors as a percentage at a given increment of the target factor were calculated during scenario modeling. The simulation results presented in Fig. 2, considering the attainable values of latent factors in the 2019–2020 interval. Correspond to the scenario of their increment, ensuring the achievement of the required target indicator (functional) with minimal resource costs for the increment of latent factors.

In Fig. 2, it can be seen that the greatest intensity of impact will be required for the following latent factors: Joint research projects, Number of publications in the Scopus database, Demand for graduates from employers, and the smallest for latent factors: Number of academic staff and Areas for educational activities.

The results of calculating the predicted values of growth indicators of basic indicators and functional in 2020, considering the chosen scenario, are presented in Fig. 3. The obtained results of scenario forecasting show that the largest increase should be obtained by the PP indicator (reputation with the employer) - 52%, and the smallest - MP (international teachers) - 8%. An increase in the values of basic indicators should lead to an increase in the value of the functional by 12%, i.e. achieving its planned value. Thus, the goal of the study has been achieved.

As a result of the calculations performed, the following conclusions can be drawn:

To guarantee the achievement of the functional value F = 26.89 by 2025, corresponding to the 435 (451) place in the QS rating, it is necessary to stepwise increase the indicator values, taking into account the factors influencing them, identified on the basis of factor analysis methods, in the interval 2020–2025. According to the following scheme.

Based on the calculation performed using the cognitive modeling methodology, it is necessary to perform impulse influences (to provide an increment) on the set of factors identified as of the present in accordance with the results presented in Fig. 3.

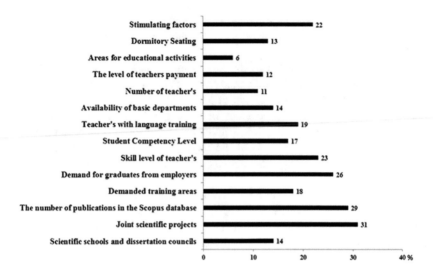

Fig. 2. Calculated values of the intensities of control actions on latent factors in the form of increments in% necessary to achieve the target increase in baseline indicators and functional in 2020.

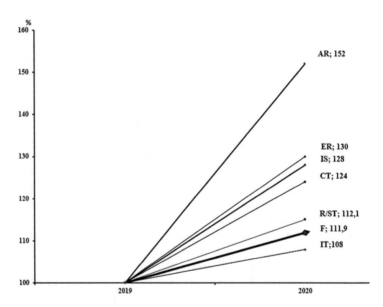

Fig. 3. Forecast of indicators growth and growth of functional in 2020 compared to 2019 (2019–100%).

At the end of the first control interval (after 1 year), after specifying the actual values of the basic indicators and functional in the QS rating, assess the change in the influence and interaction of factors on the basic indicators based on the methods of factor analysis, correct the data presented in Table 1, identify latent factors 3 - level, to update the developed model of scenario forecasting based on the cognitive map, taking into account the additional (third) level of latent factors, which allows to build a more adequate and reliable model of the cognitive map. Based on the updated model, perform calculations to justify a set of new impulse influences on factors of the 3rd, 2nd, and 1st levels of the cognitive model for the period 2020–2021.

6 Conclusion

During the research, an iterative learning cognitive model of scenario forecasting of activities to achieve the required values of the target performance indicators of the university in the QS international institutional ranking was developed. On the basis of the developed model, the calculation of the most preferable from the point of view of cost minimization of the scenario variant of the set of required intensities of influence on the control variables (latent factors) for a given planned increment of the target factor value was carried out.

The novelty of the proposed approach lies in the use of factor analysis methods to determine the relationships between latent factors, which ensure the identification of the most significant latent factors for building an adequate cognitive model. The proposed approach allows training the developed model by adjusting the structure of the cognitive map and its parameters based on the results of updating the initial data for its construction.

As a direction for further research, it is advisable to solve the problem of identifying latent factors of the second level, which should ensure an increase in the reliability and accuracy of the cognitive model.

Acknowledgments. The article was prepared with the support of the Russian Foundation for Basic Research, grants No. 18-07-00918, 19-07-01137 and 20-07-00926.

References

1. Maksimov, V.I., Kachaev, S.V.: Technologies of the Information Society in Action: Application of Cognitive Methods in Business Management. https://www.rfbr.ru/default.asp?doc_id=5222.
2. Axelrod, R.: The Structure of Decision: Cognitive Maps of Political Elites. University Press, Princeton (1976)
3. Kuznetsov, O.P.: Cognitive modeling of semi-structured situations. https://posp.raai.org/data/posp2005/Kuznetsov/kuznetsov.html
4. Roberts, F.S.: Discrete mathematical models with applications to social, biological and environmental problems. Int. J. Man-Mach. Stud. **24**, 65–75 (1986)
5. Kosko, B.: Fuzzy Cognitive Maps, vol. 184, p. 13. Nauka (1986)
6. Abramova, N.A., Kovriga, S.V.: Cognitive approach to decision-making in - structured situation control and the problem of risks. In: International Conference on Human System Interaction, Krakow (2008)

7. Frey, W.: Socio-Technical Systems in Professional Decision Making. https://cnx.org/content/m14025/latest/
8. Green, D.: Socio-technical Systems in Global Markets. https://nuleadership.wordpress.com/2010/08/23/sociotechnical-systems-in-global-markets/.
9. Karabutov, N.: Structural identifiability of nonlinear dynamic systems under uncertainty. Int. J. Intell. Syst. Appl. (IJISA) **12**(1), 12–22 (2020). https://doi.org/10.5815/ijisa.2020.01.02
10. Bodyanskiy, Y.V., Tyshchenko, O.K., Deineko, A.O.: An evolving neuro-fuzzy system with online learning/self-learning. IJMECS **7**(2), 1–7 (2014). https://doi.org/10.5815/ijmecs.2015.02.01
11. Vijayalakshmi, V., Venkatachalapathy, K.: Comparison of predicting student's performance using machine learning algorithms. Int. J. Intell. Syst. Appl. (IJISA) **11**(12), 34–45 (2019). https://doi.org/10.5815/ijisa.2019.12.04
12. Gorelova, G.V., Zakharova, E.N., Rodchenko, S.A.: Study of Semi-Structured Problems of Socio-Economic Systems: A Cognitive Approach, p. 332. Publishing house of the Russian State University, Rostov (2006)
13. Avdeeva, Z.K., Kovriga, S.V., Makarenko, D.I., Maksimov, V.I.: Cognitive approach in management. Probl. Upr. **3**, 2–8 (2007)
14. Bolotova, L.S.: Artificial intelligence systems: models and technologies based on knowledge. In: FSBEI RSUITP; FGAU GNII ITT `Informatics'. - M.: Finance and statistics,p. 664 (2012)
15. Goridko, N.P., Nizhegorodtsev, R.M.: Modern economic growth: theory and regression analysis: Monograph, Infra, p. 444 (2017)
16. Draper, N. Applied regression analysis. Williams I.D., p. 912 (2019)
17. Sokolov, G.A., Sagiots, R.V.: Introduction to regression analysis and planning of regression experiments in economics, p. 352. Infra-M (2016)
18. Mikryukov, A.A., Gasparian, M.S., Karpov, D.S.: Development of proposals for promoting the university in the international institutional ranking QS based on statistical analysis methods. Stat. Econ. **17**(1), 35–43 (2020)
19. Trakhtengerts, E.A.: Computer support for the formation of goals and strategies, p. 224. SINTEG (2005)
20. Kornoushenko, E.K., Maksimov, V.I.: Situation control using the structural properties of the cognitive map. Tr. IPU RAS, pp. 85–90 (2000)
21. Information and analytical system QS - analytics. https://analytics.qs.com/#/signin.
22. Kuznetsov, O.P.: Intellectualization of control decision support and creation of intelligent systems. Probl. Upr. **3**(1), 64–72 (2009)
23. Saati, T.: Decision-making: method of analysis of hierarchies. In: Radio and communication, p. 320 (1993)
24. Kulba, V.V., Kononov, D.A., Kovaleskiy, S.S.: Scenario analysis of dynamics of behavior of socio-economic systems, p. 122. IPU RAN, Moscow (2002)
25. Maksimov, V.I.: Structural-target analysis of the development of socio-economic situations. Manag. Prob. (3) (2005)
26. Gorelova, G.V., Melnik, E.V., Korovin, Y.S.: Cognitive analysis, synthesis, forecasting the development of large systems in intelligent RIUS. Artif. Intell. **3**, 61–72 (2010)
27. Kulba, V.V., Kononov, D.A., Kosyachenko, S.A., Shubin, A.N.: Methods for the Formation of Scenarios for the Development of Socio-Economic Systems, p. 296. SINTEG, Moscow (2004)
28. Simon, H.: The structure of Ill-structured problems. Artif. Intell. **4**, 181–202 (1973)

29. Cognitive analysis and management of the development of situations: materials of the 1st international conference, p. 196. IPU, Moscow (2001)
30. Huff, A.S.: Mapping Strategic Thought, pp. 11–49. Wiley, Chichester (1990)
31. Kizim, N.A., Khaustova,V.E.: Features of testing models based on cognitive maps for stability and reliability. In: Modern Approaches to Modeling Complex Socio - Economic Systems, p. 280. Publishing House INZHEK (2011)

Advances in Medical Approaches

Temperature Reaction of a Person with a Contact Method of Exposure to a Thermal Signal

A. A. Shul'zhenko$^{(\boxtimes)}$ and M. B. Modestov

Federal State Budgetary Institution of Science Institute of Mechanical
Engineering, A.A. Blagonravov RAS, Moscow, Russia
aa_shulzhenko.01@mail.ru

Abstract. In this work, we studied the possibility of using the contact method of thermal influence on a person to obtain a human reaction, in the form of a change in temperature over time on the surface of a woven electric heater, which simultaneously serves as a generator of a heat signal, which is a novelty. During the work, a patented program was used and modified, which describes the thermal processes occurring in the human system- the gap between the human body and the heater - woven electric heater - thermal insulation layer - the external environment, which is also a novelty. The performed mathematical modeling confirmed the results of field experiments. This work may be of interest to medical professionals.

Keywords: Woven · Electric heater · Human heat generation · Reaction · Thermal effect

1 Introduction

The human response to a heat signal has been of interest to researchers for a long time. So in work [1], to study the reaction to a heat signal, a thermode was used, a device that allows a thermal effect on thermoreceptors, and to receive an impulse of thermoreceptors depending on the level of temperature exposure. This method of measuring heat response is invasive. Using this method, results were obtained that indicate that thermoreceptors are located on the human body at different depths. For quite a long time, the method of temperature exposure to humans from external infrared emitters has been used. However, emitters located at a fairly significant distance from a person allow obtaining only a picture of the temperature distribution over the surface of a person's body practically in statics. Close to the previous method is immersing a person in water heated to a certain temperature, while simultaneously measuring his internal temperatures: rectally, orally, etc. The main feature of these methods is the transfer of heat through the coolant, which makes them inert [2].

A more modern method of heat exposure and simultaneous measurement is near infrared spectroscopy (NIRS). With this method, the concentration of several compounds

© The Author(s), under exclusive license to Springer Nature Switzerland AG 2021
Z. Hu et al. (Eds.): AIMEE 2020, AISC 1315, pp. 149–158, 2021.
https://doi.org/10.1007/978-3-030-67133-4_14

in tissues can be measured in a non-invasive way. For example, chromophores, oxyhe-moglobin, deoxyhemoglobin, myoglobin, etc. The wavelengths of monochromatic laser radiation sources are $\sim 0.7 \div 2.5$ microns [3–5]. There are still quite a large number of methods that allow studying the thermal state of a person and the operation of his indi-vidual systems [6]. In the considered method, woven electric heaters are used, providing a thermal effect in the range of temperatures familiar to humans. This approach allows for thermal provoking of thermoreceptors and includes the mechanisms of human ther-moregulation. This is done due to the effect of elastic woven heaters directly on the human body in the absence of intermediate heat carriers, which allows using a precision thermal signal, and then in the same area (the area where the heater is located) to receive a human response to this signal. The proximity of the location, and the absence of intermediate heat carriers, allows one to obtain dynamic characteristics of temperature changes in the measurement area (on the surface of the heater). This method is non-invasive, which provides a comfortable environment for specialists during research. It is also convenient that the temperature acts as an input parameter, and temperature also acts as an output parameter. The use of a small gap between the human body and the heater allows you to get a more complete picture, including the thermal changes in this gap arising from perspiration. Moreover, not only a qualitative fixation of perspiration is possible, as it happens when using the methods of U. Fere and R. Tarkhanov [8], but also a quantitative measurement of the secreted sweat,

This method was patented [9].

Now let's look at what reactions this method allows you to call. In works [9, 10], experimental data were obtained and a mathematical description of a person's reaction to thermal exposure, carried out using an elastic woven electric heater located on the surface of the human body, was carried out. A program was written and patented [11], which allows simulating thermal processes occurring within the system: a person - a gap between a person and a woven electric heater (hereinafter referred to as a gap) - a woven electric heater - a layer of thermal protection of a woven electric heater - the external environment. The study revealed that when a person is exposed to heat, sweat is released, which affects the thermal parameters of the gap between the human body and the woven electric heater (hereinafter referred to as the heater), which causes temperature changes associated with this phenomenon, and not only in the interval, and in the area where the heater is located.

It has also been determined that the heater need not be woven. Heaters of other types can also be used for these purposes, but the thermal parameters of all elements included in the system must be appropriately selected. A heater powered by a direct current source was used as a source of thermal effect on humans. The smooth nature of the change in the heat signal generated by the heater located on the table surface, that is, in the absence of various extraneous signals affecting the heater, is shown in Fig. 1.

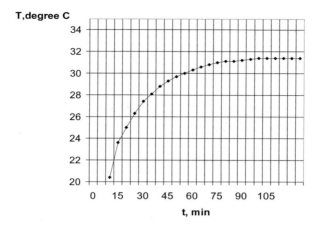

Fig. 1. Change in temperature over time of a woven electric heater located on the table surface.

Based on the measurements of the temperature on the surface of the heater, carried out during the experiments, when a person was exposed to a heat signal from the heater, the results were obtained (Fig. 2), which show that with an increase in the level of thermal exposure to a person, the response time of a person to this thermal effect decreases. That is, the formation of an elevated temperature, causing a reaction of the sweat glands, takes some time. The higher the level of the influencing signal, the shorter the waiting time for the reaction.

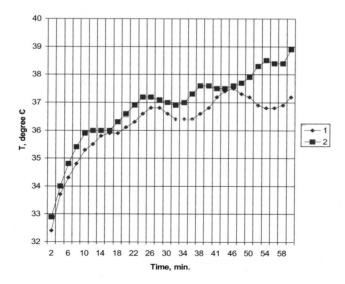

Fig. 2. Temperature changes on the surface of a woven electric heater over time when heat signals of different levels and relatively short duration are applied to the human body.

The above results gave grounds to make the assumption that with a longer exposure to heat on a person, the resulting front of elevated temperature in time will move further, deep into the human body and new reactions to the heat signal will arise. Thus, the purpose of this study is to obtain new information using the method used.

2 Results of Experiments with Laboratory Samples and Mathematical Modeling

2.1 Results of Experiments with Prototypes

A number of experiments were carried out to solve the target problem. The experimental results are shown in Fig. 3. For the sake of comparison, the results of experiments have been demonstrated, both at the level of the supplied heat signal sufficient only for the manifestation of sweating (Fig. 3, curve 3), and at the level and duration of the supplied heat signal sufficient for the manifestation of additional temperature changes. Investigations of the response to the heat signal were carried out both in the presence of a gap between the human body and the heater (hereinafter referred to as the gap) (Fig. 3, curve 2), and in its absence (Fig. 3, curve 1).

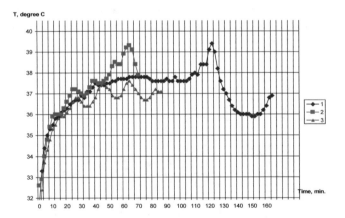

Fig. 3. Temporal distribution of temperature on the surface of a woven electric heater when a heat signal of a longer duration is applied: 1 - the level of the heat signal is sufficient for the occurrence of thermal changes caused by the human body (conditions for temperature measurement due to perspiration are excluded (there is no gap)); 2 - the level of the heat signal is sufficient for the occurrence of thermal changes caused by the human body (conditions for measuring temperature due to sweating are present (the gap is present)); 3 - the level of the heat signal of the woven electric heater is sufficient only for the occurrence of temperature fluctuations due to perspiration (there is a gap).

2.1.1   Results of Experiments with Prototypes at a Signal Level Sufficient for Perspiration, But Insufficient for the Manifestation of Additional Temperature Changes in a Person's Response to a Heat Signal

When zoomed to 86 min. The time of heat exposure, the level of which did not exceed ~ 37 °C, the total temperature changes in the area of the heater were caused by: the heat signal of the heater affecting a person; fluctuations in temperature due to a person's reaction to this heat signal, which manifests itself in the form of a change in thermal resistance in the area of the gap due to the release and evaporation of sweat. No other reasons for the temperature change appeared (Fig. 3, curve 3) [9, 10].

2.1.2   Results of Experiments with Prototypes with a Signal Level Sufficient for Sweating and the Manifestation of Additional Temperature Changes in a Person's Response to a Heat Signal and in the Presence of a Gap

Let us increase the level of the heat signal acting on a person to a temperature slightly over 38.4 °C.

At the initial stage of the curve, from 4 min. and up to 54 min, fluctuations associated with the release and evaporation of sweat are clearly visible, similar to the fluctuations shown in Fig. 2 and in Fig. 3, curve 3. Having reached 54 min, the temperature begins to rise even more. To determine the reasons for this rise, let us consider what the temperature is composed of at this moment in time. First, there is the temperature of the external environment, but it does not change during the entire time of the experiment. Secondly, the temperature generated by the heater - its change over time is smooth (Fig. 1). Thirdly, the reaction of a person in the form of the release and evaporation of sweat associated with the effect of a heat signal, but the nature of this signal also differs from that shown in Fig. 3, curve 2. Moreover, the rise in temperature occurs during the period of sweating. That is, when the temperature should drop, but it rises. It remains only to assume that this rise in temperature occurs due to the accumulation of heat due to the limitation of the outflow of heat from the person himself due to the operation of the heater, which prevents this outflow. Of course, this is not a proof yet, but only a guess.

2.1.3   The Results of Experiments with Prototypes with a Signal Level Sufficient for Sweating and the Manifestation of Additional Temperature Changes in a Person's Response to an Acting Thermal Signal and in the Absence of a Gap

Let's do some research that will shed more light on the ongoing processes. To do this, we will remove the gap between the human body and the heater, that is, we will place the heater directly on the surface of the human body. Then the temperature meter, located simultaneously on the surface of both the human body and the heater, will record, against the background of the ambient temperature, temperature changes caused only by heating of the heater and changes in the temperature of the human body (Fig. 3, curve 1). As you can see, changes reflecting the effects on the thermal field associated with the release of sweat were not observed during the entire experiment. The heater itself has only a smooth characteristic, shown in Fig. 1, and cannot generate intermittent heat in this mode of operation. There is no other reason for a rise in temperature, except for the accumulation of heat at this temperature level due to the limitations of its outflow due to the operation of the heater. Thus, the temperature rise starting on this curve from 112 min. occurs only due to the transfer of heat by the human body, which is a reaction to a heat signal.

Let us now compare the rise in temperatures in Fig. 3, on curve 1 with 112 min. and on curve 2 with 58 min. Some discrepancies in time are due to the fact that the test results, recorded in Fig. 3, curve 2, were performed in the area of the spine, where there are few muscles and the time spent by the passage of heat to the heat receptor is small (54 min.), And the tests, the results of which were recorded in Fig. 3, curve 1, were produced in the region of the kidney, where the muscles are much larger and the heat receptors are located deeper [1]. Therefore, it takes a longer time for the heat signal to travel to reach the interoreceptor (112 min). Another reason for the difference in the time the temperature rise occurs is the slight difference in the heating signal levels of the heater.

Although the tests were carried out in different areas of the human body, however, the levels of signals received in response to heat exposure, their shapes and levels are absolutely identical, which also indicates the same nature of their occurrence (Fig. 2, curves 1, 2).

After raising the temperature for 64 min. Figure 3, curve 2 and at 122 min Fig. 3, curve 1, there is a drop in temperatures caused by the switching on of the cooling mechanisms of thermoregulation with the help of interoreceptors [1, 6] (continuation of perspiration, as well as the activation of the mechanism of expansion of the vascular bed, etc.).

2.2   Results of Mathematical Modeling

For more proof that under the influence of the heat signal of the heater on a person, not only temperature changes associated with the release and evaporation of sweat occur [9, 10], but also an increase in temperature due to the transfer of heat by the human body and a drop in temperature due to the inclusion of cooling mechanisms of thermoregulation, mathematical modeling was carried out. During the simulation, the propagation of a heat signal in the area of the thermal system was considered, formed by: the human body - the gap between the human body and the heater - heater - an insulating layer on the outer surface of the heater - the external environment.

When forming the thermal model, the possibility of changing the temperature field in the system not only in space, but also in time was taken into account [12, 13].

In the area of the human body, which was presented as the environment, possible temperature changes were envisaged.

Boundary conditions of the third kind were used at the boundary between the human body and the gap [14–16].

Further, the thermal system was considered in the form of a three-layer thin wall.

The first layer was the gap between the human body and the heater. Changes in thermal parameters occurred in it due to the release and evaporation of sweat.

The second layer was a heater, where smooth heating was carried out.

The third layer is the outer heater insulation layer.

Outside the heat-insulating layer was a constant temperature environment. Boundary conditions of the third kind also operated on the border with it.

Boundary conditions of the fourth kind were applied between the layers of the thin wall.

The program [11] took into account the possibility of changing the thermal parameters in the gap between the human body and the heater. The introduction of an alternating heat signal from the side of the human body made it possible to simulate the transfer of excess heat from the human body and the activation of the cooling mechanism of thermoregulation in response to the heat signal of the heater. This addition made it possible to take into account, within the framework of the thermal system under consideration, temperature changes on the part of the human body, temperature changes as a result of changes in thermal parameters due to sweating, temperature changes due to the operation of the heater, and the temperature of the external environment.

The operation of the program is shown in Fig. 4, 5, 6.

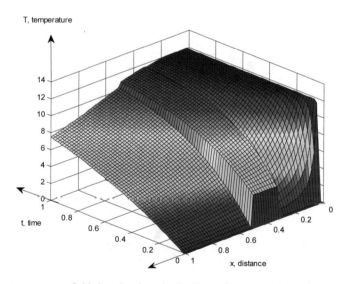

Fig. 4. The temperature field that develops in the thermal system during the operation of the heater and in the absence of a human reaction to thermal effects in the form of sweating and additional heat.

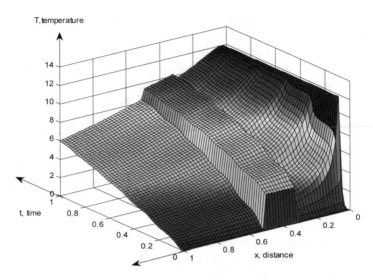

Fig. 5. The temperature field that develops in the thermal system during the operation of the heater, in the presence of a reaction to the thermal effect only in the form of perspiration and its evaporation.

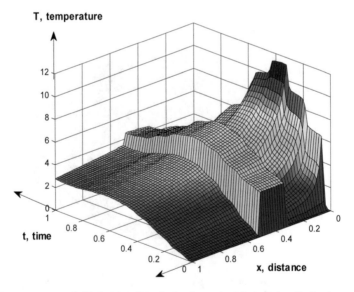

Fig. 6. The temperature field that develops in the thermal system during the heater operation, in the presence of a reaction to the thermal effect of the heater in the form of sweating and its evaporation, and for the manifestation of additional temperature changes in the human response to the thermal signal.

2.2.1   The Result of Simulating the Action of a Signal that is Insufficient Even for Perspiration

In Fig. 4 shows the simulation result under the condition when a person is exposed to the heat of a heater, the level of the heat signal of which is insufficient for the appearance of perspiration and the formation of excess heat by a person. On the right side along the X axis, in the area x = 0, the human body is located, and on the left, in the area x = 1, the external environment, in the middle of the X axis, between the points 0.4 <x <0.6 there is a heater. In the figure, we see only a smooth change in temperature over time within the considered thermal system.

2.2.2   The Result of Simulating the Effect of a Heat Signal, Sufficient Only for the Manifestation of Perspiration

In Fig. 5 shows the same thermal system and its thermal field, formed under the influence of the thermal signal of a higher level heater, and the human reaction to this heat - sweating and its evaporation, causing oscillatory changes in temperature over time. The figure shows the nature of the oscillations in time and their propagation over the space of the heating system. As can be clearly seen from the figure, temperature fluctuations due to changes in thermal parameters due to perspiration and evaporation occur not only in the area of the gap between the human body, but also in the area of the woven heater. This simulation result corresponds to the results of field tests shown in Fig. 3, curve 3. The heater signal level is insufficient for the manifestation of additional temperature changes associated with an increase in the heat signal from the human body.

2.2.3   Simulation of the Action of a Signal Sufficient for Perspiration and for the Manifestation of Additional Temperature Changes in a Person's Response to a Heat Signal from the Body

In Fig. 6 shows the thermal field formed within the considered thermal system under the influence of three temperature components against the background of the ambient temperature: the thermal field of the heater, the reaction to this heat in the form of sweating and its evaporation, in the form of temperature changes on the part of the human body. This model correlates well with the experimental data shown in Fig. 3, curve 2.

3   Conclusions

Thus, in the course of natural and mathematical experiments, results were obtained that confirm that when woven electric heaters are located on the surface of the body, having temperature meters on their surface and generating a certain level of a thermal signal that affects a person, it is possible to obtain information in the form of temperature changes in time: on the work of the sweat glands, on the increase in human body temperature due to a decrease in heat transfer, on the inclusion of cooling mechanisms: expansion of the bed of blood vessels, etc. [1]), preventing further increase in body temperature.

The proposed method provides a precise supply of a thermal signal to the human body, which makes it possible to obtain comparable results of a person's response to this signal. This opportunity is provided through the use of contact heating.

The results obtained can serve as a basis for the development of non-invasive, thermal methods for studying human physiological reactions.

References

1. Ivanov, K.P., Minut-Sorokhtina, O.P., Maistrakh, E.V.: Physiology of thermoregulation. L. Nauka, p. 470
2. Craig, A.B., Dvorak, M.: Thermal regulation of man exercising during water immersion. J. Appl. Physiol. **25**, 28–35 (1968)
3. Bonshel, R., Langberg, H., Olesen, J., Novak, M., Simonsen, L., Buulow, J., Kyar, M.: Regional blood flow during exercise in humans measured by near-infrared spectroscopy and indocyanine green. J. Appl. Physiol. **89**, 1868–1878 (2010)
4. Davis, S.L., Fadel, P.J., Cui, J., Thomas, G.D.: Skin blood flow influences near-infrared spectroscopy - derived measurements of tissue oxygenation during heat stress. J. Appl. Physiol. **100**, 221–224 (2006)
5. Binzoni, T., Cooper, C.E., Wittekind, A.L., Beneke, R., Elwell, C.E., Van De Ville, D., Leung, T.S.: A new method to measure local oxygen consumption in human skeletal muscle during dynamic exercise using near-infrared spectroscopy. Physiol. Measur. **31**, 1257–1269 (2010)
6. Eugene, H.: Vissler Human. Temperature. Control. B, vol. 7, p. 425. Springer (2018)
7 Sukhodoev, V.V.: A modified technique for measuring and evaluating galvanic skin reactions. IP RAS, p. 84 (1990)
8. Shul'zhenko, A.A., Modestov, M.B., Modestov, B.M.: A method for measuring the reaction of human sweat glands in the presence of heat. Patent No. 2578864, Bulletin No. 9 (2016)
9. Shul'zhenko, A.A., Modestov, M.B.: Simulation of human reaction to heat exposure. Bull. Sci. Tech. Dev. **5**, 23–33 (2017)
10. Shul'zhenko, A.A., Modestov, M.B.: A program for simulating the manifestation of new capabilities of a woven electric heater. Certificate of State Registration of Computer Programs No. 2020618044 (2020)
11. Marchuk, G.I.: Computational mathematics methods. Main edition of physical and mathematical literature of the publishing house "Nauka", p. 456 (1977)
12. Samarskiy, A.A., Vabishevich, P.N.: Computational heat transfer. Book House "LIBRO-KOM", p. 784 (2009)
13. Falade, K.I.: A numerical approach for solving high-order boundary value problems. Int. J. Math. Sci. Comput. (IJMSC) 1–16 (2019). https://doi.org/10.5815/ijmsc.2019.03.01
14. Falade, K.I., Tiamiyu, A.T.: Numerical solution of partial differential equations with fractional variable coefficients using new iterative method (NIM). J. Math. Sci. Comput. (IJMSC) 12–21 (2020). https://doi.org/10.5815/ijmsc.2020.03.02
15. Chowdhury, S., Hashan, T., Rahman, A.A., Saif, A.F.M.: Category specific prediction modules for visual relation recognition. J. Math. Sci. Comput. (IJMSC), 19–29 (2019) https://doi.org/10.5815/ijmsc.2019.02.02
16. Jassbi, S.J., Agha, A.E.A.: A new method for image encryption using chaotic permutation. Int. J. Image Graph. Sig. Process. (IJIGSP), 42–49 (2020). https://doi.org/10.5815/ijigsp.2020.02.05

Study of the Force-Moment Sensing System of a Manipulative Robot in Contact Situations with Tenzoalgometry of Soft Biological Tissues

Maksim V. Arkhipov[1(✉)], Mikhail A. Eremushkin[2],
and Ekaterina A. Khmyachina[1]

[1] Moscow Polytech University,
B. Semenovskaya 38, 107023 Moscow, Russian Federation
maksim_av@mail.ru, lusybagrova@gmail.com
[2] FSBI "National Medical Research Center for Rehabilitation and Resortology"
MH RF, N. Arbat 32, 121099 Moscow, Russian Federation
medmassage@mail.ru

Abstract. The active development of robotics, which is taking place today within the framework of the new economic paradigm "4th technological revolution", opens up new opportunities in many areas, including medicine. One of the topical directions for the health care system of economically developed countries is the improvement and clinical application of robotic manipulation mechanotherapy complexes. The Russian medical manipulation robot presented in the article, previously developed and patented, has not only a priority for domestic medicine, but also raises unresolved issues of intelligent situational behavior of the robot, intelligent control of complex dynamic objects, the operator-robot interface, its safety for the patient and the doctor. The article provides models for the development of software for a robotic manipulative mechanotherapy complex based on technologies for providing a biological feedback. Consideration of the development of original parameters for assessing the biomechanical characteristics of soft tissues. An algorithm for tuning the diagnostic unit of a manipulation robot for measuring the biomechanical characteristics of soft tissues is shown. The description and results of the development of a software data processing system using a diagnostic unit of a manipulation robot are given.

Keywords: Manipulation robot · Robotics · Power sensing · Supply · Control · Tensometry · Algometry · Soft tissue · Dynamic model

1 Introduction

In clinical practice, a well-known method for assessing the level of pain in trigger points, proposed by A.M. Vasilenko. called "tensoalgometry". The method of topographic tensoalgometry for assessing the biomechanical state of soft tissues, including tensometry and algometry on symmetrical areas of the patient's back surface, followed by a visual analysis of the results obtained is used to diagnose and evaluate the effectiveness of treatment of pathology of the musculoskeletal system [18].

© The Author(s), under exclusive license to Springer Nature Switzerland AG 2021
Z. Hu et al. (Eds.): AIMEE 2020, AISC 1315, pp. 159–172, 2021.
https://doi.org/10.1007/978-3-030-67133-4_15

To mitigate these shortcomings, we proposed a topographic tensoalgometry method for assessing the biomechanical state of soft tissues, including tensometry and algorithms on symmetrical areas of the patient's back surface, followed by a visual analysis of the results obtained and serving to diagnose and evaluate the effectiveness of treatment of pathology of the musculoskeletal system. Classical methods of diagnostic manual palpation of the soft tissue surface are characterized by subjectivity and lack of quantitative assessments. The lack of quantitative assessments in assessing the state of the musculoskeletal system leads to inaccurate localization of the areas of the pain sources, and, consequently, to inaccurate assignment of area and dosage of the therapeutic effect. Evaluation parameters of the soft biological tissues of various groups of patients provides a large amount of information that is necessary to determine the tactics of the prescribing process from methods of therapeutic intervention to procedures for restorative medicine. The method of automatic topographic tensoalgometry with the use of manipulative robotics makes it possible to obtain a complete diagnostic picture by the parameters of soft tissue with the localization of areas of seals. This information is necessary for a number of specialists (neurologist, orthopedist-traumatologist, chiropractor, exercise therapy doctor, massage therapist) form the tactics of restorative medicine methods. This article describes a new method of topographic tensoalgometry, which allows to expand knowledge about the state and parameters of soft biological tissue of a wide contingent of subjects - from patients undergoing restorative medicine procedures to athletes who undergo various functional tests. It will be possible to apply the proposed method in wide practice after performing technical and medical tests, which are currently continuing to be performed at the Russian Scientific Center for Rehabilitation and Balneology. The introduction of collaborative manipulators into the practice of restorative medicine will provide this area with a new non-medicinal and secure method of diagnostics that will be highly informative for specialists in this area.

2 Traditional Methods and Their Disadvantages

Traditionally, the biomechanical characteristics of soft tissues are determined by palpation: for the skin, this is a rotational compression test (to determine skin tone) and a skin fold test (elasticity test), and to assess muscle tone, it is used to determine the lateral hardness (resistance) of the muscle and depth of immersion (indentation) [1]. Thus, the deformation of pressure, displacement, stretching, twisting of soft tissues (skin, subcutaneous tissue, muscles) is assessed [20]. The obvious disadvantages of the palpation method for assessing the biomechanical characteristics of the state of soft tissues is that it is indicative, does not have clear criteria and is extremely subjective. In recent years, hardware technologies for assessing the condition of the skin and muscles have been increasingly used in clinical practice. These include the PIERA HC-220 diagnostic device (TANITA CORP., Japan), the Cutometer MPA 580 (Courage +Khazaka electronic GmbH, Germany), which characterize the elastic and elastic properties of the skin, the Pat. Dr. SZIRMAI (ELEKTROIMPEX, Hungary), used for the purpose of myotonometry [2, 3]. However, the use of these devices is limited to any one interested area (the skin of the frontal region, the projection of an individual muscle, etc.) [19].

In addition, the research results obtained with their help are not connected in any way and are considered separately. With regard to recording pain threshold and pain tolerance, mechanical algometers (such as Wagner Force Ten Digital Force Gage FPX 50, Wagner instruments, USA) are currently actively used.

A simpler instrument for measuring pain threshold is described in the patent for invention of the Russian Federation No. 2342063 "Method for quantitative assessment of individual pain thresholds."

However, these instruments are not intended to assess areas of localized tenderness, and when used for this purpose, either the determination of a statistical norm or comparison of indicators in similar symmetrical areas of the body is required. In clinical practice, a well-known method for assessing the level of pain in trigger points (painful muscle seals) proposed by Vasilenko A.M. under the name "tenzoalgometry" [4], which consists in using a tensoalgometer, made on the basis of household spring scales. At the same time, depending on the localization of measurements, replaceable nozzles are used (for measurements carried out in the head and distal parts of the limbs, the diameter of the working surface is 1.5 mm, and in the area of massive skeletal muscles −5 mm). Direct tensoalgometry is carried out by smoothly or stepwise increase in pressure on the tested area of the body until the subject's command to stop the pressure [20].

However, this method has the following disadvantages:

− doesn't assess the condition of the skin and subcutaneous tissue;
− doesn't evaluate large areas of surface tissues;
− not calibrated or calibrated instruments are used;
− doesn't give recommendations on the use of specific manual techniques, as a result of which the method of tensoalgometry does not fully reflect the functional state of superficial soft tissues, and also does not provide an opportunity to evaluate the results of treatment and choose the most optimal method for correcting existing disorders.

Quantitative assessments of the proposed method of tensoalgometry make it possible to identify areas of pain localization, to determine numerical indicators in physical units of measuring the state of the musculoskeletal system - an indicator of soft tissue elasticity, measured as the amount of force applied to deform the tissue at a certain distance (N/m). The proposed approach to the use of a manipulation robot that automatically conducts the diagnostic procedure of topographic tensoalgometry will provide a number of advantages. In comparison with manual techniques, the proposed method will allow to introduce quantitative estimates, comparison of progress over time, instant digitization of the obtained measurement results (processing, storage, output, transmission), and higher sensitivity of the obtained measurement data.

3 Automation of Tensoalgometric Topography Procedures

The proposed device relates to medical technology and can be used as a sensor system when controlling a manipulator, in particular, when automating the procedures of tensoalgometric topography and massage techniques. A device is known that contains an anthropomorphic robot manipulator with electromechanical drives, the final link of which is equipped with a massage tool [5]. The device is capable of carrying out a wide range of massage operations, however, it does not provide a high objectification of massage due to the lack of control of the manipulator force and the tensometric topography mode [21].

The closest to the invention is a device containing a force sensor associated with a manipulator and with a diagnostic unit containing means for recording muscle tone parameters at diagnosed points [6]. The device is capable of operating in the mode of muscle tone diagnostics, allowing to judge the tonification or relaxation of the massaged areas, but has limited functionality to control the force of interaction of the end link of the manipulator with soft tissues and elastic materials, as well as tensoalgometric topography at diagnosed points due to the lack of the ability to form control commands when a patient feels pain during control. The tensoalgometric topography is a sequential manual scan of an array of points, while verbal commands from the patient are monitored about the appearance of pain in the diagnosed point. In addition to waiting for a verbal command from the patient, the diagnostician monitors the collection of information on effort and tries to maintain the uniformity of the immersion rate of the diagnostic device. Such multivariate control creates certain difficulties for a specialist conducting diagnostics associated with constant stress along the auditory, tactile and snapping-motor channels, which introduces subjectivity into the process of tensoalgometric topography and entails inaccuracies and errors affecting the description of the results and the appointment of subsequent rehabilitation and treatment [2–4].

Thus, on the basis of the above, the essence of the hardware seems to be the use of an indenter based on the MES 9000 system, which is closely connected with a manipulator and with a strain-gauge unit containing a power sensor and a patient's limit switch for his algorithmic examination. The diagnostic unit is made in the form of a pushbutton switch, the signal from which is sequentially transmitted through the digital input of the I/O module to the central processor using a line through which the converted signal is transmitted to the digital input of the controller, which generates a signal to the first input of the serial port of the personal computer. The second input of the serial port of a personal computer receives a signal from the output of an analog-to-digital converter, which receives a signal from the output of the force sensor of the indenter of the MES 9000 system installed on the end link of the manipulator, and the output of the force sensor is connected to the analog input of the controller, and the controller has output digital channel through which data is transmitted to the trunk connected with the manipulator control system.

The technical result is expressed in increasing the efficiency of control of the interaction force of the indenter plate of the tensoalgometric system installed on the final link of the manipulator with soft tissues at the diagnosed points, as well as increasing the objectification of the tensoalgometric topography due to the use of a limit switch for the patient.

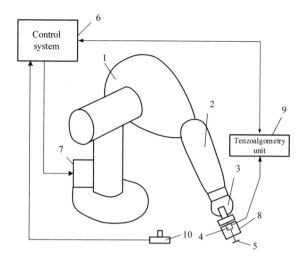

Fig. 1. Tenzoalgometry unit of manipulator control system

Figure 1 shows a diagram of the installation of the proposed device on the robot arm. Robot 1 has an anthropomorphic kinematics of links in the form of a forearm 2 and a hand 3, equipped with a force sensor 4 and a contact plate 5, as well as a control system 6 associated with electromechanical drives 7, an indenter 8, a tensoalgometry unit 9 Fig. 2 and a normally open push-button switch for the patient.

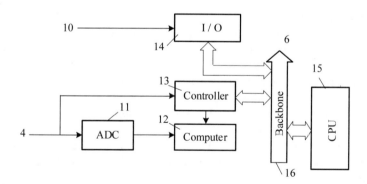

Fig. 2. Tenzoalgometry unit of manipulator control system

The tensoalgometry unit 9 contains a push-button closure 10, the signal from which is sequentially transmitted through the digital input of the input-output module 14 to the central processor 15 using the line 16, through which the converted signal is also transmitted to the digital input of the controller 13 which generates a signal to the first input of the serial port of the personal computer 12. The second input of the serial port of the personal computer 12 receives a signal from the output of the analog-to-digital converter 11, the input of which receives the signal from the output of the force sensor

4, and the output of the force sensor is connected to the analog input of the controller 13, while the controller has a digital channel through which data to highway 16 included in the control system 6.

The design of the hardware part of the invention in a static state contains an indenter rigidly connected to the manipulator and to a strain-gauge unit containing a power sensor and a limit switch for a patient's algometric examination. The tensometric topography unit is made in the form of a push-button closure, the signal from which is sequentially transmitted through the digital input of the input/output module to the central processor by means of a line, through which the converted signal is also transmitted to the digital input of the controller, which generates a signal to the first input of the serial port of a personal computer, to the second input the serial port of a personal computer receives a signal from the output of an analog-to-digital converter, to the input of which a signal is received from the output of the force sensor of the indenter of the MES 9000 system installed on the final link of the manipulator. The output of the force sensor is connected to the analog input of the controller, while the controller has a digital channel through which data is transmitted to the line connected to the manipulator control system.

The operation of the hardware device is based on the fact that the contract platform 5 of the indenter 8 passes sequentially between the deformable diagnosed points, performing pressure at each point, the coordinates of which are calculated in the control system 6, which generates control actions on the electromechanical drives 7 of the links of the shoulder 1, forearm 2 and hand 3 an anthropomorphic manipulator installed by the operator over the projection of the first diagnosed point in a semi-automatic mode, if necessary with a manual adjusting the direction of the contact platform 5 [6–9].

The procedure for controlling the force from the patient's side occurs through a push-button switch 10, and when the maximum value of the limit of the force range is reached, if the command from the patient has not been received, the manipulator 3 exerting pressure through the indenter 8 on the soft tissue at the point to be diagnosed ensures uniform growth force, which is simultaneously read by the force sensor 4 and the controller 13, the speed and accuracy of indentation are formed by commands from the central processor 15 using software-set steps of the lateral displacement.

After transformation by means of software, sections of the array of the tensoalgometric diagram are formed.

The resumption of the recording of a new volume of force information for the next diagnosed point occurs when the platform 5 mechanically contacts the soft tissue of the next diagnosed point.

Sequential interrogation of signals from the power sensor 4 and the pushbutton switch 10 at the diagnosed points through the input/output module 14, the analog-to-digital converter 11, the controller 13, the highway 16 and the central processor 15 make it possible to form a data array in a personal computer 12 for constructing tensoalgometric diagnostic diagrams that clearly reflect objective power information. The sequence of passage of the manipulator 3 trajectories consisting of diagnosed points is determined by a given array of coordinates recorded in the control system 6.

The indenter 8 installed on the manipulator flange 3 sequentially bypasses through the given coordinates of the points and in each of them controls the efforts by the built-in force sensor 4 on those paths through which the passage is carried out. When power information appears within the limits of the power range at each diagnosed point, the force corresponding to the patient's command from the push-button switch 10 is fixed, according to the programmed condition for displacement of the manipulator 3 deforming soft tissue at the diagnosed point with the force ΔF, which is given by the inequalities:

$$\text{If } F_{min} < \Delta F \leftarrow F_k, \text{ то } \Delta z_{i+1} = -\Delta,$$
$$\text{If } F_{min} = \Delta F \leftarrow F_k, F_{max} = \Delta F \leftarrow F_k, \text{ то } \Delta z_{i+1} = 0,$$

where $F_{min} \div F_{max}$ is the range of controlled effort set by software; ΔF is the current effort developed by the manipulator during soft tissue deformation at the diagnosed points; F_k - force corresponding to the achieved value of ΔF when the patient activates the limit switch; Δ is the step of immersion of the contact platform of the indenter when the manipulator is moved in the direction of the z axis.

Thus, an increase in the efficiency and objectivity of the topographic tensoalgometry method is achieved when the manipulator interacts through the indenter with soft tissues and a push-button switch used for patient self-control in the process of constructing level diagnostic diagrams.

The unit of tensoalgometry of the manipulator control system, containing an indenter connected to the manipulator, containing a power sensor and a limit switch for the patient's algometric examination, characterized in that the unit of tensoalgometric topography is implemented in the form of a push-button closure [12].

The signal from the contactor is sequentially transmitted through the digital input of the input/output module to the central processor by means of a line, through which the converted signal is also transmitted to the digital input of the controller generating the signal to the first input of the serial port of the personal computer. The second input of the serial port of a personal computer receives a signal from the output of an analog-to-digital converter, which receives a signal from the output of the force sensor of the indenter of the MES 9000 system installed on the end link of the manipulator, and the output of the force sensor is connected to the analog input of the controller, and the controller has digital channel through which data is transmitted to the trunk connected to the manipulator control system.

4   Clinical Example of the Method

Subject 28 years old, engaged in manual labor. Complaints of pain in the lower back and the right scapular region. History of the present disease. Considers himself sick for more than one year. Not actively treated. According to the results of the visual analogue scale of pain, pain syndrome was 6 points. According to X-ray data, there are signs of thoracic and lumbar osteochondrosis 2 tbsp. Topographic tensoalgometry reveals an increased density of subcutaneous tissue in the lumbar region (Fig. 3) and painful areas in the right scapula and lumbar region (Fig. 4).

Tensoalgometry

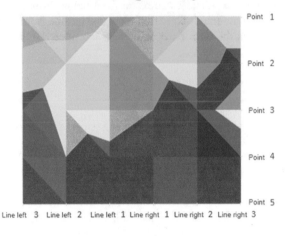

Point 1

Point 2

Point 3

Point 4

Point 5

Line left 3   Line left 2   Line left 1 Line right 1 Line right 2 Line right 3

■ 20-40   ■ 40-60   ■ 60-80

Fig. 3. The tensometry results (diagram and table with the obtained data in gf/cm^2) of patient, 28 years old, before the start of the massage course

Below is the result of the study as the obtained data in gf/cm^2. The research results are needed for further comparison of indicators.

	Line 3 left	Line 2 left	Line 1 left	Line 1 right	Line 2 right	Line 3 right
Point 1	30	30	35	43	45	45
Point 2	45	50	45	56	52	67
Point 3	69	55	50	60	67	50
Point 4	72	60	65	70	65	75
Point 5	70	65	70	75	70	72

After carrying out tensoalgorithm, it is necessary to conduct an analysis Algometry.

When prescribing a course of massage procedures for the back area, using classical massage techniques, 30% of rubbing techniques on the lumbar region and 60% of kneading techniques on the lumbar region and right scapular region were included in the massage technique.

After 12 procedures performed daily, the patient's condition improved significantly. He doesn't actively complain. According to the results of repeated topographic tensoalgometry, the area of the zone of compaction of the subcutaneous tissue of the lumbar region decreased (Fig. 5), and the zones of tenderness associated with increased muscle tone in the lumbar and right scapular regions disappeared (Fig. 6).

Algometry

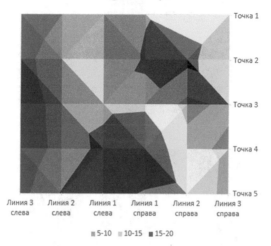

Fig. 4. The algometry results (diagram and table with the obtained data in lb) of the patient, 28 years old, before the start of the massage course

	Line 3 left	Line 2 left	Line 1 left	Line 1 right	Line 2 right	Line 3 right
Point 1	6	10	12,2	14,8	13,4	12,1
Point 2	8,3	13,1	12	15,4	16,5	10,6
Point 3	6,1	12,5	13,9	14,2	8,7	11
Point 4	6	12	16,3	18,9	11,6	8,9
Point 5	8,3	15,2	16,7	14,6	13,7	8,6

Tensoalgometry

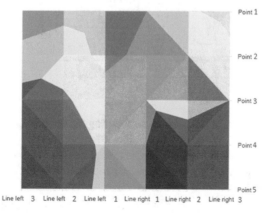

Fig. 5. The tensometry results (diagram and table with the obtained data in gf/cm^2) of a patient, 28 years old, at the end of the massage course

	Line 3 left	Line 2 left	Line 1 left	Line 1 right	Line 2 right	Line 3 right
Point 1	30	35	30	40	42	35
Point 2	45	50	35	45	56	45
Point 3	69	63	48	56	50	60
Point 4	72	70	57	60	73	70
Point 5	70	72	55	60	75	70

Algometry

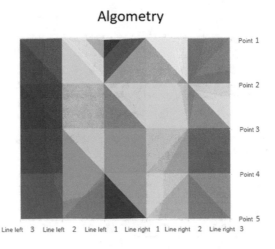

Fig. 6. The algometry results (diagram and table with the obtained data in lb) of a patient, 28 years old, at the end of the massage course

	Line 3 left	Line 2 left	Line 1 left	Line 1 right	Line 2 right	Line 3 right
Point 1	7,1	11,8	11,1	12	8,9	11,6
Point 2	7,4	12,5	11,7	10,8	9,7	9,3
Point 3	6,9	12,1	10,6	11,5	10	11,5
Point 4	7,1	10,8	10,5	11,1	8,4	8,4
Point 5	7	9,5	6,7	11,1	7,2	7,8

The graphical application is written in the programming language C ++, using the Qt library, GPL v3. When plotting the distribution of measured values of force/displacement over the body surface, standard gradient classes from the Qt library are used. The continuity of the gradient filling between adjacent lines is carried out by interpolating the values of the measured values at intermediate points and then displaying them according to the color legend (Fig. 7). Thus, the graphic display of the obtained tensoalgometry data clearly and in an accessible form shows the results of the study for a specialist, which can have a positive effect on the performance of a dosed massage effect.

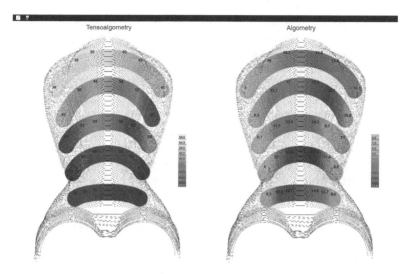

Fig. 7. Graphic display of tensoalgometry results

Thus, we have shown the advantages of the proposed method of topographic tensoalgometry, its practical feasibility and clinical utility.

For the "Method of topographic tensoalgometry" at the Federal Institute of Industrial Property, a notification was received on the acceptance and registration of an application for an invention (registration No. 2018120906 dated 06.06.2018).

The academic value of the presented part of the research is in the fact that the results obtained make it possible to fill the gap in the scientific and practical work of medical specialists, increasing the diagnostic significance of the results obtained, accumulating observation statistics. The primary results obtained during the approbation of this scientific and practical method are currently being introduced into the practice of the exercise therapy department and clinical biomechanics of the Russian Scientific Center for Rehabilitation and Balneology. Since the purpose research were to test a new diagnostic method of topographic tensoalgometry and to increase the significance and value of empirically obtained diagnostic data, a series of clinical experiments has been carried out with the involvement of volunteers and with their consent. A series of clinical experiments let us be accumulate statistical data to assess the objectivity and significance of the proposed method and to draw a conclusion about the need to continue its introduction into the widespread practice of restorative medicine.

5   Conclusions

The reliability of the results obtained contains at the series of the research work of the department of exercise therapy and clinical biomechanics of the National Medical Research Center for Rehabilitation and Balneology of the Ministry of Health of Russia.

The tables and graphs presented in the article are a sample from the massive series of scientific and practical experiments stored in the information database of the center.

For the engineering and technical component of the system, with the help of which the research was carried out, a Russian patent for an invention. The technique of the new method is also protected by a patent. Currently, work is underway to obtain copyright for the developed software product included in this system. The analysis of literature sources showed that manual methods of tensoalgomeric diagnostics, as well as manual diagnostics automated by diagnostic devices, were less precise diagnostics. It has been established that the reason for the low accuracy of measuring the elasticity of soft biological tissue from its true (actual) value is the inconstancy of the diagnostic movements performed by a person, in various applied efforts, the impossibility of ensuring the accuracy of the installation of the sensitive element of the device. Manipulative robotics has the greatest capabilities in interacting with soft tissues [Golovin V., Zhuravlev V., Arkhipov M. Robotics in restorative medicine. Robots for mechanotherapy, Lambert Academic Publishung, CmbH&Co. KG, Europe.280 p.]. This monograph examines the use of manipulation robots to perform a variety of massage therapy procedures. To implement the method of topographic thesoalgometry, a modern collaborative manipulation robot (cobot) was also chosen, which allows performing movements using algorithms of soft techniques (impedance control), as well as implementing the developed algorithm for tensoalgorithmic diagnostics of parameters of soft biological tissues. The capabilities of manipulative robotics make it possible to increase the diagnostic accuracy of measurements, automatically process the results obtained, and also store and transmit data to information systems. The tabular and graphical results presented in the experimental part, obtained using the developed information system, make it possible to obtain a visual integral assessment of the results of the tensoalgorithmic diagnostic procedure. These results are a sample of a series of experiments carried out at the Department of Physical Therapy and Clinical Biomechanics of the Russian Scientific Center of Balneology and Rehabilitation. Each illustration gives the opportunity to detail information about the parameters of soft tissues at each point of the study area. The visual representation information about the localization of problem areas in specific patients, allows adjusting the course of recovery measures. The purpose of the work was to increase the efficiency of restorative medicine procedures by improving the quality of diagnostic information obtained using a software and hardware system based on a collaborative manipulation robot (cobot) equipped with a tensoalgometric module. To solve this goal, the article described the solved tasks for the presentation of the tensoalgometry methodology, the demonstration of the engineering solution for its implementation, the informativeness of the reflection of diagnostic information. The objectives of the study were confirmed by scientific and practical experimental studies carried out in the Department of Physical Therapy and Clinical Biomechanics of the Russian Scientific Center of Balneology and Rehabilitation using specialized equipment including the MES9000 diagnostic system, the manipulation collaborative robot (cobot) Omron TM5-700 model (Fig. 10). The experiments carried out made it possible to carry out a statistical analysis of the data obtained and to confirm the high sensitivity of the new proposed method of topographic tensoalgometry. The research results showed the greatest capabilities of collaborative manipulators in terms of sensitivity, coverage area of the tested areas, fatigue, and absence of errors, compared with manual testing.

Acknowledgements. The scientific work described in this article was implemented on the meringue of the National Medical Research Center for Rehabilitation and Balneology and Moscow Polytechnic University.

References

1. Eremushkin, M.A., Kolyshenkov, V.A., Arkhipov, M.V., Vzhesnevsky, E.A.: Patent RF No. 2695020. Tensoalgometric block of the robotic manipulator control system. Patent of Russia No. 2695020, 13 Aug 2018
2. Arkhipov, M., Leskov, A., Golovin, V., Gercik, Y., Kocherevskaya, L.: Prospects of robotics development for restorative medicine. In: Proceedings of the 25th Conference on Robotics in Alpe-Adria-Danube Region, p. 499–506. Springer, Heidelberg (2016)
3. Arkhipov, M., Orlov, I., Golovin, V., Kocherevskaya, L.: Bio-mechatronic modules for robotic massage. In: Proceedings of the 25th Conference on Robotics in Italy (RAAD16), pp. 499–506. Springer, Heidelberg (2016)
4. Arkhipov, M.V., Vzhesnevsky, E.A., Eremushkin, M.A., Samorukov, E.A., Kocherevskaya, L.B.: Tensometry of soft biological tissues with manipulation robot. In: IOP Conference Series: Materials Science and Engineering, vol. 489, p. 012051 (2019). https://doi.org/10.1088/1757-899x/489/1/012051 /[Электронный ресурс]
5. Gerasimenko, M.Y., Eremushkin, M.A., Arkhipov, M.V., Kolyagin, Y., Antonovich, I.V.: Prospects for the development of robotic manipulation mechanotherapy complexes. Physiotherapy, Balneology Rehabilitation **16**(2), 65–69 (2017)
6. Golovin, V., Zhuravlev, V., Arkhipov, M.: Robotics in Restorative Medicine. LAPLAMBERT Academic Publishing, GmbH & Co. KG, Saarbrücken, p. 270 (2012)
7. Golovin, V., Samorukov, A., Arkhipov, M., Kocherevskaya, L.: Robotic restorative massage to increase working capacity. Altern. Integr. Med. **7**(2), 209–215 (2018)
8. Golovin, V.F., Arkhipov, M.V., Kocherevskaya, L.B.:Robotics to improve the combat effectiveness of military personnel. Sat. In: Abstracts of the International Scientific and Technical Conference Extreme Robotics and Conversion Technologies, pp. 54–57 (2018)
9. Golovin, V.F., Samorukov, A.E., Arkhipov, M.V., Zhuravlev, V.V.: Review of the state of robotics in restorative medicine. Mechatron. Autom. Manage. **8**, 42–50 (2011)
10. Timofeev, G.: Methods of apparatus research of human skin. Cosmet. Med. (4) (2005)
11. Khmyachin, D.V., Khmyachina, E.A., Archipov, M.V., Shumeiko, G.A., Panov, U.P., Ippolitov, S.A.: Distributed control system for mobile medical complexes. In: IOP Conference Series: Materials Science and Engineering, vol. 489, p. 012047 (2019). https://doi.org/10.1088/1757-899x/489/1/012047 [Электронный ресурс]
12. Eremushkin, M.A.: Manual methods of treatment in a complex of rehabilitation measures for pathology of the musculoskeletal system: abstract of thesis. Doctors of medical sciences: 14.00.22, 14.00.51/Center Scientific Research Institute of Traumatology and Orthopedics, N. N. Priorova. - Moscow (2006)
13. Korolkova, T.N., Soghomonyan, A.V.: Methods of functional diagnostics of the skin to assess the effectiveness of treatment of purulent lipodystrophy. Russ. J. Skin Venereal Dis. (6) (2015)

14. Vasilenko, A.M., Zhukolenko, L.V., Popkova, A.M.: Tensoalgometry (clinical use). Russ. Med. J. (1–S), 51–56 (1998)
15. Method of massage and device for its implementation, R. Patent No. 2145833, A 61 H 7/00 (1998)
16. Bio-controlled robot, RU 105588, A61N 15/00 (2011)
17. Rather, N.N., Patel, C.O., Khan, S.A.: Using deep learning towards biomedical knowledge discovery. Int. J. Math. Sci. Comput. 2, 1–10 (2017). Published Online April 2017 in MECS , (http://www.mecs-press.net), https://doi.org/10.5815/ijmsc.2017.02.01
18. Uçar, M.K., Bozkurt, M.R., Bozkurt, F.: Sympathetic skin response: a new biological signal that can be used in diagnosis of fibromyalgia instead of beck depression inventory. Int. J. Image, Graph. Sig. Process. 7, 32–40 (2016). Published Online July 2016 in MECS, (http://www.mecs-press.org/), https://doi.org/10.5815/ijigsp.2016.07.04
19. Uçar, M.K., Bozkurt, M.R., Bozkurt, F.: Determination of new bio signal and tests alternative to verbal pain scale for diagnosing fibromyalgia syndrome. Int. J. Image Graph. Sig. Process. 3, 1–8 (2016). Published Online March 2016 in MECS, (http://www.mecs-press.org/) https://doi.org/10.5815/ijigsp.2016.03.01
20. Sridevi, S., Vijayakuymar, V.R., Anuja, R.: A survey on various compression methods for medical images. Inter. J. Intell. Syst. Appl. 3, 13–19 (2012). Published Online April 2012 in MECS (http://www.mecs-press.org/) https://doi.org/10.5815/ijisa.2012.03.02
21. Mohammed, R.H., Elnaghi, B.E., Bendary, F.A., Elserfi, K.: Trajectory tracking control and robustness analysis of a robotic manipulator using advanced control techniques. Int. J. Eng. Manuf. 6, 42–54 (2018). Published Online November 2018

The Influence of Hydroplasma
on the Proliferative and Secretory Activity
of Human Mesenchymal Stromal Cells

P. S. Eremin[1(✉)], I. R. Gilmutdinova[1], E. Kostromina[1],
I. G. Vorobyova[1], M. Yu. Yakovlev[1], F. G. Gilmutdinova[2],
S. V. Alekseeva[3], E. I. Kanovsky[4], and V. M. Inyushin[3]

[1] National Medical Research Center for Rehabilitation and Balneology
of the Ministry of Health of the Russian Federation,
32 Novy Arbat Street, Moscow 121099, Russian Federation
ereminps@gmail.com
[2] Family Health Center "MiR",
62 Mira Street, Orenburg 460040, Russian Federation
[3] Al-Farabi Kazakh National University, 71/19 Al-Farabi Avenue,
Bostandyk District, Almaty City 950040/A15E3B4, Kazakhstan
[4] TOO "Organic Production",
304 Tole Bi Str., Almaty City 050031/A10X2K1, Kazakhstan

Abstract. Here we review the influence of the biologically active supplement "Hydroplasma" on the proliferative potential of cultured human mesenchymal stromal cells (hMSCs). The following parameters were assessed: cytotoxicity of the studied drug, cell population doubling time, cellular migration, cell cooperation (wound healing assay), and secretory activity of the cell culture. We found that in "Water for Life" Hydroplasma (in concentrations recommended by the manufacturer) and in "Absolute Energy" Hydroplasma (in extremely low concentration) there were no statistically significant differences in the proliferative and secretory activities of the cell cultures relative to the control condition. When estimating the total impact of AE Hydroplasma in standard concentrations, hMSCs death and a low content of cytokines was observed.

Keywords: Human mesenchymal stromal cells · Proliferative potential · Oxidative stress · Hydroplasma

1 Introduction

Damage to the cell membranes and other cell structures by free oxygen radicals is one of the principal reasons for pathological processes that lead to various somatic diseases. Also, as organisms age, the activity of free radicals increases, resulting in oxidative stress of cell structures [1, 2].

Oxidative stress is an intracellular chemical condition that reflects an imbalance between ROI formation (free radicals and peroxides) and the capability of cells to clear them through the antioxidant system (AOS) (e.g., glutathione, ascorbic acid, tocopherol, carnitine, pentoxifylline, etc.) [3]. To correct the processes of free-radical

Z. Hu et al. (Eds.): AIMEE 2020, AISC 1315, pp. 173–179, 2021.
https://doi.org/10.1007/978-3-030-67133-4_16

oxidation (FRO) in various diseases, drugs with antioxidant activity are used. All antioxidants (AO) are classified as indirect and direct-acting drugs. Indirect-acting AO exhibit activity *in vivo* and are ineffective *in vitro*. They stimulate AOS and can reduce FRO intensity. Direct-acting AO have pronounced antiradical properties, which are determined during *in vitro* testing. The majority of drugs with antioxidant effect belong to this group [4–6].

Lately, there has been a significant increase in interest in herbal, natural drugs (nutraceuticals, parapharmaceuticals) as they are safer and more adjusted to human physiology. One of the most common natural compounds is bioflavonoids, the largest class of plant polyphenols (7). Among the numerous biological supplements with bioflavonoids, we can single out the so-called "Hydroplasma". According to the manufacturer, "Hydroplasma" is a concentrate for the preparation of biogenic water, which increases the degree of biological activity of any fluid, streamlining its structure, creating the effect of structuring bioplasm with energy-intensive biogenic memory of "living water". This product neutralises the impact of radicals; restores the central and nervous system and muscular system; improves brain function by 20–25%; increases strength under physical activity; increases endurance and work capacity; fills the body with energy; and improves concentration and memory. This product neutralises all sorts of dependencies and is beneficial during high mental buoyancy, physical training, mountain climbing, and in all areas where human efforts are needed. This concentrate is required for treating a wide range of diseases, and for recovery from severe conditions, and promotes early recovery, exhibits preventive properties, provides health, youth, energy, stress resistance and longevity. The purpose of our work was to investigate the effect of hydroplasma on the proliferative and secretory activity of MMSCs to assess the safety of its use in humans.

2 Materials and Methods

2.1 Cell Culture

The influence of hydroplasma on the proliferative potential of human mesenchymal stromal cells (hMSCs) culture was studied using a commercial hMSC cell line (Sigma-Aldrich, Lot.: 492-05A, 2nd passage). Cell mass growth was conducted according to the manufacturer's recommendations using the growth medium intended for hMSCs cultivation. The test products, Absolute Energy (AE) Hydroplasma and Water for Life (WoL) Hydroplasma, were diluted in a growth medium according to the instructions to the concentration of $\times 1$, $\times 0.5$ and $\times 2$. For AE Hydroplasma, a maximum dilution of $\times 10^{-5}$ was included in the study.

2.2 Assessment of Proliferative Activity of HMSCs

To estimate the proliferative potential, cell culture was subcultured onto 6-well plates with 10^4 cells/cm^2 density. The studied drugs were added at different concentrations and cultivated during 5 days. Cell growth was monitored every 24 h using Zeiss Axio Observer A1 inverted microscope with an AxioCam MRC5 digital camera (Germany).

Upon completion of cultivation, cells were removed from the plastic surface with a trypsin solution with 0.25% EDTA (Stem Cell Technology, USA). The calculation was performed with viability assessment using a Bio-Rad TC-20 automated cell counter (Bio-Rad, USA) according to the manufacturer's methodology and the time of cell population doubling (DT) was calculated using the formula: $TD = (\log_2 2) \times t / [\log_2(N/N0)]$, where t is population growth time, N is the number of cells over time t, and N0 is the initial number of cells.

The wound healing assay method was used to study cell migration and cell cooperation. Cell culture was subcultured onto 6-well plates with 10^4 cells/cm^2 density. Further, it was cultivated until a single layer formation was formed. Next, a mark was made using a sterile serological pipette tip. Then, the studied drugs were added to the culture medium, and the changes were then visualised under a microscope.

2.3 Evaluation of Secretory Activity of HMSCs

To assess the secretory activity at the end of the proliferative potential experiment, hMSCs culture medium was examined. The specimens of the resultant conditioned media were carefully sampled in separate labelled vials and frozen at-20 °C. After obtaining samples of the conditioned media, the secretory activity of cells in the culture medium was evaluated using the Vector-Best ELISA kits according to the manufacturer's method.

2.4 Statistical Analysis

Statistical data processing was performed using IBM SPSS Statistics 19 package. A pair-wise comparison of groups was analysed using Mann-Whitney non-parametric criterion. The data are presented as Me [Q1; Q3].

3 Results and Discussion

3.1 Cytotoxic Effects of Hydroplasma in Cultured HMSCs

In our visual assessments, signs of toxicity of the WoL drug were not observed. Cells grow well in the presence of the drug. During coculturing with WoL, the control group had an advanced growth rate on day 3. There were no visual differences between the × 0.5, × 1, and × 2 dilutions. In this regard, it was decided to continue further experiments with the × 1 dilution. When assessing the cell culture growth with the AE drug in × 0.5, × 1, × 2 dilutions, the cultured hMSCs died. This was probably due to the presence of active photodynamic components having blue-greenish tint in the drug composition. When using the maximum dilution of the AE drug (× 10^{-5}), no toxic effect on cells was observed (Fig. 1).

3.2 Proliferative Activity of HMSCs

During the study of the cells' proliferative activity, the dynamics of growth of experimental and control hMSCs cell cultures were similar. The number of cells

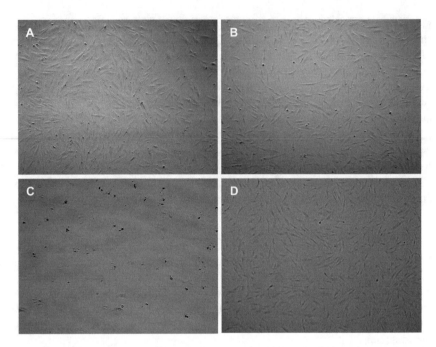

Fig. 1. hMSC cells cultured in the medium (5th day of cultivation). A) Native culture of hMSCs cells. B) hMSCs cell culture in a medium containing the studied WoL drug at a concentration of × 1. C) hMSCs cell culture in a medium containing the studied drug AE at a concentration of × 1. D) hMSCs cell culture in a medium containing the studied AE drug at a concentration of × 10^{-5}. Magnification: × 50.

obtained from one cultural glassware in the control group was 221×10^3 [197×10^3; 237.5×10^3], compared to 210×10^3 [193.5×10^3; 223.5×10^3] cells in the experimental WoL group at a dilution rate of × 1 and 212.5×10^3 [190.75×10^3; 242×10^3] in the AE group at a concentration of 10^{-5}. The time of cell number doubling (DT) for the hMSCs control group was 54.77 h [52.62; 64.23], which corresponds to the literature data [8]. When cultivating cells in media containing the WoL drug, DT was 56.57 h [54.67; 63.61]. Because no viable cells were found in the cell culture of hMSCs, cultivated in the medium containing AE drug in standard concentrations, the cell population doubling time was not calculated. With AE concentration of × 10^{-5}, the proliferation rate of hMSCs cells was 55.9 h [52.23; 65.33]. There was no statistically significant difference between the control cell culture and cells cultured in media containing WoL, which presupposes the same rate of cell culture proliferation. To visually demonstrate the proliferative activity of hMSCs cell culture, the wound healing assay was conducted, which assesses cellular migration and cell cooperation (Fig. 2).

The control cells were actively migrating and completely overlapped the "defect" on the single-layer formation. For the hMSCs that were cultivated in the medium containing WoL, the cells overlapped the defect by 80%, as well as in the medium with the addition of AE at a concentration of × 10^{-5}. Complete overlap of the defect was

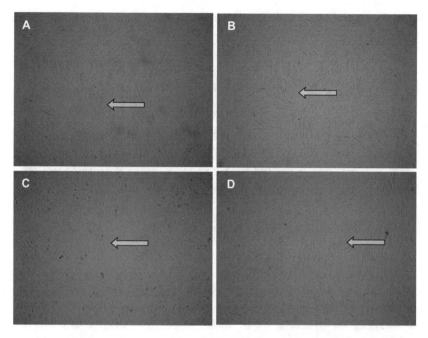

Fig. 2. Wound healing assay. Cultivation (18 h) after forming a mark on the single-layer formation. A) hMSCs cultivated in the standard growth medium, B) hMSCs cultivated in medium containing WoL diluted at a concentration of × 1. C) hMSCs cultivated in a medium containing AE drug at a concentration of × 1. D) hMSCs cultivated in a medium containing AE drug at a concentration of × 10^{-5}. Magnification: × 50. The grey arrows indicates the scratch location.

observed 24 h after the start of the experiment. When cultivated in a medium containing AE Hydroplasma at a concentration of ×1, the cells did not migrate. Instead, they shrivelled-up; single dead cells were observed in the growth medium.

3.3 Secretory Activity of HMSCs

The data on the secretory activity of the conditioned medium in the control group (Table 1).

The data are presented the median and the quartiles: Me [Q1; Q3]. Statistical analysis was performed using Mann-Whitney method. * $p < 0.05$.

The secretory activity of hMSCs cultivated in a medium containing WoL or AE at extremely low concentrations is comparable with the control value. The level of secretion by TNFα cells was slightly higher. No statistically significant differences were found. The secretory activity of IL-6/-8 in hMSCs, cultivated in a medium with the addition of AE at normal concentration, was 5-times lower; cells did not secrete TNF-α, and the secretion of IL-10 was 2–3 times higher. Apparently, this was due to early cell death.

Table 1. Results of the study of the secretory activity of hMSCs in the control and experimental groups.

	TNFα (pg/ml)	IL-6 (pg/ml)	IL-8 (pg/ml)	IL-10 (pg/ml)
Control	2.4 [2.2; 2.85]	71.8 [71.7; 74.6]	65.1 [61.9; 67.7]	2.8 [1.7; 3.4]
WoL × 0.5	3.5 [2.1; 3.8]	68.6 [62.4; 71.9]	67.1 [60.5; 70.4]	1.8 [1.3; 2.2]
WoL × 1	4.8 [1.8; 5.4]	70.4 [58.7; 72.4]	58.7 [54.4; 61.1]	1.6 [1.1; 2.3]
WoL × 2	1.8 [1.7; 2.1]	75.9 [75.1; 78.9]	68.1 [67.1; 70.6]	1.3 [1.1; 1.6]
AE × 10^{-5}	2.3 [1.9; 2.6]	70.3 [65.3; 78.7]	68.3 [63.9; 71.1]	2.5 [1.2; 3.1]
AE × 0.5	0	22.5 [16.5; 27.6]*	17.3 [16.2; 18.5]*	5.8 [4.6; 6.7]*
AE × 1	0	19.1 [10.5; 25.2]*	15.7 [15.3; 16.4]*	6.2 [4.9;7.1]*
AE × 2	0	11.8 [10.3; 12.8]*	12.9 [11.3; 13.8]*	5.0 [4.3; 5.5]*

4   Conclusions

In this study, two samples of hydroplasma were analysed, WoL and AE. Each sample was analysed under three different concentrations, and each analysis was repeated three times. The average value and standard deviation were calculated from the data obtained. When assessing the overall effect of WoL and AE Hydroplasma at extremely low concentrations on the proliferative and secretory activity of hMSCs, no statistically significant differences were found compared to the hMSCs control culture. When estimating the total impact of AE Hydroplasma, the death of individual hMSCs was observed from the first day of cultivation. By the fifth day after the beginning of the experiment, all of the cells were dead. It was associated with a low content of TNFα and IL-6/-8 in the conditioned medium, as well as a minor increase in the amount of IL-10.

References

1. Kalikinskaya, E.Y.: Antioxidants: protection against aging and diseases. Nauka i Zhizn **8**, 90–94 (2000). (in Russian)
2. Badzhinyan, S.A.: Antioxidant therapy. Protection of brain from free radicals. Medical Science of Armenia. V. LVII **1**, 35–44 (2017). (in Russian)
3. Novikov, A.I., Fesenko, V.N., Korenkov, D.G., et al.: New approach to correction of oxidative stress in seminal plasma in men with pathospermia in infertile marriage. Vop. Urol. Andrologii **4**(3), 11–17 (2016). (in Russian)
4. Shakhmardanova, S.A., Gulevskaya, O.N., Seletskaya, V.V., et al.: Antioxidants: classification, pharmacological properties the use in the practice of medicine. Fundam. Med. Biol. **3**, 4–15 (2016). (in Russian)
5. Ushkalova, E.A.: Antioxidant and anti hypoxic properties of Actovegin in cardiac patients. Trudny Patient **3**(3), 22–26 (2005). (in Russian)
6. Halliwell, B.: The antioxidant paradox. Lancet **355**, 1179–1180 (2000)

7. Tregubova, I.A., Kosolapov, V.A.: Spasov AA Antioxidants: current situation and perspectives. Usp. Fiziol. Nauk **43**(1), 75–94 (2012). (in Russian)
8. Aisenstadt, A.A., Enukashvili, N.I., Zolina, T.L., Alexandrov, L.V., Smoljaninov, A.B.: Comparison of proliferation and immunophenotype of MSC, obtained from bone marrow, adipose tissue and umbilical cord. Bull. North-West State Med. Univ. Named After I.I. Mechnicov. **7**(2), 14–22 (2015). (in Russian)

The Biotechnological Method for Constructing Acoustic and Vibration Sequences Based on Genetic DNA Sequences

Roman V. Oleynikov[1], Sergey V. Petoukhov[1,2(✉)], and Ivan S. Soshinsky[2]

[1] Mechanical Engineering Research Institute, Russian Academy of Sciences, M. Kharitonievsky pereulok, 4, Moscow, Russia
spetoukhov@gmail.com
[2] Moscow State Tchaikovsky Conservatory, Bolshaya Nikitskaya, 13/6, Moscow, Russia

Abstract. The paper is devoted to the usage of binary-oppositional molecular features of DNA nucleotides for presentation of DNA sequences as interrelated parallel sets of binary-numeric sequences, which can be converted into corresponding sequences of polyphonic incentives of different physical nature: acoustical, vibrational, electrical, magnetic, and optical. Such interrelated sets of binary-numeric sequences reproduce in some degree a complex informational content of genetic sequences and are interesting for their applications in different biotechnologies concerning cell cultures, regenerative medicine and also musical therapy. The binary-numeric representations of eukaryotic and prokaryotic genomes are described, which have universal algebra-harmonic properties, connected with the harmonic progression known long ago in musicology and other scientific fields. These genomic properties were discovered due to a presented method of oligomer sums.

Keywords: DNA sequences · Genomes · Binary numbers · Rhythm · Dyadic group

1 Introduction

A living organism is a huge chorus of coordinated oscillatory processes, the interaction of which sometimes determines the complex general rhythm of oscillations of biological subsystems. The study of these oscillations and their rhythmic features is an urgent scientific problem, the development of which leads to practical applications in a number of areas: medical diagnostics and physiotherapy, cellular biotechnology, regenerative medicine, ergonomics, music therapy, plant growth stimulation [1–3].

As indicative examples of oscillatory bioprocesses with a complex rhythm can be called some cardiac arrhythmias, as well as pulsations of individual cells, which are accompanied by the generation of complex sounds in the acoustic range, recorded by sensitive devices [4]. The study of such cellular phenomena is carried out within the framework of a scientific area called "sonocytology" and is considered particularly promising for the diagnosis of cancer and other diseases.

Z. Hu et al. (Eds.): AIMEE 2020, AISC 1315, pp. 180–190, 2021.
https://doi.org/10.1007/978-3-030-67133-4_17

Many oscillatory processes are hereditary in nature, that is, they are associated with the genetic system and are transmitted from generation to generation. All inherited physiological systems and processes must be structurally consistent with the genetic coding system in order to be genetically encoded for transmission to future generations. Therefore, a deep understanding of inherited oscillatory processes should be associated with the study of the molecular genetic system, including the complexly organized nucleotide sequences of DNA molecules that carry genetic information. These sequences are not random and have many repeats and complementary palindromes in the genomes of organisms. For example, in the human genome, about a third of DNA sequences are represented by complementary palindromes [5, 6].

Mendel discovered in experiments on the crossing of organisms that the inheritance of traits occurs according to algebraic rules, despite the colossal heterogeneity of the molecular structure of bodies. According to Mendel's law of independent inheritance of traits, information from the level of DNA molecules dictates the macrostructure of living bodies through many independent channels, despite strong noises and interference. For example, hair, eye, and skin colors are inherited independently of each other. This determinism is provided by unknown algorithms of multichannel noise-immune coding. Accordingly, each organism is a multichannel machine of noise-immune coding.

Modern technologies of noise-immune communication and computer informatics are built on the basis of binary numbers, implemented by systems of two-position switches or indicators. The principles of functioning of a living organism are reminiscent of these binary-numerical principles of computers in connection with the well-known universal law "all-or-none" for different types of excitable cells of the organism. All excitable cells, for example, neurons and muscle units, respond to external stimuli either "yes" or "no": the cell does not respond at all to stimuli of a subthreshold value, and to stimuli of any supra-threshold value it reacts in full amplitude.

In connection with the importance of cognition of cellular and other complex biorhythms, which seem to be inherited in a significant degree and connected with genetic nucleotide sequences, we study rhythmic-like features in DNA sequences for a possible technological applications. For this, a convenient and substantiated mathematical language is required that would link the analysis of complex biorhythms with the mathematics of signal processing theory, structural features of genetic informatics and the principles of computers. Such a language should make it possible to record and store samples of complex rhythms in the form of mathematical records for their analytical processing and reproduction in applied problems in the form of sequences of mechanical, electrical, light and other physical influences. Besides, it can be additionally used to simulate rhythmic spatial configurations realized in numerous inherited morphological patterns.

This article describes an approach to the initial development of such a language based on binary-numerical sequences, dyadic analysis, and characteristics of DNA molecular alphabets. It has some connection with the methodology for representing complex musical rhythms with binary numbers, which was previously developed in musicology [7–10]. The possibilities of representing information nucleotide texts of DNA in the form of binary-numerical sequences are considered. Such a representation allows not only to study the principles of noise immunity of these texts, but also, using computer programs, it is easy to transform the binary-numerical representations of genetic texts into the corresponding sequences of physical influences of various nature for their use in biotechnology and regenerative medicine. DNA nucleotide sequences can be represented in the form of binary-numeric sequences in many voluntary ways. For example, in a sequence of nucleotides, you can designate three out of four nucleotides with the digit 0, and the fourth nucleotide with the digit 1. Our represented approach differs from many possible arbitrary approaches in that the transformation of nucleotide sequences into binary-numerical sequences is based on a natural system of the binary-oppositional molecular characteristics in the alphabet of 4 DNA nucleotides. These binary-oppositional characteristics allow transforming any nucleotide sequences in three kinds of the parallel and different binary-numeric sequences mathematically interrelated to each other by the known logical operation of modulo-2 addition. These three parallel and different binary-numerical sequences can be simultaneously converted into acoustic or other physical stimuli, resulting in a single "polyphonic" physical sequence of complex rhythmic character for various biotechnological and medical applications.

At the same time, the article describes some of the results of the study of binary-numerical representations of DNA in the genomes of eukaryotes and prokaryotes. These results reveal the connection of these representations of genomes with harmonic progression, known in the theory of musical harmony since the time of Pythagoras. The paper contains two sections, concluding remarks, acknowledgments, and references.

2   Binary-Numeric Representations of Genetic DNA Sequences

All living organisms have the same molecular basis for genetic coding based on the nucleotide sequences of DNA and RNA. Genetic information in DNA is recorded as long sequences of four types of nucleotides: adenine A, cytosine C, guanine G, and thymine T. This information determines the inherited physiological structures of the body and has the properties of noise immunity, which is necessary in connection with the reproduction of DNA molecules during cell division and transmission to future generations. Modern science doesn't know the following:

1) Why the nucleotide DNA-alphabet is constructed by Nature on the base of 4 letters C, G, A and T? (Why not, for example, 30 or 50 letters?);
2) Why the alphabet is constructed on the base of these simple molecules though millions other molecules exist?

But science knows that this set of 4 molecules possesses symmetrical properties since it has 3 pairs of binary-oppositional traits or indicators:

1) Two letters are purines (A and G), and other two are pyrimidines (C and T). From the standpoint of these traits one can denote $G = A = 0, C = T = 1$. In this case any DNA-text is represented by a corresponding binary sequence. For example, GCATGAAGT.... is 010100001...

2) Two complementary letters (A and T) have two hydrogen bonds and other complementary letters (C and G) have three bonds. From the standpoint of these traits A = T = 0, C = G = 1, and the same DNA-text GCATGAAGT.... is 110010010...
3) Two letters are keto (G and T) and two other letters are amino (A and C). From the standpoint of these traits A = C = 0, G = T = 1, and the text GCATGAAGT.... is 100110011...

So, any DNA-text is a bunch of 3 parallel messages on 3 different binary languages. In other words, the described existence of binary-oppositional indicators in the DNA alphabet dismembers the 4-letter DNA alphabet into three specified sub-alphabets. The use of these sub-alphabets allows each gene and, in general, each fragment of the nucleotide sequence to be represented by three parallel and different binary sequences, which form a family that is closed relative to the known logical operation of modulo-2 addition described below: the logic sum of any two of these binary messages gives the third message (for example, 010100001 and 110010010 gives 100110011).

One can convert each of these logically interrelated and different binary sequences into a corresponding sequence of incentives of different physical nature: acoustical, vibrational, electrical, magnetic, or optical. If you play any two of these three sequences of physical stimuli at the same time and the same way (or play all three sequences at the same time), then you will get a whole "polyphony" of physical stimuli sequences having a complex rhythmic character. These "genetic" sequences of physical stimuli reflect the logical features of the modeled nucleotide DNA sequences and can be used for various applications.

The connection of the DNA alphabet of nucleotides with the logical operation of modulo-2 addition shows a deep conjugation of DNA texts with binary numbers and dyadic analysis of binary sequences. Modulo-2 addition is utilized broadly in the theory of discrete signal processing as a fundamental operation for binary variables. By definition, the modulo-2 addition of two numbers written in binary notation is made in a bitwise manner in accordance with the following rules:

$$0 + 0 = 0, 0 + 1 = 1, 1 + 0 = 1, 1 + 1 = 0 \tag{1}$$

The set of binary numbers 000, 001, 010, 011, 100, 101, 110, 111 forms a dyadic group with 8 members, in which modulo-2 addition serves as the group operation [11]. By analogy dyadic groups of binary numbers with 2^n members can be presented. The distance in this symmetry group is known as the Hamming distance. Since the Hamming distance satisfies the conditions of a metric group, the dyadic group is a metric group. The modulo-2 addition of any two binary numbers from (2) always gives a new number from the same series. The number 000 serves as the unit element of this group.

The representation of DNA texts by binary numbers is especially interesting because it is on binary numbers that many methods of noise-immune transmission of information, Boolean algebra of logic, logical holography, spectral logic, and sequency analysis are built, the concept of geno-logical coding is developed [12]. Besides, the study of complex musical rhythms, that is, sequences of various durations of musical notes in musical pieces gave rise to the method of representing these rhythms in binary-numerical sequences, which is currently used in interdisciplinary studies of the

Moscow State Conservatory. The study of various musical rhythms is important because, as is known, it is the rhythms that are the most or one of the most influencing factors in musical works. In particular, this study is important for the development of music therapy, which is very popular all over the world [13].

Let us turn to the binary-numerical representation of DNA nucleotide sequences based on, for example, the first of the above-mentioned binary DNA sub-alphabets where purines $G = A = 0$ and pyrimidines $C = T = 1$. If a DNA sequence is represented as a sequences of its fragments of length n, then these fragments are called n-plets or oligomers, as is customary in genetics. It turns out that the corresponding binary-numerical representations of the genomes of eukaryotic and prokaryotic genomes obey some algebraic hyperbolic rules reflected features of the cooperative organization of binary-numerical oligomers. Let us explain this phenomenological fact and hyperbolic rules.

Each of long binary-numeric sequences, which presents a genomic DNA, can be considered as a sequence of binary monomers (like as 0-1-0-0-1-0-...), or a sequence of binary doublets (like as 01-00-01-11-...), or a sequence of binary triplets (like as 010-001-111-...), etc. We focus at the numerical analysis of sets of binary n-plets, which belong to the equivalence classes (or cooperative groupings) of 0-oligomers or 1-oligomers, where all oligomers start from the same binary number 0 or 1 correspondingly. For example, the class of the 1-oligomers contains the following n-plets: 2 binary doublets 11, 10; 4 binary triplets 111, 110, 101, 100; etc. The total amount of different kinds of binary n-plets, which start with the same binary number, under fixed n is equal to 2^{n-1}.

To simplify a theoretical explanation, let us consider the example of an analysis of the oligomer cooperative organization of human chromosome №1 by the method of oligomer sums (abbreviation, the OS-method). The totality of data obtained by analyzing a binary-numeric sequence by the OS-method is called its OS-representations.

The application of the OS-method to the analysis of the binary-numeric representation of DNA in human chromosome № 1 includes the following steps, which are typical also for the analysis of other DNA sequences:

- Firstly, in the considered binary sequence, one should calculate phenomenological quantities S_0 and S_1 of digits 0 and 1 correspondingly. In the human chromosome № 1, the following quantities exist: $S_0 = 115125320$, $S_1 = 115355692$;
- Secondly, to construct the oligomer sums sequences, one should calculate the total amounts $\Sigma_{0,n}$ and $\Sigma_{1,n}$ of n-plets in equivalence classes of 0-oligomers and 1-oligomers under $n = 2, 3, 4, \ldots$ (here, for example, the symbol $\Sigma_{0,3}$ refers to the total amount of binary triplets, which start with 0). These total amounts regarding each of the classes are members of the appropriate OS-sequence of the class.

For human chromosome № 1, phenomenological values of the total amounts of n-plets from the class of 0-oligomers and 1-oligomers are shown in the graphical forms for $n = 1, 2, 3, \ldots, 20$ in Fig. 1. Here the abscissa axis represents the values of n, and the ordinate axis represents the values of the total amounts $\Sigma_{0,n}$ and $\Sigma_{1,n}$ of n-plets, which start with 0 and 1. The amazing result is that all 20 phenomenological points with coordinates $[n, \Sigma_{0,n}]$ and $[n, \Sigma_{1,n}]$ lie along with the hyperbolas $S_0/n = 115125320/n$ and $S_1/n = 115355692/n$ correspondingly with a high level of accuracy, whose range is $-0.030\% \div 0.03\%$. Initial data on this chromosome were taken in the GenBank: https://www.ncbi.nlm.nih.gov/nuccore/NC_000001.11.

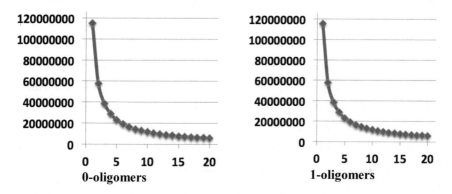

Fig. 1. The graphs of the OS-sequences of binary n-plets from the classes of 0-oligomers and 1-oligomers for the human chromosome № 1. In these graphs, the abscissa axis represents the values $n = 1, 2, 3, \ldots, 20$. The ordinate axis represents the set of phenomenological total amounts $\Sigma_{0,n}$ and $\Sigma_{1,n}$ of n-plets from these classes.

Figure 2 shows appropriate numeric data, which were used as the basis for the graphs in Fig. 1.

This result is striking because it shows that knowing only the sums of binary number 0 (or 1), that is, only one member of the OS-series shown in Fig. 1, one can predict all other 19 members with the high accuracy. Similar results were received for all human 22 autosomes and 2 sex chromosomes X and Y.

These results indicate, at least for human chromosomes, that the following hyperbolic rule exists:

- For any of classes of 0-, and 1-oligomers in individual chromosomes, the total amounts $\Sigma_{0,n}(n)$ and $\Sigma_{1,n}(n)$ of their n-plets, corresponding different n, are inter-related each other through the general hyperbolic expression $\Sigma_{N,n} \approx S_N/n$ with a high level of accuracy (here N refers to any of binary numbers 0 and 1; S_N refers to the sum of binary monomers N; $n = 1, 2, 3, 4, \ldots$ is not too large compared to the full length of the nucleotide sequence).

Similar results were received for many eukaryotic and prokaryotic genomes, whose list is in the preprint [14]. The same oligomer sums method is applicable for analyzing not only complete genomes but also for analyzing – with interesting and unexpected results – of long genes, giant viruses, etc. where long nucleotide sequences are analogically presented as binary-numeric sequences. These genetic binary sequences can be study by known mathematical tools of the dyadic analysis: dyadic shifts, dyadic convolutions, dyadic derivative, and so on.

In various genomes, the revealed and described hyperbolic progressions S_0/n and S_1/n differ from each other only in the magnitude of numerators in their expressions. By the corresponding compression of the ordinate axis in these Cartesian coordinate

n	1	2	3	4	5
0-olig.	115125320	57562404	38373124	28782469	23025759
1-olig.	115355692	57678104	38453882	28837786	23070446

n	6	7	8	9	10
0-olig.	19184994	16449982	14391676	12790520	11510591
1-olig.	19228509	16475880	14418452	12818482	11537515

n	11	12	13	14	15
0-olig.	10466323	9591526	8854157	8225714	7675075
1-olig.	10486488	9615226	8875154	8237219	7690331

n	16	17	18	19	20
0-olig.	7195528	6772990	6395359	6062515	5756293
1-olig.	7209536	6784719	6409142	6068060	5767760

Fig. 2. The total amounts of binary n-plets from the classes of 0-oligomers and 1-oligomers in the case of the human chromosome № 1.

systems (that is by appropriate scaling of numerators S_0, and S_1), each of these hyperbolic sequences is transformed into the canonical hyperbolic sequence (2):

$$y = 1/n : 1/1, 1/2, 1/3, 1/4, 1/5, \ldots, 1/n \qquad (2)$$

The series (2) is known long ago as the harmonic progression (or the harmonic sequence) where each term is the harmonic mean of the neighboring terms. For this reason, the revealed hyperbolic sequences in genomes can be also called genomic harmonic progressions, and, in this mathematical sense, one can talk about the harmonic rules and the harmonious organization of genomes. The historically famous name "the harmonic progression" comes from the connection (2) with the series of harmonics in music. The sums of the first members of the harmonic progression (2) are called harmonic numbers. The rich centuries-old history of the study of the harmonic progression and harmonic numbers is associated with the names of Pythagoras, Orem (d'Oresme), Leibniz, Newton, Euler, Fourier, Dirichlet, Riemann, and other researchers. Using musical terminology, where the term "timbre" refers to the totality of the set of sound frequencies in a prolonged sound, one can conditionally say that the oligomer sums method represents the analyzed nucleotide sequence as some "oligomer timbre". The series of harmonic numbers serves as the discrete analog of the continuous function of natural logarithm $\ln(n)$ [15]; that, in particular, connects the harmonic progression (2) with the Weber-Fechner logarithmic law, which is the main psychophysical law and dictates information peculiarities for all inherited sensory channels - vision, hearing, smell, etc., whose organs (eyes, ears, nose, etc.) very differ each other in appearance. It testifies that genetic and different psychophysical levels of inherited biological informatics structurally correlated on the algebra-harmonic basis [14, 16].

3 Transformations of Binary Sequences into Sequences of Physical Incentives

Let us explain one of possible ways to transform long binary-numeric sequences into sequences of possible physical influences of difference nature - acoustical, vibrational, and electrical, etc. – for various applications in biotechnologies.

Since the DNA alphabet contains three binary sub-alphabets described above, the nucleotide sequence of any of the genes can be represented by three different binary-numerical sequences. Each of these binary sequences can be converted, for example, into a sequence of acoustic sounds according to one rule or another. With the simultaneous sounding of all these genetic binary sequences, a whole polyphony of sounds with complex rhythms arises, which can be used for various applied purposes.

By transforming these binary-numerical sequences by dyadic analysis operations, for example, by the dyadic shift operation, you get extended sets of genetic binary sequences for applied purposes.

In the genetic coding system, 64 triplets, each of which consists of three nucleotides (AAA, ATC, CGT, ...), play a particularly important role as they encode amino acids and stop-signals of protein synthesis. When using the mentioned binary sub-alphabet of nucleotides $A = C = 0$ and $T = G = 1$, a set of these 16 triplets is transformed into a set of eight 3-bit binary numbers, forming a dyadic group (3):

$$000, 001, 010, 011, 100, 101, 110, 111 \qquad (3)$$

If you have a presentation of any of genes in a form of a corresponding long binary-numerical sequence of members of the dyadic group (3), you can convert this binary sequence to a sequence of durations of acoustic sounds by a few ways. For example, you can use a set of 8 different durations of musical note sounds (or other sounds), each of which is related only to one of 8 members of the dyadic group (2). With this approach to acoustical representations of genes, the length of binary oligomers in a fragmentary representation of a binary sequence is essential: if you represent a binary sequence as a sequence of binary doublets (making up a 2-bit dyadic group 00, 01, 10, and 11), then 4 types of durations will be enough. In a general case, if you represent a binary sequence as a sequence of n-plets, then 2^n types of durations of sounds are needed. This way has some relation with a method of binary-numeric representations of musical rhythms, which was developed in the field of musicology [7–10] and is now used in interdisciplinary research at the Moscow State Conservatory. For developing this direction of researches, methods described in [17–22] can be useful.

It should be noted that the concept "musical rhythm" here means a sequence of different durations of note sounds, where all relative durations of sounds belong to a fragment of a special geometric progression 1, $2^{-1}, \ldots, 2^{-k}$ (usually $k < 8$). For this reason, each of such durations can be expressed by a sum of the shortest duration. In musical works, all sequences of notes are usually divided into typical repeated fragments and their associations. In a considered musical piece, one can usually choice the most minimal duration of its notes and use it for an expression of a duration of each of other notes by an appropriate sum of this minimal duration. In this case, a rhythm of the whole musical piece is represented as a sequence of rhythms of its separate fragments represented by binary sequences. In a concrete musical piece, its repeated fragments can have a duration expressed by a sum of 2, 4, 8, ..., 2^k minimal durations. Sequences of the durations of note sounds in musical pieces can provide very complex musical rhythms; but such musical rhythms are only particular cases of rhythms possible in the genetic sequences, when these genetic sequences are represented by binary sequences to be converted into corresponding sequences of physical events, like sequences of acoustical sounds.

Acoustic conversations of binary representations of the genes are also a useful way to study the characteristics of genetic texts by people with a developed sense of rhythm, which is so essential in music. As is known in musicology, « *interaction and the struggle of elements of regularity (rhythmic "consonances") with elements of irregularity (rhythmic "dissonances") is one of the most important factors of musical expressiveness and shaping, due to which moments of tension and resolution appear in a work, a dynamic current is created in the musical form* » [23], [p. 20]. In this regard, experimental studies were carried out [7–10], which made it possible to put forward and substantiate the hypothesis that people have some physiological predispositions to emotionally perceive different rhythms. Simultaneously these experimental studies revealed that most people have a predisposition to perceive music played at a certain pace, which corresponds to the pace of fast running - an average of about 180 steps per minute. The predisposition to this pace may have an evolutionary origin, since it corresponds to the fast running of a person, who fleeing a predator or in pursuit of prey. Additional interesting thematic questions are described in [24].

4 Some Concluding Remarks

The more we know about genetics, the more we understand in what high degree genetics determines many aspects of the life of each organism. For the authors of this article, the question of the relationship between genetic informatics and the inherited biorhythmic characteristics of living organisms is of particular interest. The system of molecular genetic coding has certain logic, presented in the concept of geno-logical coding, and associated with the binary oppositional features of the DNA nucleotide system [12]. The approach described in our article uses these binary oppositional features to represent any of the DNA nucleotide sequences in the form of three different and parallel binary-numerical sequences interrelated on the basis of the logical operation of modulo-2 addition. For the first time, attention is drawn to the fact that the simultaneous conversion of these binary-numerical sequences into sequences of physical stimuli generates a poly-modal (or "polyphonic") complex-rhythmic sequences, reflected hidden logical features of DNA-texts. Considering the close relationship of inherited physiological structures with the genetic coding system, it seems justified to investigate in the future possible biotechnological, medical and musicological applications of these "genetic" complex-rhythmical sequences of physical stimuli of different natures: acoustic, vibrational, electrical, optical and others. Such complex rhythmic sequences obtained from DNA nucleotide sequences should be investigated for their possible connection with complex rhythmic processes in organisms.

In the course of the described study, attention was drawn to the connection between the binary-numerical representations of DNA sequences of the eukaryotic and prokaryotic genomes with certain hyperbolic rules and the so-called harmonic progression (2), widely and long known in science, including musicology. In further study of genetic and musical sequences, a special attention should be paid to their relations to the harmonic progression (2) and harmonic numbers, which are connected with the genetic coding system. On this way, we are waiting for new interesting results and applied solutions. The described approach and results are connected with the quantum information modeling long DNA sequences [14, 25] and understanding biological bodies as quantum information entities. The described results are parts of the development of algebraic and quantum biology.

The authors not only developed the described approach to mathematical modeling of genetic and musical rhythms, but also created original computer programs and devices for transforming binary-numeric sequences in sequences of low-intensity physical events of various natures: acoustic, vibrational, electrical, magnetic, and optical. These devices are designed for applied tasks and are been preparing to be patented.

Acknowledgments. The authors are grateful to their colleagues M. He, Z. Hu, A. Koblyakov, K. Zenkin, I. Stepanyan, and V. Svirin for research assistance.

References

1. Winfree, A.T.: Timing of Biological Clocks. W H Freeman & Co, N.Y (1987). ISBN-10: 9780716750185; ISBN-13: 978-0716750185
2. Aschoff, J. (ed.): Biological Rhythms. Plenum Press, N-Y (1981)
3. Glass, L., Mackey, M.C.: From Clocks to Chaos. The Rhythms of Life. Princeton University Press, New Jersey (1988). ISBN 0-691-08496-3
4. Pelling, A.E., Sehati, S., Gralla, E.B., Valentine, J.S., Gimzewski, J.R.: Local nano mechanical motion of the cell wall of saccharomyces cerevisiae. Science **305**(5687), 1147–1150 (2004). https://doi.org/10.1126/science.1097640
5. Gusfield, D.: Algorithms on Strings, Trees, and Sequences: Computer Science and Computational Biology. Cambridge University Press, Cambridge, p. 556 (1997)
6. McConkey, E.: Human Genetics: The Molecular Revolution. Jones and Barlett, Boston (1993)
7. Soshinsky, I.S., Oleynikov, R.V.: The phenomenon of rhythm and its mathematical representation. Binary code (2019). https://youtu.be/L_y3fuWbRKY
8. Soshinsky, I.S., Oleynikov, R.V.: Tempo in music. Evolutionary factor (2020). https://youtu.be/QCQHY8VJTgc

9. Soshinsky, I.S., Oleynikov, R.V.: How the brain perceives rhythm. Study (2020). https://youtu.be/GqMV20CuQmA
10. Soshinsky, I.S., Oleynikov, R.V.: Rhythm in Music (2020). https://youtu.be/Kkz78ffdVGw
11. Harmuth, H.F.: Information Theory Applied to Space-Time Physics. The Catholic University of America, DC, Washington (1989)
12. Petoukhov, S.V., Petukhova, E.S.: Symmetries in genetic systems and the concept of genological coding. Information 8(1), 2 (2017). https://doi.org/10.3390/info8010002
13. Shushardzhan, S.V., Petoukhov, S.V.: Engineering in the scientific music therapy and acoustic biotechnologies. In: Hu, Z., Petoukhov, S., He, M. (eds) Advances in Artificial Systems for Medicine and Education III. AIMEE 2019. Advances in Intelligent Systems and Computing, Springer, Cham, vol. 1126, pp. 273–282 (2020). https://doi.org/10.1007/978-3-030-39162-1_25
14. Petoukhov, S.V.: Hyperbolic rules of the oligomer cooperative organization of eukaryotic and prokaryotic genomes. Preprints 2020050471, 95 (2020). https://doi.org/10.20944/preprints202005.0471.v2, https://www.preprints.org/manuscript/202005.0471/v2
15. Graham, R.L., Knuth, D.E., Parashnik, O.: Concrete Mathematics. A Foundations for Computer Science. Addison-Wesley, Massachusetts (1994). ISBN 0-201-55802-5
16. Petoukhov, S.V.: The system-resonance approach in modeling genetic structures. Biosystems 139, 1–11 (2016)
17. Awadalla, M.H.: Spiking neural network and bull genetic algorithm for active vibration control. IJISA 10(2), 17–26 (2018). https://doi.org/10.5815/ijisa.2018.02.02
18. Hata, R., Akhand, M.A.H., Islam, M.M., Murase, K.: Simplified real-, com plex-, and quaternion-valued neuro-fuzzy learning algorithms. IJISA 10(5), 1–13 (2018). https://doi.org/10.5815/ijisa.2018.05.01
19. Babichev, S., Korobchynskyi, M., Mieshkov, S., Korchomnyi, O.: An effectiveness evaluation of information technology of gene expression profiles processing for gene networks reconstruction. IJISA 10(7), 1–10 (2018). https://doi.org/10.5815/ijisa.2018.07.01
20. Hussain, A., Muhammad, Y.S., Sajid, M.N.: An efficient genetic algorithm for numerical function optimization with two new crossover operators. IJMSC 4(4), 42–55 (2018). https://doi.org/10.5815/ijmsc.2018.04.04
21. Wani, A.A., Badshah, V.H.: On the relations between lucas sequence and fibonacci-like sequence by matrix methods. IJMSC 3(4), 20–36 (2017). https://doi.org/10.5815/ijmsc.2017.04.03
22. Hamd, M.H., Ahmed, S.K.: Biometric system design for iris recognition using intelligent algorithms. IJMECS 10(3), 9–16 (2018). https://doi.org/10.5815/ijmecs.2018.03.02
23. Holopova, V.N.: Musical Rhythm. Muzyka, Moscow (1980)
24. Oleynikov, R.V.: Musical System Building: The Scientific Explanation of Intervals, Chords, Major and Minor, Keys and Modes. URSS, Moscow (2020). ISBN 978-5-9710-6977-5
25. Petoukhov, S.V., Petukhova, E.S., Svirin, V.I.: Symmetries of DNA alphabets and quantum informational formalisms. Symmetry Cult. Sci. 30(2), 161–179 (2019). https://doi.org/10.26830/symmetry_2019_2_161, http://petoukhov.com/PETOUKHOV%20GENETIC%20QUANTUM%20INFORMATIONAL%20MODEL%202019.pdf

Modelling of Piezoelectric Disk Transducers Operated on Non-axisymmetric Oscillations for Biomedical Devices

Constantine Bazilo[(✉)] and Liudmyla Usyk

Cherkasy State Technological University,
Shevchenko Blvd, 460, Cherkasy 18006, Ukraine
b_constantine@ukr.net, luda.usyk@gmail.com

Abstract. This paper will review the research conducted on the mathematical description of the non-axisymmetric oscillations of a piezoelectric element with separated electrodes. The authors have devised a measurement scheme for physical and mechanical constants of piezoceramics to obtain meaningful and reliable quantitative results describing the important parameters of the physical state of piezoelectric transducers. As one of the outcomes, the research offers a construction scheme for the mathematical model which allows incorporating a set of geometrical, physical, mechanical, and electric parameters of a real-life design of the piezoelectric transducer with sector-electroded working surface.

Keywords: Piezoelectric disk · Non-axisymmetric oscillations · Physical processes · Mathematical description

1 Introduction

Over past decades, innovational industrial technologies have been extensively penetrating various spheres of human activity all over the world. This happens primarily due to continuously expanding implementation of such areas as nanotechnology, bionics, functional medicine, nano electronics, and other fields, as well as with significant investments into these areas made by the world's leading research and production companies [1].

A distinctive feature of the material, possessing the ability to convert energy in both directions, enables piezoelectric materials to become highly suitable for designing intelligent systems [2]. Such systems are capable of regulating their behavior basing on information received from the environment in real time [3, 4]. As an example of a successful implementation of intelligent systems, we can mention here industrial and medical robots [5].

Sensory and executive properties, ability to generate and detect ultrasonic frequencies (that is, ultrasound), and overall high potential of piezoelectric materials in intelligent systems justify their high suitability for medical diagnosis and therapy [5–7]. Tactile sensors, which are devices that detect contact or touch, find medical applications in minimally invasive surgery. Their sensitivity is sufficient to detect the closeness of soft tissues, which helps to operate surgical instruments gently and safely during

Z. Hu et al. (Eds.): AIMEE 2020, AISC 1315, pp. 191–200, 2021.
https://doi.org/10.1007/978-3-030-67133-4_18

surgery [8]. Piezoelectric accelerometers function as overly sensitive vibration sensors. In medical applications, they are aimed at detecting the inertia forces created by the body's movement [9]. The pressure sensor can be employed for continuous measuring arterial pressure, for example, in the course of heart surgeries [10].

Nowadays, elements of electronic devices often comprise micro electro mechanical systems (MEMS). In MEMS, partial electroding is most commonly used for working surfaces. It therefore becomes possible to control the energy of oscillatory processes in MEMS by manipulating the geometric parameters of the electroded surfaces. When partial electroding is performed with a violation of axial symmetry of the entire structure, additional opportunities for controlling stress-strain state parameters of the piezoelectric disk and transmission characteristics of the electromechanical system as a whole are enabled [11].

Diversity and variation of practical applications of disks with sector electroding naturally stimulate theoretical research, the purpose of which is to predict character-istics and technical parameters of piezoelectric devices, designed on their basis. The forecast is devised based on a mathematical model, which is the main result of the theoretical description of a real device. Studying the mathematical model of the real device permits us to determine the set of geometrical, physical, mechanical, and electric parameters of the real object, which will ensure the implementation of technical indicators of the functional element of the piezoelectronics specified in the technical task. This significantly reduces the time and the cost required to develop new functional elements of piezo electronics. The cost of resources saved thus constitutes the com-mercial price of the mathematical model [11].

The aim of this research paper will be to construct a mathematical model of piezoceramic disk transducers with separated electrodes operating on non-axisymmetric oscillations.

2   Literature Review

Systems used in critical applications, such as health care, robotics, etc., must be exceptionally reliable [12]. A failure of "critical systems" may lead to personal injury, environmental damage, or significant economic losses [13]. Developing systems of critical application, as well as increasing their efficiency, are both impossible without prior solving the problem of their mathematical modeling [14].

In practice, disks with sector electroded surfaces become a basic element in many microelectromechanical structures [15]. On the question of modern disk piezoelectric elements design, findings reveal that most of them have an asymmetrical arrangement of electrodes on their surface [16–19]. This arrangement of the electrodes, as well as the method of fastening the elements, induces non-axisymmetric oscillations within the disk elements.

Several authors [20–22] have attempted to describe modeling non-axisymmetric oscillations of thin disks covered with asymmetric electrodes; however, the models of elements mentioned in the above studies differ from those used in modern devices. A prior study [23] offers a simulation of non-axisymmetric oscillations of a disk resonator; nonetheless, the mentioned analysis lacks ат exhaustive consideration of the piezoelectric properties of the device.

Theoretical and experimental studies of the kinematic characteristics of non-axisymmetric planar oscillations in piezoelectric thin disks and disks with partial electroding of one or both surfaces launched in the second half of the twentieth century. The monograph [23] attempts to summarize the results of these studies. Currently, a considerable amount of literature is being published on various aspects of the theory of non-axisymmetric planar oscillations in thin piezoelectric disks [24–26]. The above works adopt, so to speak, a kinematic approach to describing non-axisymmetric oscillations, that is, an approach that takes into account the sole fact of absence of the axial symmetry (a series of trigonometric cos $m\varphi$ or sin $m\varphi$, in which $m = 1, 2, 3, \ldots$; φ represents the circular coordinate in the cylindrical coordinate system), but real-life conditions under which these oscillations excite and exist are not taken into account. In the monograph [23], while attempting to construct the theory of planar non-axisymmetric oscillations in a thin piezoceramic disk, the author wrongly uses Helmholtz's representation of the vector of displacement of material particles in an oscillating piezoceramic disk through scalar and vector potentials. The discrepancy lies in the fact that in anisotropic elastic environments, the Helmholtz's representation does not lead to separating the equation of motion into two Helmholtz equations for scalar and vector potentials. Meanwhile, scalar and vector potentials in the monograph [23] are described as solutions of the corresponding Helmholtz equations. The latter leads to constructing expressions for calculating the components of the displacement vector of the material particles of the piezoceramic disk, in which cylindrical functions of integer orders are included, which does not correspond to the real situation. We would also like to point out that the above studies lack consideration of the actual electrical state of the volume elements of the piezoceramic disk under different electrodes, which excludes the very possibility of using published theoretical constructions as a calculation method for defining transfer characteristics of the functional piezoelectric elements based on disks with separated electrodes [11].

Hence, the development of principles and methods for calculating the characteristics and parameters of the mechanical component of MEMS poses a relevant and practically significant problem. The latter inspires a search for new approaches to the procedure of calculating the parameters of the stress–strain state and transmission characteristics of piezoelectric transducers with separated electrodes [27].

3   Methodology

First of all, let us consider the construction, which is shown in Fig. 1.

The lower surface of the disk $z = 0$ (z is the coordinate line of the cylindrical coordinate system (ρ, φ, z), connected to the axis x_3 of the right-handed Cartesian coordinate system (x_1, x_2, x_3) (Fig. 1)) is covered with a thin layer of metal and is grounded; that is, its electric potential equals to zero at any time. The upper surface of the disk $z = \alpha$ is covered with two sector electrodes (positions 1 and 2). Sector electrodes 1 and 2 are separated by a non-electrode gap, the width of which is significantly less than the thickness of the piezoceramic disk.

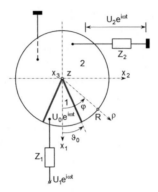

Fig. 1. Design of a piezoelectric transducer with sector electrodes

A harmonical time-varying difference of electric potentials $U_0 e^{i\omega t}$ is being applied to the sector electrode 1. Symbol Z_1 indicates the complex output resistance of the electrical signal source. Obviously, $U_0 = U_1 Z_{el}^{(1)}(\omega) \Big/ \Big[Z_1 + Z_{el}^{(1)}(\omega) \Big]$ where U_1 marks the amplitude value of the potential difference at the output of the source of the electrical signals; $Z_{el}^{(1)}(\omega)$ represents electrical impedance of the piezoceramic disk element under the sector electrode 1. The electric potential forms an alternating electric field under the electrode 1, which generates harmonical time-varying elastic deformations within the volume of the piezoceramic disk. Harmonic oscillations of the material particles of the disk generate a polarization charge on the sector electrode 2 $q_2(t) = Q_2 e^{i\omega t}$, where Q_2 defines the amplitude value of the polarizing charge. With its electric field, the polarizing charge $q_2(t)$ moves the free carriers of electricity (electrons) within the electric current conductor that connects the electrical load Z_2 to the sector electrode 2. When an electric current passes through the complex resistance, a potential difference $U_2 e^{i\omega t}$ is released on it, where U_2 marks the amplitude value of the potential difference in the secondary electric circuit of piezoelectric transducer.

Therefore, we can present the equation that expresses calculations for the input electrical impedance as follows:

$$Z_{el}^{(1)} = \frac{1}{i\omega C_1^* \Psi^{(0)}(\gamma, \lambda, R)}, \qquad (1)$$

where

$$\Psi^{(0)}(\gamma, \lambda, R) = \frac{4 K_{31}^2}{\pi} \sum_{m=0}^{\infty} \frac{\sin^2 m\vartheta_0}{m^2 \vartheta_0} \frac{\Psi_z^{(m,0)}(\gamma, \lambda, R)}{\Psi_0^{(m)}(\gamma, \lambda, R)} + 1;$$

the symbol $\Psi_z^{(m,0)}(\gamma, \lambda, R)$ designates here a dimensionless analytical design

$$\Psi_z^{(m,0)}(\gamma, \lambda, R) = \frac{1}{\gamma R} \int_0^{\gamma R} \gamma \rho \, W_z^{(m,0)}(\gamma, \lambda, \rho) \, d(\gamma \rho); \quad \text{while} \quad W_z^{(m,0)}(\gamma, \lambda, \rho) \quad \text{marks}$$

zero approximation to the exact value of the linear invariant of the strain tensor (volume deformation) in the mode of planar oscillations;

$$\Psi_0^{(m)}(\gamma, \lambda, R) = F_C^{(m,0)}(\lambda, R)[\gamma R J_{v_m-1}(\gamma R) - (v_m - k) J_{v_m}(\gamma R)]$$

$$+ km\left[F_A^{(m,0)}(\gamma, \lambda, R) J_{\mu_m}(\lambda R) + \frac{\pi}{2} N_{\mu_m}(\lambda R) F_C^{(m,0)}(\lambda, R) \int_0^R \frac{1}{x} F_{v_m}^{(0)}(\gamma x) J_{\mu_m}(\lambda x)dx\right];$$

$$F_C^{(m,0)}(\lambda, R) = \lambda R J_{\mu_m-1}(\lambda R) - (\mu_m + 1) J_{\mu_m}(\lambda R);$$

$$F_A^{(m,0)}(\gamma, \lambda, R) = mJ_{v_m}(\gamma R) + \frac{\pi}{2}\left[\lambda R N_{\mu_m-1}(\lambda R) - (\mu_m - 1) N_{\mu_m}(\lambda R)\right] \int_0^R \frac{1}{x} F_{v_m}^{(0)}(\gamma x) J_{\mu_m}(\lambda x)dx;$$

$$F_{v_m}^{(0)}(\gamma\rho) = \frac{(3-k)m}{1-k}\left\{J_{v_m}(\gamma\rho) + \left(\frac{k+1}{3-k}\right)\gamma\rho\left[J_{v_m-1}(\gamma\rho) - \frac{v_m}{\gamma\rho}J_{v_m}(\gamma\rho)\right]\right\};$$

$\mu_m = \sqrt{1 + 2m^2/(1-k)}$ is the fractional order of the Bessel function $J_{\mu_m}(\lambda\rho)$ and Neumann function $N_{\mu_m}(\lambda\rho)$; $v_m = \sqrt{1 + (1-k)m^2/2}$ is the fractional order of the Bessel function; $k = c_{12}^*/c_{11}^*$ is the piezoelectric parameter.

The function $\Psi^{(0)}(\gamma, \lambda, R)$ is a zero approximation to the exact value of the function $\Psi(\gamma, \lambda, R)$, which determines the frequency-dependent change of the input electrical impedance of a degenerate piezoelectric transducer. If we accept that $\Psi^{(0)}(\gamma, \lambda, R) \rightarrow \infty$, which is observed when $\Psi_0^{(m)}(\gamma, \lambda, R) = 0$, then electrical impedance will constitute $Z_{el}^{(1)} \rightarrow 0$, which corresponds to the physical state of the electromechanical system, which is called electromechanical resonance. In the immediate vicinity of the electromechanical resonance frequency, function $\Psi^{(0)}(\gamma, \lambda, R) = 0$ and $Z_{el}^{(1)} \rightarrow \infty$, which corresponds to the electromechanical antiresonance.

Figure 2 demonstrates the results of calculations according to formula (1) of the electrical impedance modulus $Z_{el}^{(1)}$ for different values of the angular size ϑ_0 of the sector electrode. Numerical values ϑ_0 in arc degrees are indicated in the figures field of the graph. The values were calculated for PZT piezoceramics with the following physical and mechanical parameters: $c_{33}^E = 108, 6$ GPa; $c_{11}^E = 94, 8$ GPa; $c_{12}^E = 44$ GPa; $\chi_{33}^\varepsilon = 7, 0227 \cdot 10^{-9}$ F/m; $e_{33} = 13, 17$ C/m^2; $e_{31} = -5, 89$ C/m^2; $\rho_0 = 7428$ kg/m^3. The dimensions of the piezoceramic disk constituted: radius $R = 33 \cdot 10^{-3}$ m, thickness $\alpha = 3 \cdot 10^{-3}$ m. The calculations were performed for two piezoceramic quality factors Q_p: 80 units (solid line) and 160 units (dashed line).

The values of the modulus $Z_{el}^{(1)}$ were plotted on the ordinate axes, while the frequency values f were plotted on the abscissa axes. When performing the calculations, we added the first three circumferential harmonics, that is, expressions with index values $m = 0, 1, 2$. The graphs clearly demonstrate that along with increasing angular size ϑ_0 of the sector electrode within the frequency range of $(5 \div 100)$ kHz,

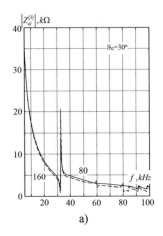
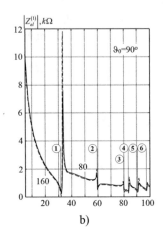

a)   b)

Fig. 2. The electrical impedance modulus $Z_{el}^{(1)}$ of the part of the piezoceramic disk under the sector electrode 1.

electromechanical resonances are manifested which are caused by the zero circumferential harmonic (resonances 1 and 4 in Fig. 2, b), the first circumferential harmonic (resonances 2 and 5) and the second harmonic (resonances 3 and 6).

Next, we calculated the zero approximation to the exact value of the frequency equation of non-axisymmetri mode m of the planar oscillations of a thin piezoceramic disk, which constitute $\Psi_0^{(m)}(\gamma, \lambda, R)$, while $\gamma = \omega / \sqrt{c_{11}^*/\rho_0}$ indicates the wave number of radial oscillations of the thin disk; $\lambda = \omega / \sqrt{(c_{11}^* - c_{12}^*)/(2\rho_0)}$ marks the wave number of circumferential planar oscillations of the material particles of a thin piezoceramic disk, where ρ_0 is the piezoceramics' density; c_{11}^* denotes the elasticity modulus of a thin piezoceramic disk for the mode of planar oscillations; $c_{12}^* = c_{12}^E(1 - c_{12}^E/c_{33}^E)$; c_{12}^E and c_{33}^E are the reference values of the elasticity modulus for piezoceramics polarized by the thickness of the disk.

It should be noted that either the first root $\chi_1^{(1)}$ or, what is equal to that, the frequency of the first electromechanical resonance for the first ($m = 1$) circular mode, have a lesser value as compared to the frequency of the first electromechanical resonance of axisymmetric radial oscillations. As the number of the circular mode m increases, the numerical values of the dimensionless frequency increase monotonically; accordingly, the total number of roots of the equation $\Psi_0^{(m)}(\gamma_n, \lambda_n, R) = 0$ within any fixed range of numerical values of dimensionless wave numbers (frequencies) has a fixed (limited) value.

4   Experiments and Results

A recent study [28] considers a method of experimental determination of the electrical impedance of a piezoceramic disk. The scheme employed in our experimental research is demonstrated in Fig. 3. Prior to determining the electrical impedance, the piezoceramic disk is weighed and its thickness α and radius R are to be measured.

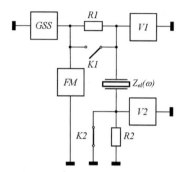

Fig. 3. Wiring diagram for measuring electrical impedance of a piezoceramic disk

A general recommendation for experimental studies of this kind may be to use massive disks, which minimize the effects associated with the added mass, that arise in the process of soldering conductors to the electroded surfaces [29] of the disk.

One more recommendation for measuring the electrical impedance would be to employ a model, the electrical circuit of which is shown in Fig. 3, where *GSS* is a generator of sinusoidal signals, *FM* marks a frequency meter, *V1* and *V2* represent voltmeters, *K1* and *K2* are mechanical keys, *R1* and *R2* designate load resistors. The sample under investigation is indicated as $Z_{el}(\omega)$. The positioning of the keys *K1* and *K2*, as the diagram demonstrates, corresponds to the mode of electrical impedance measurement in the vicinity of the electromechanical resonance frequency. In case of measurements within the vicinity of the frequencies of electromechanical antiresonance, the key *K1* is be closed and key *K2* is to be open [30].

Figure 4 presents the data obtained from measuring the electrical impedance of the PZT piezoelectric disk Ø66 × 3 mm with different values of the opening angle ϑ_0 of the sector electrode in the primary electric circuit.

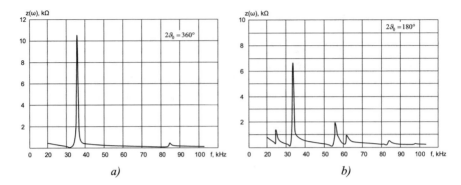

Fig. 4. Electrical impedance of the piezoelectric disk at different values of the angle of opening of the sector electrode in the primary electric circuit: *a)* $2\vartheta_0 = 360°$; b) $2\vartheta_0 = 180°$

5   Discussions

The results of the experiment suggest that, as the opening angle of the sector decreases, an increasing number of non-axisymmetric modes participate in the formation of the stress-strain state of the disk, and the number of resonances after the first axisymmetric radial resonance increases. The numerical values of resonance frequencies are therefore determined by the size, physical and mechanical parameters of the disk element material. The ratios of the resonance frequencies of the same disk are practically determined solely by its size. Thus, we observed a fairly satisfactory coincidence of theoretically and experimentally determined relations of resonance frequencies. The discrepancy between the absolute values of the resonance frequencies can be explained by the disparity between the piezoceramic physical and mechanical parameters, on which the calculations were based, and those that were inherent in the experimentally studied object. Comparing the curves, we concluded that the difference between the quality factor of the experimentally studied sample and the quality factor on which the calculations were based constituted at least 1.2 times.

Thus, we can assert that the nature of the change demonstrated by the curves in a fairly wide frequency range coincides. The latter implies that expression (1), which is a mathematical model of a piezoelectric disk with sector electroding, is sufficiently adequate to the real object and the processes occurring in it. In its turn, it suggests that the mathematical description of the stress-strain state of the disk element also corresponds satisfactorily to the real state of affairs.

6   Conclusions

This study set out to determine the features of a mathematical description of non-axisymmetric oscillations of a piezoelectric element with separated electrodes. We developed a set of computational procedures based on the fundamental principles of mechanics and electrodynamics, the sequential implementation of which allows constructing a mathematical model of a disk piezoelectric transducer with sector electrodes. The results of this investigation prove that the frequency of the first elec-tromechanical resonance for the first circumferential mode is characterized by lesser values than the frequency of the first electromechanical resonance of axisymmetric radial oscillations. Another important finding was that with a decreasing angle of the sector opening, an increasing number of non-axisymmetric modes participate in the formation of the stress-strain state of the disk. Correspondingly, the number of reso-nances after the first axisymmetric radial resonance increases as well. The mathematical model, constructed in this research, considered the case with a short-circuited electrode of the output section of the converter. The following research may possibly be aimed at studying and building a mathematical model of multi-section transducers with sector electroding.

Acknowledgements. The research was conducted within the framework of the state-funded research topic "Development of highly efficient intellectual complex for creation and research of piezoelectric components for instrumentation, medicine and robotics".

References

1. Sharapov, V.: Piezoceramic Sensors. Springer (2011)
2. Safari, A., Akdogan, E.K.: Piezoelectric and Acoustic Materials for Transducer Applica-tions. Springer, New York (2008)
3. Olakanmi, O.O.: A multidimensional walking aid for visually impaired using ultrasonic sensors network with voice guidance. Int. J. Intell. Syst. Appl. (IJISA) **6**(8), 53–59 (2014). https://doi.org/10.5815/ijisa.2014.08.06
4. Li, J.: Design of active vibration control system for piezoelectric intelligent structures. Int. J. Educ. Manag. Eng. (IJEME) **2**(7), 22–28 (2012). https://doi.org/10.5815/ijeme.2012.07.04
5. Sienkiewicz, M.L.K.: Concept, Implementation and Analysis of the Piezoelectric Resonant Sensor/Actuator for Measuring the Aging Process of Human Skin. Institute National Polytechnique de Toulouse, Toulouse (2016)
6. Zhu, X.: Piezoelectric ceramic materials: processing, properties, characterization, and applications. In: Piezoelectric Materials: Structure, Properties, and Applications. Nova Science Publishers, Inc. (2009)
7. Vives, A.A.: Piezoelectric Transducers and Applications. Springer (2008)
8. Jalkanen, V., Andersson, B.M., Bergh, A., Ljungberg, B., Lindahl, O.A.: Resonance sensor measurements of stiffness variations in prostate tissue in vitro–a weighted tissue proportion model. Physiol. Meas. **27**(12), 1373–1386 (2006)
9. Okuno, R., Yokoe, M., Akazawa, K., Abe, K., Sakoda, S.: Finger taps movement acceleration measurement system for quantitative diagnosis of Parkinson's disease. In: 2006 International Conference of the IEEE Engineering in Medicine and Biology Society, New York, pp. 6623–6626 (2006)
10. Vijaya, M.S.: Piezoelectric Materials and Devices. CRC Press, Boca Raton (2013)
11. Petrishchev, O.N., Bazilo, C.V.: Principles and Methods of Mathematical Modelling of Oscillating Piezoelectric Elements. Gordienko Publ., Cherkasy (2019). (in Russian)
12. Koren, I., Krishna, C.M.: Fault-Tolerant Systems. Elsevier/Morgan Kaufmann, Burlington (2007)
13. Sommerville, I.: Software Engineering. Pearson, Boston (2016)

14. Poroshin, S.M., Semenov, S.G.: Development and research of the mathematical model of computerized information-measuring control system of critical application, taking into account the factor of external influences. Inform. Process. Syst. **2**(109), 226–234 (2013). (in Russian)
15. Varadan, V., Vinoy, K., Jose, K.: RF MEMS and Their Application. Technosphere, Moscow (2004). (in Russian)
16. Yan, L., Wu, J., Tang, W.C.: Piezoelectric micromechanical disk resonators towards UHF band, vol. 2, pp. 926–929 (2004)
17. Wang, J., Ren, Z., Nguyen, C.T.-C.: 1.156-GHz self-aligned vibrating micromechanical disk resonator. IEEE Trans. Ultrason. Ferroelectr. Freq. Contr. **51**(12), 1607–1628 (2004)
18. Nguyen, C.T.-C.: Integrated micromechanical circuits fueled by vibrating RF MEMS technology, pp. 957–966 (2006)
19. Nguyen, C.: MEMS technology for timing and frequency control. IEEE Trans. Ultrason. Ferroelectr. Freq. Contr. **54**(2), 251–270 (2007)
20. Huang, C.-H., Lin, Y.-C., Ma, C.-C.: Theoretical analysis and experimental measurement for resonant vibration of piezoceramic circular plates. IEEE Trans. Ultrason. Ferroelectr. Freq. Contr. **51**(1), 12–24 (2004)
21. Lin, Y.-C., Ma, C.-C.: Experimental measurement and numerical analysis on resonant characteristics of piezoelectric disks with partial electrode designs. IEEE Trans. Ultrason. Ferroelectr. Freq. Contr. **51**(8), 937–947 (2004)
22. Huang, C.-H.: Resonant vibration investigations for piezoceramic disks and annuli by using the equivalent constant method. IEEE Trans. Ultrason. Ferroelectr. Freq. Contr. **52**(8), 1217–1228 (2005)
23. Shulga, M.O., Karlash, V.L.: Resonant electromechanical vibrations of piezoelectric plates. Naukova dumka (2008). (in Ukrainian)
24. Naciri, I.: Modeling of MEMS resonator piezoelectric disc partially covered with electrodes. Am. J. Mech. Appl. **4**(1), 1–9 (2016)
25. Staworko, M., Uhl, T.: Modeling and simulation of piezoelectric elements – comparison of available methods. Mechanics **27**(4), 161–171 (2008)
26. Leinonen, M., Palosaari, J., Juuti, J., Jantunen, H.: Combined electrical and electromechanical simulations of a piezoelectric cymbal harvester for energy harvesting from walking. J. Intell. Mater. Syst. Struct. **25**(4), 391–400 (2014)
27. Bazilo, C.: Modelling of bimorph piezoelectric elements for biomedical devices. In: Hu, Z., Petoukhov, S., He, M. (eds) Advances in Artificial Systems for Medicine and Education III. AIMEE 2019. Advances in Intelligent Systems and Computing, vol. 1126. Springer, Cham, pp. 151–160 (2020). https://doi.org/10.1007/978-3-030-39162-1_14
28. Bazilo, C., Zagorskis, A., Petrishchev, O., Bondarenko, Y., Zaika, V., Petrushko, Y.: Modelling of piezoelectric transducers for environmental monitoring. In: Proceedings of 10th International Conference "Environmental Engineering", Vilnius Gediminas Technical University, Lithuania (2017). https://doi.org/10.3846/enviro.2017.008
29. Medianyk, V.V., Bondarenko, Y., Bazilo, C.V., Bondarenko, M.O.: Research of current-conducting electrodes of elements from piezoelectric ceramics modified by the low-energy ribbon-shaped electron stream. J. Nano- Electron. Phys. **10**(6), 06012-1–06012-6 (2017). https://doi.org/10.21272/jnep.10(6).06012
30. Petrishchev, O.N., Bazilo, C.V.: Methodology of determination of physical and mechanical parameters of piezoelectric ceramics. J. Nano- Electron. Phys. **9**(3), 03022-1–03022-6 (2017). https://doi.org/10.21272/jnep.9(3).03022

Inherited Bio-symmetries and Algebraic Harmony in Genomes of Higher and Lower Organisms

Sergey V. Petoukhov[1,2(✉)], Elena S. Petukhova[1], Vitaly I. Svirin[1],
and Vladimir V. Verevkin[1]

[1] Mechanical Engineering Research Institute, Russian Academy of Sciences,
M.Kharitonievsky pereulok, 4, Moscow, Russia
spetoukhov@gmail.com
[2] Moscow State Tchaikovsky Conservatory,
Bolshaya Nikitskaya, 13/6, Moscow, Russia

Abstract. The article is devoted to a new class of biological symmetries related to the oligomer organization of eukaryotic and prokaryotic genomes and genetically inherited physiological systems. The discovery of these symmetries is based on a new analytic method of oligomer sums connected with quantum information modeling of long DNA nucleotide sequences. The focus is on the previously formulated hyperbolic rules for the oligomer cooperative organization of a wide set of eukaryotic and prokaryotic genomes. These rules show the connection of genomic structures with the historically famous harmonic progression $1, 1/2, \ldots, 1/n$. The universality of these hyperbolic rules is evidenced by the presented results of a special study of the genomes of species living in extreme conditions of high radiation, high and low temperatures, and so on. Mathematical features of the harmonic progression are described, which are related to the harmonic mean, harmonic numbers, and the basic invariant of projective geometry. These features allow developing not only new model approaches to inherited biological structures related to the genetic code system but also some new methods in biotechnologies and problems of artificial intelligence connected with understanding living bodies as quantum-information entities.

Keywords: Genomes · Symmetry · Extremophiles · Harmonic progression · Harmonic mean · Harmonic numbers · Cross-ratio · Projective transformation

1 Introduction

In their development of artificial intelligence systems, medical engineering, and biotechnology, engineers take into account the properties of living organisms, whose physiological systems are genetically inherited from generation to generation and are often endowed with certain relationships of symmetry. Inherited physiological systems are often structurally consistent with the genetic coding system and bear the imprint of its symmetric and other features. Therefore, for a deeper understanding of physiological systems, it is important for engineers, physicians, and biologists to know the structural

Z. Hu et al. (Eds.): AIMEE 2020, AISC 1315, pp. 201–211, 2021.
https://doi.org/10.1007/978-3-030-67133-4_19

laws of genetic informatics and features of biological symmetries. The disclosure by the science of bio-informational patents used by all living organisms in genetic informatics can provide a lot of useful information for technology, medicine, and science, including communication technologies, artificial intelligence, personal pharmacology, and others.

The fundamental difference between living bodies and inanimate objects is that genetic DNA and RNA molecules play a dictatorial role in living bodies (inanimate objects have no such dictatorial molecules in principle since they are governed by the average random motion of millions of particles) [1]. The study and use of the properties of DNA molecules with their nucleotide sequences, which carry genetic information, are some of the key tasks of modern science. Of cardinal importance is the fact that all living organisms have the same molecular basis for their genetic coding system, associated with DNA and RNA molecules, which can be considered as a fact of some molecular "symmetry" between all living bodies. In living organisms, hereditary information is recorded in DNA molecules in the form of texts using four nucleobases: adenine A, guanine G, cytosine C, and thymine T. The genomic nucleotide sequences of a great number of biological species are freely available in the international Gen-Bank (https://www.ncbi.nlm.nih.gov/genbank/).

A study of the principles of symmetry in the structures of DNA and its informational nucleotide sequences in the last century led to important results: for example, a DNA double helix model was created by Watson and Crick, two Chargaff's rules, which are related with "a grammar of biology" [2], were formulated concerning the equalities of percentages of different nucleotides in DNA filaments, etc. Currently, the totality of the results of studies of symmetries in the genetic system allows us to think that the genetic system is a whole multi-level tree of coordinated symmetry relationships.

This article presents and analyzes some new biological symmetries, which are revealed under a study of the oligomer organization of eukaryotic and prokaryotic genomes. This study is based on a new analytic method of oligomer sums, which is connected with a quantum algorithmic modeling of long DNA sequences [3, 4]. Research objectives in this paper are related to checking a universal status of the mentioned hyperbolic rules, connected with the harmonic progression, by studying genomes of extremophiles. They are also related to studying a connection of some inherited morphogenetic symmetries with the mathematical features of the harmonic progression. The term "oligomer" refers to a molecular complex of chemical that consists of a few repeating units. In DNA, oligomers can have different lengths and are also called n-plets, where n refers to the oligomer length. The described results open a new chapter in the study of biological symmetry and support the understanding of living bodies as quantum information entities. They differ from all known results publications on biological symmetries.

By a logical structure of the paper, it initially describes the oligomer sums method and results to its application to the analysis of extremophiles genomes; after this, it shows that the received results are connected with the harmonic progression, whose mathematical features are related to known non-Euclidean symmetries in inherited morphogenetic structures. The conclusion part of the article gives some remarks on the results.

2   Genomic Hyperbolic Rules and the Harmonic Progression

Each of the long DNA sequences of nucleotides can be represented as a sequence of monomers (like as A-C-A-T-G-T-…), or a sequence of doublets (like as AC-AT-GT-GG-…), or a sequence of triplets (like as ACA-TGT-GGA-…), etc. In each of such fragmented representations, one can calculate and compare total amounts of oligomers (or n-plets) in the equivalence classes of A-oligomers, or T-oligomers, or C-oligomers, or G-oligomers. Each of these classes contains all oligomers, which start with the noted nucleotide. For example, the class of the A-oligomers contains the following sub-sets of n-plets: 4 doublets AA, AT, AC, AG; 16 triplets AAA, AAT, AAC, …, AGG; etc. This method of analysis is termed as the oligomer sums method.

Its applications to the analysis of DNA sequences in eukaryotic and prokaryotic genomes have revealed the following general hyperbolic rule, speaking about the existence of algebraic invariants in full genomic DNA sequences [3]:

- For any of the named classes of A-, T-, C-, or G-oligomers in an individual genomic DNA sequence, the total amounts $\Sigma_{N,n}(n)$ of their n-plets, corresponding different n, are interrelated each other through the general expression $\Sigma_{N,n} \approx S_N/n$ with a high level of accuracy (here N refers to any of nucleotides A, T, C, or G; S_N refers to the number of monomers N; $n = 1, 2, 3, 4, …$ is not too large compared to the full length of the nucleotide sequence). The phenomenological points with coordinates $[n, \Sigma_{N,n}]$ practically lie on the hyperbola having points $H_N = S_N/n$.

This rule is true for many analyzed genomes. But is it universal? In particular, is it fulfilled for organisms living in extreme conditions (they are termed extremophiles)? To study this, we turned, in particular, to the genome of the microorganism Deinococcus radiodurans, which is a champion in survival in extreme conditions and is listed in the *Guinness Book of World Records* as "the world's toughest bacterium" [5]. This extremophile is the most radiation-resistant organism known: it survives not only in conditions of lack of nutrition and drought but also under the influence of radiation thousands of times higher than the level of radiation withstand by humans. Scientists develop technologies of the usage of this organism in cleaning up toxic waste. It is also seen as an object for the storage of information, which is capable of surviving a nuclear disaster. Figure 1 shows the results of our analysis of the genomic DNA sequence of this extremophile.

In our Laboratory, we analyze the genetic characteristics of other extremophiles, the study of which is useful for many practical and theoretical problems. The https://en.wikipedia.org/wiki/Extremophile website contains a table of extremophiles. For the analysis of their genomes by the oligomer sums method, we used 1–2 organisms from each category of the table. The initial data on the genomes were taken from the GenBank. In all these genomes, confirmations of the noted hyperbolic rule were received without any exclusion (see detailed numeric results in the preprint [3]). In particular, such confirmations were received for the following extremophiles:

- *Pyrolobus fumarii 1A*, complete genome, 1843267 bp (this extremophile lives in submarine hydrothermal vents), https://www.ncbi.nlm.nih.gov/nuccore/NC_015931.1;

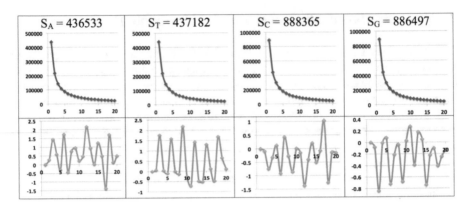

Fig. 1. The results of the analysis - by the oligomer sums method – the extremophile *Deinococcus radiodurans R1* chromosome 1, complete sequence, 2648638 bp (this extremophile lives in conditions of cosmic rays, X-rays, radioactive decay), https://www.ncbi.nlm.nih.gov/nuccore/NC_001263.1. All abscissa axes show the values $n = 1, 2, \ldots, 20$. The top row demonstrates that model hyperbolic progressions S_A/n, S_T/n, S_C/n, S_G/n (red lines) almost completely cover the oligomer sums sequences of phenomenological values (the ordinate axes show appropriate values). The bottom row shows in percent slight alternating deviations of phenomenological values of the oligomer sums sequences from model values (the ordinate axes show fractions of percent %).

- *Pyrococcus furiosus DSM 3638*, complete genome, 1908256 (this extremophile lives in submarine hydrothermal vents), https://www.ncbi.nlm.nih.gov/nuccore/NC_003413.1;
- *Synechococcus lividus PCC 6715* chromosome, complete genome, 2659739 bp (this extremophile lives in low-temperature conditions), https://www.ncbi.nlm.nih.gov/nuccore/NZ_CP018092.1);
- *Psychrobacter alimentarius strain PAMC 27889* chromosome, complete genome, 3332539 bp (this extremophile lives in soda lakes), https://www.ncbi.nlm.nih.gov/nuccore/NZ_CP014945.1;
- *Clostridium paradoxum* JW-YL-7 = DSM 7308 strain JW-YL-7 ctg1, whole genome shotgun sequence, 1855173 bp (this extremophile lives in volcanic springs, acid mine drainage), https://www.ncbi.nlm.nih.gov/nuccore/LSFY01000001.1;
- *Halobacterium sp. NRC-1*, complete genome, 2014239 bp (this extremophile lives in conditions of high salt concentration), https://www.ncbi.nlm.nih.gov/nuccore/NC_002607.1;
- *Chroococcidiopsis thermalis PCC 7203*, complete genome, 6315792 bp, (this extremophile lives in conditions of desiccation), https://www.ncbi.nlm.nih.gov/nuccore/NC_019695.1.

In various genomes, the revealed and described hyperbolic sequences S_A/n, S_T/n, S_C/n, and S_G/n differ from each other only in the magnitude of numerators in their expressions. By appropriate scaling of numerators S_A, S_T, S_C, and S_G, each of these hyperbolic sequences is transformed into the historically famous hyperbolic progression (1):

$$y = 1/n : 1/1, 1/2, 1/3, 1/4, 1/5, \ldots, 1/n \tag{1}$$

The harmonic progression (1) was studied Pythagoras, Leibniz, Newton, Euler, Fourier, Dirichlet, Riemann, and other researchers. It has interesting mathematical features, which are described in the next Sections and which can be useful for developing new classes of models of genetically inherited physiological structures.

3 The Harmonic Progression, the Harmonic Mean and Harmonic Numbers

The harmonic progression (1) is a recurrent sequence related to the harmonic mean. Let us show this. By definition, the harmonic mean H of the positive real numbers x_1, x_2, ..., x_n is defined to be

$$H = \frac{n}{\frac{1}{x_1} + \frac{1}{x_2} + \ldots + \frac{1}{x_n}} \tag{2}$$

In the case of three positive real numbers x_n, x_{n+1}, x_{n+2}, which are interrelated by the harmonic mean, the following expressions hold:

$$x_{n+1} = 2/(1/x_n + 1/x_{n+2}); \, x_{n+2} = x_n x_{n+1}/(2x_n - x_{n+1});$$

$$x_n = x_{n+1}x_{n+2}/(2x_{n+2} - x_{n+1}) \tag{3}$$

In the harmonic progression (1), any three adjusting members $1/n$, $1/(n+1)$, and $1/(n+2)$ are interrelated by the harmonic mean expressions (3). For example, the expression (4) shows that the harmonic mean of two these members $1/n$ and $1/(n+2)$ is equal to the third member $1/(n+1)$:

$$2/\{1/(1/n) + 1/(1/(n+2))\} = 2/\{n + n + 2\} = 1/(n+1) \tag{4}$$

Knowing any two adjacent members $1/n$ and $1/(n+1)$, one can recurrently construct all the harmonic progression (1) using the expressions (3).

One can consider a sparse harmonic progression (harmonic epi-progression), in which only every mth term of the original progression remains (m is a positive integer number):

$$1/1, 1/(1+m), 1/(1+2m), 1/(1+3m), 1/(1+4m), \ldots \tag{5}$$

Any three adjacent members $1/(1+km)$, $1/(1+(k+1)m))$, and $1/(1+(k+2)m)$ of such harmonic epi-progression are also interrelated by the harmonic mean:

$$2/\{1/(1/(1+km) + 1/1/(1+(k+2)m)\} = 2/(2+2km+2m) = 1/(1+(k+1)m) \tag{6}$$

This similarity of properties of the sparse harmonic epi-progressions (5) and the harmonic progression (1) resembles a similarity of properties of long DNA nucleotide sequences and their sparse subsequences called epi-chains and described in [6].

The harmonic progression $1/n$ (1) and hyperbola $y = 1/x$ have also deep relations with the function of the natural logarithm ln (x). The natural logarithm can be defined for any positive real number "a" as the area under the hyperbola $y = 1/x$ from 1 to a [7]. The function of the natural logarithm is widely used in mathematical biology and relates to inherited biological phenomena. For example, different types of inherited sensory perceptions are subordinated to the main psychophysical law of Weber-Fechner: sight, hearing, smell, touch, taste, etc. The law states that the intensity of the perception is proportional to the natural logarithm of stimulus intensity. Because of this law, for example, the power of sound in engineering technologies is measured on a logarithmic scale in decibels. The logarithmic nature of perception provides the body with a huge expansion of the range of perceived stimulus. For example, hammer strikes on a steel plate sound a hundred billion times louder than the quiet rustle of leaves, and the brightness of a voltaic arc trillions of times exceeds the brightness of some faint star, barely visible in the night sky [8]. If sensations were proportional to stimuli, then the sensory organs would have an extremely difficult task to give sensations that differ billions and trillions of times. The organism with its genetically inherited sense organs bypasses this difficulty, as "logarithm" the irritations it receives. It is natural to think that this inherited ability is structurally linked with the genetic coding system, which ensures the coding of the sensory organs and their transmission to descendants.

The sums of the first members of the harmonic progression (1) are called harmonic numbers H_n [9, 10]. «*These numbers appear so often in the analysis of algorithms that computer scientists need a special notation for them*» [9, p. 273]. The series (7) presents first harmonic numbers H_n:

$$H_n = \sum_{k=1}^{n} 1/k : 1, 3/2, 11/6, 25/12, 137/60, \ldots \tag{7}$$

Harmonic numbers are connected with the function of the natural logarithm: the harmonic number H_n «*is the discrete analog of the continuous function ln(x)*» [9, p. 276]. Figure 2 shows an approximation of the function of the natural logarithm in Fig. 3 by the sum of areas 1, 1/2, 1/3, 1/4, …, $1/n$ of the shown rectangles; sums of these areas form the series of harmonic numbers H_n (1).

In 1740, Euler obtained an asymptotic expression for the harmonic numbers H_n:

$$H_n = \ln(n) + \gamma + \varepsilon \tag{8}$$

where $\gamma = 0.5772\ldots$ is the Euler-Mascheroni constant, and the value ε tends to zero with increasing n.

One should note that if you consider a hyperbolic sequence like S/n (Fig. 1), where S denotes the number of a certain kind of nucleotides, the sum of its first members is equal to an appropriate harmonic number H_n multiplied by the scale factor S (9):

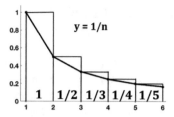 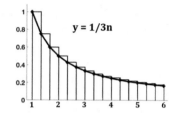

Fig. 2. Left: an approximation of the area under the hyperbola $y = 1/x$ by the sum of areas 1, 1/2, 1/3, 1/4, ..., $1/n$ of the shown rectangles, whose areas form the series of harmonic numbers H_n (7). Right: the improved approximation of the natural logarithm function by compressed harmonic numbers $H_n^{[m]}$ under $m = 3$.

$$\sum\nolimits_{k=1}^{n} s/k = S * \sum\nolimits_{k=1}^{n} 1/k \qquad (9)$$

To improve an approximation of the logarithmic function ln (x) by the total area of the rectangles, you can consider - instead of the classic harmonic progression 1, 2, ..., $1/k$ - more fractional sequences, called the "compressed harmonic progression" (or briefly, the C-harmonic progression):

$$1, 1 + 1/m, 1 + 2/m, \ldots, 1 + k/m, \qquad (10)$$

where the additional factor m can take integer values 1, 2, 3,... The sums of the first members of the C-harmonic progression (10) are called as "compressed harmonic numbers" (or C-harmonic numbers). Figure 2, at right, shows an improved approximation of the continuous function ln (x) by C-harmonic numbers $H_n^{[m]}$ at $m = 3$: the area under the hyperbola, representing the natural logarithm, is better approximated by the total area of the rectangles equal to the C-harmonic number $H_n^{[3]}$. The larger the value m, the better the approximation of the continuous function ln (x) by discrete C-harmonic numbers $H_n^{[m]}$ is provided.

Replacing ln (x) and ln (x_0) in the expression of the Weber-Fechner law by corresponding C-harmonic numbers $H_n^{[m]}$ and $H_r^{[m]}$, you get the discrete analog of the Weber-Fechner law based on the harmonic progression and C-harmonic numbers.

The authors believe that harmonic numbers H_n and compressed harmonic numbers $H_n^{[m]}$ will be useful for modeling inherited physiological systems and processes by recurrence sequences, including the modeling biological spiral systems and processes in their relation to the genetic coding system.

4 The Harmonic Progression and the Basic Invariant of Projective Geometry

Another important feature of the harmonic progression (1) is related to the cross-ratio, which is the basic invariant of the projective geometry based on the group of projective transformations. In geometry the cross-ratio (or the double ratio) is a number, which is associated with four points on the real line, having coordinates X_1, X_2, X_3, and X_4, and is defined by the expression (11):

$$(X_3 - X_1) * (X_4 - X_2) / \{(X_3 - X_2) * (X_4 - X_1)\} \tag{11}$$

The cross-ratio is preserved under the projective transformations of a projective line when lengths and simple ratios of line segments are changed (Fig. 3, at left). The cross-ratio of any 4 adjacent members of the harmonic progression (1) is equal to 4/3. The same is true for any 4 adjacent members of sparse harmonic epi-progressions (5). The cross-ratio of any 4 adjacent members of a series of natural numbers 1, 2, ..., n is also equal to 4/3 since the transformation $y = 1/x$ is one of the kinds of the projective transformations $z = (ax + b)/(cx - d)$. In other words, the harmonic progression (1) and a series of natural numbers are projective similar to each other (Fig. 3, at right).

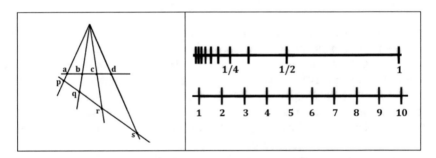

Fig. 3. Examples of projective transformation. At left: a central projective tranformation changes lengths and simple ratios of segments of projective lines but preserves the cross-ratios. At right: arrangements of numbers of the harmonic progression (1) and a series of natural numbers 1, 2, ..., n on the number line.

The connection of the harmonic progression (1) with the cross-ratio, as a basic invariant of projective geometry, leads to the following conclusions:

1. Different fragments of the harmonic progression can be considered as projective-invariant to each other;
2. In eukaryotic and prokaryotic genomes, different hyperbolic sequences of oligomer sums, corresponding to A-, T-, C-, and G-oligomer classes, are about projective similar to each other with a high level of accuracy;
3. Different eukaryotic and prokaryotic genomes, regarding their oligomer sums sequences, are about projective similar to each other with a high level of accuracy.

These data on the connection of the genetic system with the cross-ratio allow considering - in a new light – the materials known long ago on relations of geometric invariants of genetically inherited morphogenetic and growth phenomena with the cross-ratio [11–13]. For example, the cross-ratio was applied to the analysis of projective-invariant longitudinal proportions of the three-component kinematic blocks in human bodies: the phalanges of fingers, the three-membered extremities (shoulder-forearm-wrist and hip-shin-foot) and the three-membered body (in anthropology the body is subdivided into the upper, torso and lower parts). The values of the cross-ratios (wurfs) in all the blocks, at least during the entire individual post-natal development, group around the benchmark 1.3 [12]. This is especially interesting since the growth of the human body is essentiall nonlinear (Fig. 4): for instance, the upper part growth 2.4-fold, the torso 2.8-fold and the lower part 3.8-fold.

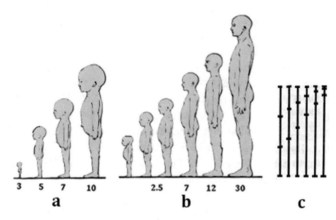

Fig. 4. Ontogenetic changes in the proportions of the human body (from [14]). a, b - antenatal and postnatal stages: a - in the lunar months; b - in years (the first on the left is a newborn); c - three-segment structures whose cross-ratios (wurfs) are equal to 1.3.

This is largely true of dwarfs and giants and a broad range of highly organized animals as well as normal human subjects. These findings are essential for an important theme on Euclidean and non-Euclidean symmetries in biological bodies, to which many books and international forums, including the Nobel Symposium, were devoted [11, 12]. One can recall that projective transformations and the cross-ratios are one of the basic mathematical tools in studying vision perception and creating methods of a computer vision, recognition of images, realistic computer graphics, and systems of artificial intelligence; a great number of publications is devoted to these topics, including, for example [15–20].

4.1   Some Concluding Remarks

The discovery of the presented new genomic symmetries gives new knowledge about the unity of the world of all living organisms, types of the biological symmetries, and the peculiarities of biological evolution. Living matter appears as an algebraic-harmonic entity whose important features are revealed by quantum-information modeling of long DNA nucleotide sequences [3, 4]. The received results support ideas of P. Jordan on quantum biology [1] and give new materials for its development using contemporary knowledge on quantum informatics and genetics. We believe that received results, which overcome limitations of previous knowledge, will be used in future works on algebraic and quantum biology, biological evolution, artificial intelligence systems, and biotechnological applications.

One can separately note the result that the cooperative system of oligomers in the genomes is associated with the harmonic progression, which is widely known in connection with a musical harmony and the frequency system of musical overtones. But the harmonic progression is important not only in music. At least from the time of the Pythagorean doctrine of the aesthetics of proportions, the following idea exists: "*the aesthetic principle is the same in every art; only the material differs*" [21]. In light of this, architecture has long been interpreted as frozen music and music as dynamic architecture. Additional speculation about the possible genetic basis of some aesthetic parallels in various arts arises though the very idea of the connection between the feeling of beauty and the genetic system is not new: it is reflected, for example, in the title of the article "*Beauty is in the genes of the beholder*" about a connection of some parameters of the DNA double helix with the golden section [22].

Researchers of the genetic system study the Nature system of storage, processing, and transmission of information, which has no direct analogies in modern science and technology, but which is studied on the basis of analogies with their achievements. The disclosure of informational patents of living nature can make an important contribution to scientific and technological progress.

Acknowledgments. The authors are grateful to their colleagues M. He, Z. Hu, Yu.I. Manin, I. Stepanyan, and G. Tolokonnikov for research assistance.

References

1. McFadden, J., Al-Khalili, J.: The origins of quantum biology. Proc. Roy. Soc. A **474**(2220), 13 (2018). https://doi.org/10.1098/rspa.2018.0674
2. Chargaff, E.: Preface to a Grammar of Biology: a hundred years of nucleic acid research. Science **172**, 637–642 (1971)
3. Petoukhov, S.V.: Hyperbolic rules of the oligomer cooperative organization of eukaryotic and prokaryotic genomes. Preprints 2020, 2020050471 (2020). https://doi.org/10.20944/preprints202005.0471.v2. https://www.preprints.org/manuscript/202005.0471/v2. 95 p.
4. Petoukhov, S.V., Petukhova, E.S., Svirin, V.I.: Symmetries of DNA alphabets and quantum informational formalisms. Symmetry Culture Sci. **30**(2), 161–179 (2019). https://doi.org/10.26830/symmetry_2019_2_161
5. DeWeerdt, S.E.: The world's toughest bacterium. Genome News Network, 5 July 2002. https://www.genomenewsnetwork.org/articles/07_02/deinococcus.shtml
6. Petoukhov, S.V.: Nucleotide Epi-Chains and New Nucleotide Probability Rules in Long DNA sequences. Preprints 2019, 2019040011 (2019). https://doi.org/10.20944/preprints201904.0011.v2, https://www.preprints.org/manuscript/201904.0011/v2

7. Conway, J.H., Guy, R.K.: The Book of Number. Copernicus, New York (1995). ISBN 0-387-97993-X
8. Vilenkin N.Ya.: Functions in nature and technology (Funktsii v prirode i tehnike). Prosveschenie, Moscow (1985). (in Russian)
9. Graham, R.L., Knuth, D.E., Patashnik, O.: Concrete mathematics: a foundation for computer science, 2nd edn. (1994). ISBN 0-201-55802-5.
10. Sondow, J., Weisstein, E.W.: Harmonic number. From MathWorld – A Wolfram Web Resourse (2020). https://mathworld.wolfram.com/HarmonicNumber.html
11. Petoukhov, S.V.: Biomechanics, Bionics and Symmetry. Nauka, Moscow (1981)
12. Petoukhov, S.V.: Non-Euclidean geometries and algorithms of living bodies. Comput. Math. Appl. **17**(4–6), 505–534 (1989). https://www.sciencedirect.com/science/article/pii/0898122189902484
13. Petoukhov, S.V., He, M.: Symmetrical Analysis Techniques for Genetic Systems and Bioinformatics: Advanced Patterns and Applications. IGI Global, USA (2010)
14. Petten, B.M.: The Human Embryology. Medgiz, Moscow (1959). (in Russian)
15. Khan, R., Debnath, R.: Human distraction detection from video stream using artificial emotional intelligence. IJIGSP **12**(2), 19–29 (2020)
16. Erwin, D.R.N.: Improving retinal image quality using the contrast stretching, histogram equalization, and CLAHE methods with median filters. IJIGSP **12**(2), 30–41 (2020)
17. Niaz Mostakim, Md., Mahmud, S., Khalid Hossain Jewel, Md., Khalilur Rahman, Md., Shahjahan Ali, Md.: Design and development of an intelligent home with automated environmental control. IJIGSP **12**(4), 14 (2020)
18. Arora, N., Ashok, A., Tiwari, S.: Efficient image retrieval through hybrid feature set and neural network. IJIGSP **11**(1), 44–53 (2019)
19. Anami, B.S., Naveen, N.M., Surendra, P.: Automated paddy variety recognition from color-related plant agro-morphological characteristics. IJIGSP **11**(1), 12–22 (2019)
20. Ahmed, M., Akhand, M.A.H., Hafizur Rahman, M.M.: Recognizing bangla handwritten numeral utilizing deep long short term memory. IJIGSP 11(1), 23–32 (2019)
21. Schumann, R.: On Music and Musicians, Wolff, K. (ed.), New York (1969)
22. Harel, D., Unger, R., Sussman, J.L.: Beauty is in the genes of the beholder. Trends Biochem. Sc. **11**, 155–156 (1986)

Telemedicine Monitoring with Artificial Intelligence Elements

Lev I. Evelson[1], Boris V. Zingerman[2], Rostislav A. Borodin[2],
Inna A. Fistul[2], Irina G. Kargalskaja[3], Alexandra M. Kremenetskaya[4],
Olga S. Kremenetskaya[5], Sergei A. Shinkarev[6],
and Nikita E. Shklovskiy-Kordi[4(✉)]

[1] Innovation Scientific Center of Information and Remote Technologies,
241050 Bryansk, Russian Federation
levelmoscow@mail.ru
[2] TelePat, 196787 Moscow, Russian Federation
boriszing@gmail.com
[3] Committee "Patient Oriented Telemedicine",
109390 Moscow, Russian Federation
[4] National Medical Research Center for Hematology,
125167 Moscow, Russian Federation
nikitashk@gmail.com
[5] Center for Theoretical Problems of Physicochemical Pharmacology,
119991 Moscow, Russian Federation
tkolga@mail.ru
[6] Lipetsk Oncological Hospital, 398005 Lipetsk, Russian Federation

Abstract. The paper is devoted to the remote monitoring health statement. The problems and ways of their solution are presented. A great attention is paid to psychological and managerial problems. Their important role here is displayed. The main aims of the monitoring are to improve medical care and simultaneously to unload the attending physician from routine work. Under the pandemic conditions, reduction of number of the face-to-face visits is also important. To reach the above targets developed telemedicine system (platform) is proposed. It includes Artificial Intelligence (AI) elements. In particular, there is a subsystem of intelligent agents (IA), intended for improvement of the efficiency and convenience of remote interaction between the patient and the doctor. Experimental application of the proposed platform conformably to the cancer patients took place in various Russian medical institutions. The results described in this article corroborate high effectiveness of the proposed approach. Ways of the perspective development of the proposed system are presented. They are further development of the IA subsystem, wide application of the IoT and analysis of the collected and depersonalized medical data to reveal general regularities.

Keywords: Intelligent agents · Medical messenger · Telemedicine · Remote monitoring health · Artificial intelligence

1 Introduction

Now remote monitoring health is very actual and important area of telemedicine and of information technologies (IT). Many scientific articles are devoted to general aspects of this field [1, 2]. Many telemedicine projects devoted to remote monitoring are developing in various countries. Perhaps all large-scale IT companies invest money into such projects. The COVID-19 pandemic has caused many challenges and some more accelerated development and application of such medical service. Every year various top-10 lists and results of the trends analysis are published [3]. Many companies dealing with remote health monitoring also present their estimation of trends and description of advantages of the company product [4, 5]. The survey of governmental institutions takes place, for example, in [6]. Usually in analogous projects the main role belongs there to technologies, first of all, "Internet of Things" (IoT) [7]. There are many various sensors recording some quantitative parameters of a patient condition (in real time or periodically) and sending them to the server [8, 9]. Many telemedicine projects now are centered to IoT. The opportunity to know always the current objective parameters is very important and probably IoT would continue wide application in telemedicine. However, there are many qualitative subjective parameters which are also important but IoT can't take them into consideration. They can be determined only by the patient and can be estimated correctly by the physician in co-operation with the patient. In spite of subjectivity and some fuzzy character, such information is not less important than quantitative parameters. It can be formed in interaction between a patient and his attending doctor. There are some facilities which can help to the doctor to solve the problems within remote monitoring more effectively. They are specific medical messenger and artificial intelligence (AI) methods [10]. Analysis of various telemedicine projects realized in Russia and other countries have displayed that the success depend on some psychology and management details. They can be disconnected with technologies but can cancel even very good technology. In spite of importance various technological means, the physician is a principal figure in medicine and his role in telemedicine monitoring is the same too. Often he is very busy and he needs useful tools which can unload him. There are things which can be done and must be done only by a doctor but there also many things which can be successfully made by IT and AI.

May be the main problem is that care of public health has to turn to new model of medicine. This model is "patient centered". Here IT and AI not only have to become good tools but to change the medicine paradigm. Now personal contacts between a patient and a doctor usually take place when the patient has some problems with his (her) health. The problems disturb him to feel well; he comes to his family doctor and complains about the problems. The doctor should be a good psychologist to understand the complaints correctly. Sometimes the patient feels shy before the doctor. There are often situations when it's not so easy to describe the problems with health. It can be especially difficult in remote interaction. Sometimes the patient doesn't know if his condition is normal or not. From other hand, he can have some individual peculiarities and the doctor must understand and take into consideration them. In normal non-emergency situations, the doctor should be, first of all, a wise adviser, who has a great knowledge and experience on the problems arising in the patient, as well as knowing the patient well. The doctor should be a guru and pilot for the patient in a confusing and dangerous labyrinth. It would help not only to overcome problems, but also to prevent their occurrence. The patient should make (guided by the advice of the doctor) his own decisions, accumulate and analyze important information concerning his health, allow the doctor with the least difficulties and most effectively perform his role.

The main objectives of this research are to bring out the current psychological and managerial problems conveying the remote monitoring health technologies, to propose and to test the methods allowing solve the problems effectively for health care and comfortably for the doctor and the patient. Also the ways of further research and practical development should be planned and discussed.

2 Methodology

The methodology of this research includes technological approaches connected with theory of information systems and AI. However, psychological and managerial aspects of the remote monitoring health also take an important role. It is based on long-term practical experience of the "medical part" of our research team. The authors had developed a telemedicine platform which would be able to solve the above problems [11]. The main concepts, feasibilities and features of the proposed system are following.

1. Conception of the proposed general approach considers a patient as the principal in everything connected with his (her) health. So, it is the patient centered system. Remote monitoring is considered as continuation of treatment initiated by the attending doctor. It should provide effective and convenient remote interaction between a patient and his attending doctor. Remote consulting and monitoring health of the patient by the doctor are considered as a remote (telemedicine) continuation of face-to-face medical care. Only the attending physician may remotely adjust the prescribed treatment. He has to become a wise adviser for the patient and the patient must trust to the doctor. The system has to be convenient for the both, to give self-confidence to the patient and to unload the doctor from routine work.
2. Asynchronous personal interaction between a doctor and a patient: the patient asks when the question arises, and the doctor answers when he has the opportunity, but necessarily within the time assigned in the contract (working day, 12 h, 24 h).
3. Monitoring the effectiveness of the medical institution, in particular, reminding the doctor to respond on time.
4. It is possible to keep correspondence in an agreed volume at the patient, at the doctor, in the medical institution. The telemedicine platform works together with Personal Electronic Medical Card (PEMC). Kept information is ordered according with the PEMC structure. The verbalization of a patient's complaints becomes also a part of the PEMC.
5. A great attention is paid to visualization of information sent from a patient. Some special tools had been developed: color diagrams reflecting a patient condition

according with noted symptoms (we calls it "heat map"), graphs displaying dynamics of the quantitative parameters in the united system of time scales, etc.

6. AI methods and technologies take important role in the proposed system. They help to patient to present his complaints in understandable form, assist to a doctor to estimate the patient condition according with information received from him. AI allows classify automatically the situations: a) urgent actions are necessary; b) urgent actions are not necessary but personal attention of the attending doctor is needed; c) urgent actions are not necessary and personal attention of the attending doctor is not needed. In the case c) AI initiates standard response including recommendations for the patient and some useful information relevant to the situation. AI methods are realized by subsystem of based on them software modules - intelligent agents (IA).

7. A patient receives periodically questionnaires appointed by the attending doctor. There many various questionnaires developed during long-term research for some classes of diseases and of patients (cancer, hematological diseases, patients after transplantation of organs, pregnant women and so on). That research part was done by scientific team headed by prof. A. Vorobiev [12].

8. There is information part including virtual Patient Library and Doctor Library. They consist of verified important information resources which are relevant to the specific areas. For example, the Patient Library integrates now 1,250 pages of cancer patient-oriented information content supporting patient teaching cases from leading experts on the specifics and treatments for life and nutrition, rehabilitation and care. It is structured on cancer diagnoses (it includes 21 types of cancers) as well as stages of the patient's life. After that the COVID-19 pandemic started, 13 video teaching cases had been added to the information part of the system. They concern the life conditions of cancer patients under the COVID-19 quarantine situation (it was done together with prof. N. V. Zhukov). Those cases present many important recommendations such as: how to control breath under the COVID-19, prophylaxis peculiarities for patients with cancer, psychological aspects during pandemic, etc.

Compared to conventional universal messengers such as WhatsApp, the following advantages can be highlighted.

1. All types of interaction between the doctor and the patient pass data through a special channel established by the medical organization. Accordingly, they are not a private matter of the doctor but they are officially regulated by the clinic. The doctor's actions are recorded and can be monitored by a specially appointed representative of the clinic, including the quality and compliance with the terms of the consultation.

2. There are no any patient private contacts with the doctor, as all interactions are carried out through the channel established by the clinic. The doctor's professional correspondence with his patients is separated from his private correspondence in messengers and is concentrated in one place.

3. There are many special means included into the system (questionnaires, general medical information, teaching cases for different diseases and classes of patient's condition, intelligent agents intended for processing various real medical situations, etc.). Certainly, there are no such means in universal messengers.

4. Presence of the relevant information part. Certainly, WhatsApp users can look for needed information in Internet but it would be difficult to find so relevant content.

The features of proposed system are corresponding with Russian peculiarities. Introduction of the system into health care practice in other countries expects efforts from multidisciplinary international team. One of the important features of the proposed system is presence of the fundamental medical knowledge and already collected information about many diseases. This knowledge was accumulated in long-term medical research. It concerns many modes of cancer, hematological diseases and others. Such knowledge would be useful in some international research.

3 Intelligent Agents in the Remote Interaction Between the Patient and the Doctor

The peculiarity of this project is that the elements of AI are embedded in the patient's personal communication channel with his doctor. It allows the doctor to adjust the recommendations given by the intelligent agent, as so as he always sees those recommendations in the tape of dialogue. Proper use of this advantage would allow carry out an easy inculcation of means of conversational AI in the system of remote monitoring health.

Subsystem of the intelligent agents (IA) includes software modules which are inclined to the patient profile by the attending physician according with the medical standards. Chat bots can ask elaborating questions; analyze automatically the patient answers, estimate and classify the patient condition and reveal alarm situations demanding urgent actions, cases when personal attention from the doctor is necessary, etc. The patient is waiting psychologically for the quick response from the doctor. So the patient's dialogue with the IA should fill the pause during which the doctor cannot answer. It also must give him useful (adequately selected) information and training content related to the asked question. Procrastinating with answers to such questions can be critical to the patient's health. Among of the questions there may be many typical ones which don't require the attention of the doctor. In addition, standardized protocols for monitoring and managing patients of certain categories may involve a large amount of information that needs to be requested from patients to assess their condition. The quick communication with the doctor is necessary only in a situation when the patient has significant deviations from the established standards of his health. In other situations the doctor's actions can be replaced by IA successfully. There are many reminders about needed procedures or examinations (from daily reminders of medication and blood pressure measurement to annual reminders of tests). The analysis of information related to those reminders and performed medical procedures can also be pre-processed by IA, involving the doctor only in cases requiring his personal attention.

In accordance with the above, criteria for choice of the dialogue scenarios of IA (chat bots) should be taken so:

1. There should be a selection of typical (non-binding) questions, which could be reasonably and adequately answered during the patient's dialogue with IA.

2. Standardized monitoring involves a significant amount of automatically generated messages (reminders and other information messages, questionnaires that need to be filled out, etc.). Preliminary automatic processing of the messages received from patients must be done to attract the doctor's attention only in the case of significant deviations from the established norm.

3. Availability of the necessary information content (high quality professional medical information) to implement the scenario has to take place.

JSM method [13] was taken as a general approach to create IA. Implementation of IA includes here 3 following directions of information exchange which can be considered at different levels.

(a) Patient – doctor: automatic analysis of messages, questions, questionnaires and current indicators of the patient's condition and decision-making based on this analysis. In particular, the identification of situations requiring urgent medical intervention should be done. The first level is based on a clear algorithm, which fulfils a cyclical check of the entire set of predetermined quantitative indicators. There are also predetermined norm intervals for all those indicators. It is checked if the fact of hit into the norm interval takes place or not. If "no" that IA sends urgent message to the doctor. It is enough if there is at least 1 indicator beyond the norm to determine dangerous situation and to send the message. For example, the set of indicators can include the upper and lower limits of blood pressure, blood glucose, body temperature, etc. There are various questionnaires filled up by a patient. The questions stipulate for only two versions of answers: "yes" and "no". Specific questionnaire for detecting the presence of COVID-19 symptoms had been implemented. The second level adds a check of the patient's textual information in the messages according with the keywords, which may be qualitative signs of a critical situation. For example, the patient reports that he recently lost consciousness. When checking the text, the keywords "lost consciousness" are identified, and an urgent message to the doctor is formed. At the third level, more complex question systems can be included, providing for the evaluation of quality indicators by the patient not only in binary form, but also in a points order scale, for example, from 1 to 10. The approach based on the theory of fuzzy sets can be proposed. The description of this approach in the practical medical example is given in [14]. Similar approach can be used for discovery of anomalies in the information flow [15]. At the following levels, some semantic search tools for key concepts can be used in relation to patient message analysis. Besides, software implementing image recognition (ultrasound, MRI, CT, histological images, etc.) should be involved [16].

(b) "Doctor - patient": automatic formation of answers to typical questions and reactions to typical situations that do not require the mandatory participation of the doctor, as well as the selection of information materials relevant to the patient's condition, medical history and questions. At the first level, a list of typical questions asked by patients is drawn up and standard answers are prepared. When real questions from the patient are received by the system, they are compared by keywords with the typical questions. If coincidence takes place, the typical response is worked out. First, a list of typical situations described in the space of quantitative and qualitative parameters is formed. In this space the criteria (procedures) of the situation to the typical class are defined. For each typical listed situation, a response procedure is specified. After implementation of the classification procedure, the response is launched.

(c) "Patient and physician - PEMC": the automatic structuring of information obtained as a result of the interaction between the patient and the doctor, and inclusion of it in the appropriate form into the PEMC. This direction is closely related to the problem of the optimal structure of PEMC [17], which is a single personal confidential electronic archive. The patient stores all his health information collected and received from various medical institutions, organizations related to health or healthy lifestyle. The patient gives the right to automatically supplement PEMC during remote monitoring. The next level provides for an automated intelligent synthesis of typical PEMC records. It should also unload the doctor. It's planned to apply the methods as like as described in [18].

4 Experiment and Results

Described above the telemedicine platform had been adapted for remote monitoring health condition of cancer patients after hard treatment. As an experiment, the pilot utilization of the adapted system (ONCONET) took place during 2018–2020 in 22 medical organizations in 10 regions of Russia. 174 doctors and 382 patients participated in the testing of the system. The main aims were: to test effectiveness of the proposed approach of the remote monitoring as compare with patients without it.

Below the most important part of the experiment and the results are described.

The largest number of patients had been tested in the Lipetsk Oncology Hospital during implementation of the ONCONET system since 2019 till April 2020. 61 doctors took part in the project. They provided remote monitoring for 206 patients. The study involved patients with breast cancer, lung, cervical and ovarian cancer, prostate, thyroid, kidney, stomach and colorectal cancer, lymphoma. The reference group of patients not involved into remote monitoring (209 persons) had been selected too. Similarity in composition of diagnoses and stages of diseases and treatment regimens, as well as on the sex-age composition was observed.

The results on two important types of the complications are seen in the Table 1. The significance level was taken as $\alpha = 0{,}05$.

Table 1. Quantity of complications

Complications	Group under remote monitoring, %	Reference group, %	Significance of Difference Yes/No
Hematological	32,5	41,6	YES
Nephrotoxic	3,6	13,6	YES

According to the results of the pilot project, patients indicated the "normal" condition of the symptom in 89.9% of cases (scattering from 8% to 100% for different symptoms); a state with certain deviations from the "norm" in 8% of cases (spread from

0% to 92%); severe and critical states in 2.1% of cases (spread from 0% to 47%). So, in the most of all cases personal attention of the attending doctor wasn't necessary. Other cases were revealed at a proper time. It allowed avoid complications in many cases.

So, the both aims, improvement of medical help and unloading of the physician, had been reached.

The feedback on the system received from doctors and involved patients were positive. Doctors marked improvement of their work conditions and good objective results. The patients noted more self-confidence and calmness than in the reference group. The quantitative difference in the frequency of treatment complications for patients participating in ONC0NET and the reference group (in addition to the Table 1) was following:

- disruptions in the gastrointestinal (0.7% vs. 2.6%);
- neurotoxic manifestations are practically reduced to a minimum (0% vs. 3.2%);
- skin complications were reduced to zero (0% in the group ONCONET vs. 1.6% in the control) Thanks to the skin photos attached to the questionnaires, the phenomenon had been detected at an early stage.

The postponements of the next chemotherapy course (very important parameter) took place significantly less (7.9% for ONCONET vs. 11.9% in the reference group).

So, the results of the pilot utilization of the proposed system were obviously good for the patients and for the doctors.

5   Discussion

Results of experimental application of the proposed approach and developed system have displayed that we have an effective and convenient tool to provide remote monitoring health. AI elements are utilized there to facilitate the doctor's work, to reduce probability of some mistakes, to reveal at a proper time undesirable deviations, to give more relevant knowledge and calm to the patient. Besides (or together with) of above noted JSM method, the method of speech analysis [19] can be used. The methods devoted to automatic images recognition [20] are also can be taken. The AI elements allow receive more exact description of qualitative indicators of the patient's health and his psychological condition. Together with objective quantitative parameters and tests results, the general health state of the patient can be estimated well. The proposed system should be developed further.

It is very important that everything is recorded and stored in the PEMC. Analysis of the PEMC should have two sides. The first one is analysis of personal data in order to reveal individual norms and important tendencies. They are needed to be taken into account for treatment correction. The second side is depersonalization of the collected data and further analysis to reveal important regularities concerning big groups of patients. Data Mining approach should be used and AI should take important role there. AI should be useful much because the data would have peculiarities typical of the AI field: not enough definite, not enough complete to be processed by algorithm software. A combination of precise and fuzzy parameters would take place and flexible approach combining clear algorithms and AI methods would be needed.

The further development of the analytical part founded on the AI elements would take into consideration individual peculiarities of the patient physiological and psychological profile (his individual norms) and would to have important general regularities analyzing collected and depersonalized data.

6   Conclusion

Developed remote monitoring approach and the proposed telemedicine platform with elements of AI form an effective way to improve much medical care. The leading role of the doctor does not reduced, but he gets the opportunity to focus on a creative assessing the patient's condition and prescribing treatment. The patient gets qualified medical advices without necessity to go to a face-to-face appointment and wait a long time for a conversation with a doctor (it is especially important under pandemic). The doctor gets responsible patient trusting his doctor.

So, we have the first level of working telemedicine system. It can already now improve medical care but transition to the following levels is needed to increase the improvement much. Further development of the system is associated with an expansion of AI functions concerning other types of diseases and analysis of the collected and depersonalized medical data to bring out important general regularities. We plan also for perspective to expand utilization of IoT technologies. It would be needed to organize (using AI) a preliminary analysis of data flow from a variety of mobile devices. It's also planned to make supplement of information messages from the devices with possibilities of some actions (here robots should play an important role). It would provide much more knowledge of the patient contained in PEMC and it would enable us to determine the individual norm in many indexes and prescribe treatment with greater consideration of the individual characteristics of a patient.

The methods and technologies are important but psychological and managerial aspects of their application are not less important. It requires close cooperation in united international team including doctors and specialists in medical informatics and AI.

Acknowledgments. This work was supported by the Ministry of Science and Higher Education of the Russian Federation (AAAA-A18-118012390247-0), by RFBR grant (project № 19-07-01235).

References

1. Topol, E.: Digital medicine: empowering both patients and clinicians. Lancet **388**(10046), 740–741 (2016)
2. Mamyrbekova, S., Nurgaliyeva, Z., Saktapov, A., Zholdasbekova, A., Kudaibergenova, A.: Medicine of the future: digital technologies in healthcare. In: E3S Web of Conferences, vol. 159, no. 04036, pp. 1–10. BTSES-2020 (2020)
3. https://evercare.ru/remote-monitoring-top10. Accessed July 2020
4. https://mlsdev.com/blog/telemedicine-app-development. Accessed June 2020

5. https://www.quanticate.com/blog/bid/70489/remote-monitoring-during-clinical-trials-a-risk-based-approach. Accessed July 2020
6. National Institute of Standards and Technology. Health IT Usability (2018)
7. Moreno, H.B.R., Ramírez, M.R., Hurtado, C., Lobato, B.Y.M.: IoT in medical context: applications, diagnostics, and health care. In: Chen, Y.W., Zimmermann, A., Howlett, R., Jain, L. (eds.) Innovation in Medicine and Healthcare Systems, and Multimedia. Smart Innovation, Systems and Technologies, vol 145. Springer, Singapore (2019)
8. Mena, L.J., Felix, V.G., Ochoa, A., Ostos, R., Gonzalez, E., Aspuru, J., Velarde, P., Maestre, G.E.: Mobile personal health monitoring for automated classification of electrocardiogram signals in elderly. Comput. Math. Methods Med. (2018)
9. Parthiban, N., Esterman, A., Mahajan, R., Twomey, D.J., Pathak, R.K., Lau, D.H., Roberts-Thomson, K.C., Young, G.D., Sanders, P., Ganesan, A.N.: Remote monitoring of implantable cardioverter defibrillators: a systematic review and meta-analysis of clinical outcomes. J. Am. Coll. Cardiol. **65**, 2591–2600 (2015)
10. Zingerman, B.V., Vorobyev, A.I., Shklovsky-Kordy, N.E.: About telemedicine: patient-doctor. Physician Inf. Technol. **1**, 61–79 (2017)
11. Shklovskiy-Kordi, N., Borodin, R., Zingerman, B., Shifrin, M., Kremenetskaya, O., Vorobiev, A.: Web-service medical messenger - intelligent algorithm of remote counseling. HIS. LNCS, vol. 11148, pp. 193–197 (2018)
12. Zingerman, B.V., Shklovsky-Kordy, N.E.: Electronic medical record and the principles of its organization. Doct. Inf. Technol. **2**, 37–57 (2013)
13. Finn, V.K.: Distributive lattices of inductive JSM procedures. Autom. Doc. Math. Linguist. **48**(6), 265–295 (2014)
14. Evelson, L.I., Dubovoy, I.I., Borisova, E.P.: Improving screening methods to determine belonging to the risk group for the consequences of alcohol effect (with information technology and mathematical modeling). Physician Inf. Technol. **3**, 6–17 (2018)
15. Sun, L., Dong, H., Liu, A.X.: Aggregation functions considering criteria interrelationships in fuzzy multi-criteria decision making: state-of-the-art. IEEE Access **6**, 68104–68136 (2018)
16. Xu, J., Luo, X., Wang, G., et al.: A deep convolutional neural network for segmenting and classifying epithelial and stromal regions in histopathological images. Neurocomputing **191**, 214–223 (2016)
17. Zingerman, B.V., Shklovsky-Kordy, N.E., Karp, V.P., Vorobyov, A.I.: Integrated electronic medical card: tasks and problems. Doc. Inf. Technol. **1**, 24–34 (2015)
18. Myo, W.W., Wettayaprasit, W., Aiyarak, P.: A cyclic attribution technique feature selection method for human activity recognition. IJISA **11**(10), 25–32 (2019)
19. Karim, R.M.: Optimization parameters of automatic speech segmenttation into syllable units. IJISA **11**(5), 9–17 (2019)
20. Hu, Z., Dychka, I., Sulema, Y., Valchuk, Y., Shkurat, O.: Method of medical images similarity estimation based on feature analysis. IJISA **10**(5), 14–22 (2018)

Modeling of the Cognitive Properties of the Brain Thalamus as a System with Self-organized Criticality

M. E. Mazurov[(✉)]

Russian University of Economics, Moscow, Russia
mazurov37@mail.ru

Abstract. A mathematical model of the cognitive properties of the "specific" structure of thalamic nuclei is proposed, which allows explaining a number of its remarkable properties. The mathematical model is presented as a nonlinear dynamic system with self-organized criticality in the form of a selective neural network with selective neurons-nuclei. The substantiation is given for the representation of the thalamus as a system with self-organized criticality due to the fact that the thalamus is a controlled self-oscillatory relaxation system near the excitation threshold of relaxation oscillations in the presence of a stabilizing nonlinear negative feedback. Computational experiments of a mathematical model of the thalamus have been carried out, which have shown that the thalamus has the properties of cognitive processing of incoming sensory information, distribution, selective and amplifying properties. The responses of the mathematical model to subthreshold stimulation were studied, which showed the relevance of the mathematical model to the responses of a neuronal cell in real physiological conditions. The comparison of the properties of the mathematical model with the real properties of the excitability of a neuron in real conditions of a physiological experiment is given. The relevance of the mathematical model for the presence of relay, selective and amplifying properties for sensor signals is shown.

Keywords: Thalamus · Thalamic nuclei · Network model · Self-organized criticality · Mathematical model

1 Introduction

The aim of the work is to develop a mathematical model of the thalamus. A mathematical model of the cognitive properties of the "specific" structure of thalamic nuclei is proposed, which allows explaining a number of its remarkable properties. The mathematical model is presented as a nonlinear dynamic system with self-organized criticality in the form of a selective neural network with selective neurons-nuclei. The substantiation is given for the representation of the thalamus as a system with self-organized criticality due to the fact that the thalamus is a controlled self-oscillatory relaxation system near the excitation threshold of relaxation oscillations in the presence of a stabilizing nonlinear negative feedback.

Z. Hu et al. (Eds.): AIMEE 2020, AISC 1315, pp. 222–233, 2021.
https://doi.org/10.1007/978-3-030-67133-4_21

In the thalamus, more than 150 paired nuclei are distinguished, which are divided into three groups: relay, associative, and nonspecific [1]. Each of the nuclei specializes in the primary relaying of information from the senses of one type or another to the cerebral cortex, receives strong feedbacks from the corresponding zone of the cerebral cortex, surrounding nuclei, hippocampus, hypothalamus, medulla oblongata and others, which regulates the activity of this nucleus and the degree of filtering the incoming stream of information. All thalamic nuclei, to varying degrees, have three common functions: relay, switching, integrative and amplifying.

Consider the anatomy and functional role of the thalamus. The location of the thalamus in the brain and some of its functions are illustrated in Fig. 1.

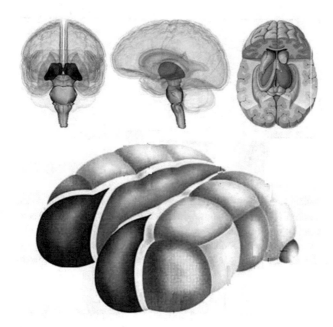

Fig. 1. Location of the thalamus in the brain and some of its functions. The thalamus in 3 projections is shown in figure from above

Consider the functional role of individual thalamic nuclei.

Relay nuclei of the thalamus are divided into sensory and non-sensory. Sensory relay cores switch the streams of afferent sensory impulses to the sensory cortex zones. They also transcode and process information. The formation of visual sensations is facilitated by the lateral geniculate body, which switches visual impulses to the occipital cortex. The relay for switching auditory impulses to the temporal cortex of the posterior part of the sylvian sulcus is the medial geniculate body. Non-sensory relay nuclei of the thalamus switch non-sensory impulses into the cortex, which enter the thalamus from different parts of the brain. In the anterior nuclei of the thalamus, impulses mainly come from the hypothalamus. The ventral nuclei are involved in the regulation of movements, performing a motor function.

The associative nuclei of the thalamus receive impulses not from the conductive pathways of the analyzers, but from a number of other nuclei of the thalamus.

Nonspecific nuclei have numerous inputs from other thalamic nuclei. In addition, information from all specific sensory systems, motor centers of the trunk, cerebellar nuclei, basal ganglia and hippocampus, as well as from the cerebral cortex enters the nonspecific thalamic nuclei. It is thanks to these many connections that the thalamic nuclei act as an integrating mediator.

An illustration of the physiological and cognitive structure of connections between the thalamus and the human cerebral cortex is shown in Fig. 2.

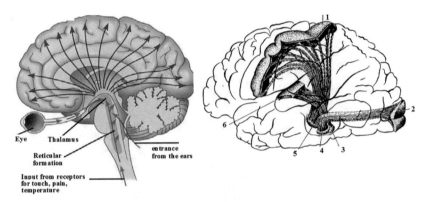

Fig. 2. Physiological and cognitive structure of the thalamus. Thalamus connections with the human cerebral cortex. Schematic model of the three main sensory systems of the left hemisphere of the brain, ascending to the cells of its cortex. 1 - area of general sensitivity of the cerebral cortex; 2 - visual area of the cerebral cortex; 3 - medial geniculate body; 4 - lateral geniculate body; 5 - lateral nucleus of the optic hillock; 6 - the auditory area of the cerebral cortex.

Thus, the thalamus is an input device to which a huge amount of sensory information is fed. This information, depending on its purpose, should be classified, redistributed to specific sections of the neocortex. This task is solved with the help of specific nuclei of the thalamus.

Mathematical models that explain the set of remarkable properties of the thalamus are currently lacking. The possibility of constructing such models is due to a set of properties: the property of self-organized criticality, the fact that the thalamus is a controlled self-oscillating relaxation system near the excitation threshold of relaxation oscillations, and the presence of a stabilizing nonlinear negative feedback.

We note right away that the set of thalamic nuclei in combination with their connections can be considered topologically equivalent to a selective neural network, in which selective neurons are used as thalamic nuclei. This presentation allows you to more effectively represent the cognitive properties of the thalamus. A series of works devoted to the description of the functioning of neural networks is given in [2–8].

2   Theory Mathematical Model of the Cognitive Properties of the Thalamus

For modeling cognitive processes, consciousness, acad. K. V. Anokhin proposed to use multilevel hyper-network models. The components of these models are elements of kogi, comma, and cognitoma. The block diagram of this simulation is shown in Fig. 3.

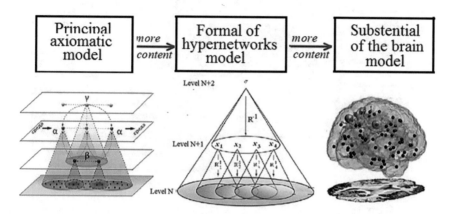

| Principal axiomatic model | more content → | Formal of hypernetworks model | more content → | Substential of the brain model |

Fig. 3. Structural modeling of a hyper-network system according to K. V. Anokhin

In order for K. V. Anokhin could describe the dynamics of electrical processes in the brain, it is necessary to provide them with physical and physiological content. A mathematical model of the cognitive properties of the "specific" structure of the thalamus in the form of a single-layer selective neural network representing a nonlinear dynamic system operating in the mode of self-organized criticality is proposed. It is topologically equivalent to a single layer neural network with selective neurons and is shown in Fig. 4. The structure of the mathematical model of the thalamus includes nuclei that represent ensembles or pools of the same type of selective impulse neurons. The selectivity of neural ensembles is ensured by their cluster organization.

Excitation of the neural pool occurs when the sum of the excitatory and inhibitory input signals is exceeded, that is, under the condition

$$\left(\sum_{i \in K_j} x_i^{(1)} - \sum_{i \in L_j} x_i^{(2)} \right) \geq U_{pj} \,,$$

where $x_i^{(1)}$, $x_i^{(2)}$ are excitatory and inhibitory signals, U_{pj} is the excitation threshold, and $(j = 1, \ldots, n)$ K_j, L_j are selective clusters containing excitatory and inhibitory neurons to which the input impulse signals belong $x_i^{(2)}$, $x_i^{(1)}$. Each neural pool corresponds to a certain cluster, the total number of clusters is equal n. Excitation of the neural pool when the threshold U_{pj} is exceeded leads to the excitation of a relaxation auto-generator contained in the impulse neuron, which produces a periodic impulse sequence, a spike

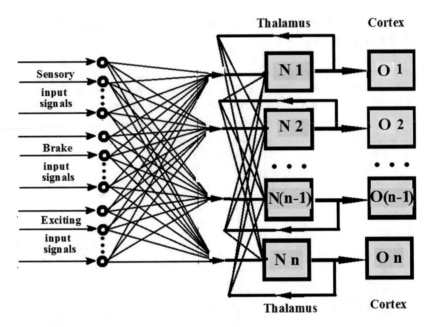

Fig. 4. Mathematical model of the cognitive properties of the "specific" structure of the thalamus in the form of a nonlinear dynamic system of a topologically equivalent single-layer neural network with selective neurons

or a single impulse entering a specific area of the cerebral cortex. Any known model of a pulsed neuron can be used as a mathematical model of a relaxation auto-generator.

The proposed mathematical model of the thalamus in the form of a set of nuclei in combination with their connections is topologically equivalent to a selective neural network in which selective neurons are used as thalamic nuclei. The mathematical model takes into account the presence of feedbacks with the neocortex. In general, the mathematical model also has similarities with the Hopfield neural network, in which there are feedbacks. This presentation allows you to more effectively reflect the cognitive properties of the thalamus.

3   Realization of Self-organized Criticality in the Thalamus

The remarkable properties of the thalamus are as follows: the presence of relay distribution properties; selectivity in relation to input sensory information; significant increase in the intensity of weak sensory signals. These properties can be realized using a system with positive feedback, which is in a near-threshold mode or at a bifurcation point, at the transition of which a self-oscillating mode arises.

Information about such capabilities of near-threshold systems, also called regenerative systems, is well known in the theory of self-orcanized criticality [9, 10], in radio engineering for the creation of regenerative radio receivers [11]. In one cascade of such a radio receiver, significant selectivity and amplification were achieved up to a million or more times. Naturally, the evolution of nature used such a remarkable property to create the brain.

There are two possible ways to significantly increase the sensitivity of systems with self-organized criticality located near the self-excitation threshold. The first way is possibly a large approximation to the self-excitation threshold. An example of such technical systems is a regenerative radio receiver with positive feedback. It allows you to increase the gain of the input signal up to a million times, while simultaneously increasing the selectivity by increasing the quality factor of the used oscillatory circuit [11]. The fundamental disadvantage of such a system is a significant increase in noise and the possibility of switching to a self-oscillating mode, in which all useful properties are lost and it becomes a source of interference.

The second possible mode of a system with self-organized criticality is the use of an additional subthreshold signal, which would provide a transition to a self-oscillating mode for a certain time, and then this mode was extinguished. Due to the emergence and suppression of the self-oscillating regime, it is possible to achieve an extremely large gain and, at the same time, to ensure the necessary stability. In technical devices, the intermittent pumping mode was called supergenerative amplification [11].

The used mathematical model of the "specific" structure of the thalamus, shown in Fig. 3, implements subthreshold synchronization of the potentials of neurons of the thalamic nuclei, which are relaxation self-oscillating systems, a subthreshold signal. This provides a significant selective amplification of weak sensory signals sent to the cortex, which ensures their better retention in memory.

Let us explain the mechanism of the possibility of achieving ultrahigh sensitivity in systems with self-organized criticality using the example of the Van der Pol - Fitzhugh equation describing a self-oscillating relaxation system

$$\frac{du}{dt} = u - \frac{u^3}{3} - v + I + U; \quad \frac{dv}{dt} = \varepsilon(a - u + bv) ,$$

where I - displacement current, a, b, ε - parameters: $a = 0.7$, $b = 0.8$. If $I = 0.142$, then the solution to the equation is a single pulse, which can be caused by external excitation. At $I = 0.4$, the equations describe relaxation oscillations [13]. The phase portrait of the Van der Pol - Fitzhue equation at $I = 0.142$ is shown in Fig. 5 on top.

The value of the OA potential is equal to the value of the threshold required for the appearance of a pulse. If the isocline is moved parallel to itself closer to point A, then the threshold value will decrease. If the isocline passes through point A, then the excitation threshold is zero and point A will be critical. With further movement of the isocline, the system will go into a self-oscillating mode. Thus, near the minimum point A, it is possible to achieve high and ultra-high sensitivity. In this case, the greater the amplitude of the subthreshold potential, the greater the frequency of spike generation. These patterns are illustrated in Fig. 5.

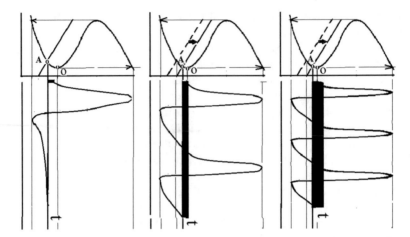

Fig. 5. Generation - of neural impulses when excited by a near-threshold signal. In figure on the left - generation of a single pulse when the OA threshold is exceeded. In figure in the center and on the right, generation of spikes from two and three pulses with subthreshold excitation, the amplitude of which is equal to the height of the rectangular pulse shown in black

4 Researching the Relevance of the Mathematical Model of the Thalamus

The structure of the proposed mathematical model of the thalamus corresponds to the available anatomical data of its nuclear structure, the functional properties also correspond to the available experimental data. According to the available experimental data, the relay selective and amplifying structure of the thalamus is specific nuclei of the thalamus, which process input sensory signals and send them to specific areas of the cortex. The control action or feedback on specific nuclei comes from various areas of the cortex, hippocampus, hypothalamus, medulla oblongata and other systems surrounding the thalamus. This is reflected in the mathematical model by the presence of inhibitory and excitation inputs. The presence of control from all feedbacks can be considered as the effect of inhibitory and exciting signals at the input. The influence of these signals can be considered, as follows from relation (1), as modulation of the excitation threshold of neurons - thalamic nuclei R1-YN.

Consider the description of the computational method. A direct method for describing the synchronization of relaxation systems was proposed, based on the modified axiomatic Wiener-Rosenbluth method and Kronecker's inequalities from the theory of uniform almost periodic functions [12–16].

The initial data are the form of the relaxation oscillation and the threshold function that describes the refractoriness of the relaxation oscillator, absolute during the fast phase and relative during the slow phase. It was shown that for the main synchronization modes the relation [12–16]_

$$-\varepsilon < (n_1 T_1 - n_2 T_2) < 0,$$

where T_1, T_2 are the periods of the synchronizing signal and the relaxation oscillator, $\varepsilon = f^{-1}(U)$; $f(T)$ is the threshold sensitivity in the phase of relative refractoriness. To describe transient processes preceding stationary synchronization, the relations can be used

$$-\varepsilon < (T_0 + n_1 T_1 - n_2 T_2) < 0$$

The modified axiomatic model of the system is specified in the form of two different functions on sequential time intervals

$$f(t) = \begin{cases} f_1(t) & 0 \le t \le t_1 \\ f_2(t) & t_1 < t \le T \end{cases}, \quad f_\partial(t) = \begin{cases} \infty \\ f_3(t) \end{cases} \tag{1}$$

where $f_1(t), f_2(t)$ are the functions characterizing the relaxation self-oscillations in the interval of "fast" and "slow" changes in the relaxation self-oscillation, $f_1(t)$ the function was specified in the form of a short pulse, in the form of a linear slowly varying function $f_2(t)$; $f_3(t)$ – a function characterizing the dynamic threshold of excitation in the interval of the "slow" phase $f_3(t) = 0$. The computational algorithm implements relation (1). In the numerical implementation, the selection ε of almost-periods of uniform almost-periodic functions obtained as a result of synchronization is carried out by separating the remainder from dividing the current time of the process by the sum of the periods of the synchronizing signal. Further, a condition is used that describes the extraordinary excitation of the relaxation self-oscillation when the dynamic threshold is exceeded by the instantaneous sum equal to the value of the relaxation self-oscillation and the amplitude of the external synchronizing signal.

5   Results of an Experimental Study of the Mathematical Model of the Thalamus

The realization of near-threshold amplification of relaxation oscillations was tested in a computational experiment using a modified axiomatic method using the properties of uniform almost periodic functions [12–16]. An effective mode is super-regenerative when using control signals from other systems: cortex, hypothalamus, hippocampus and others. The theta rhythm of the hippocampus can be one of the control signals that implements the process of attention and better assimilation of information. When testing the model, control signals of various shapes were used: 1) rectangular with different amplitudes; 2) sinusoidal; 3) linearly increasing and linearly decreasing. The obtained results of the computational experiment are illustrated in Fig. 6 from above. It is interesting to compare the results obtained with the experimental results obtained by Ming-Li Zhao and Chun-Fang Wu when the neuron of Drosophila was exposed to signals of different shapes: rectangular with different amplitudes; linearly increasing and linearly decreasing [17]. These experimental results are shown in Fig. 6e–h. The authors note that the effect of a similar effect on brain neurons is almost identical.

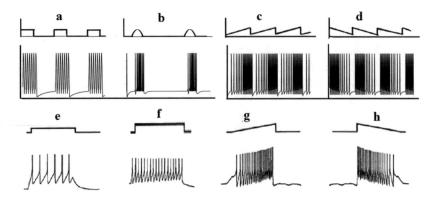

Fig. 6. The responses of the mathematical model of the thalamus to testing signals of various shapes in a computational experiment are shown in Figs. a, b, c, d. The responses of a neuronal cell in a physiological experiment to influences of various forms are shown in Figs. e, f, g, h. As control signals, a - rectangular signals, b - sinusoidal signals, c - signals in the form of a linearly increasing function, d - linearly decreasing function were used. The corresponding experimental results obtained by Ming-Li Zhao and Chun-Fang Wu are shown at the bottom of Fig. e, f, g, h

Experimental studies of the proposed mathematical model in a computational experiment are unique and performed for the first time in the world. The uniqueness of the results is explained by the significant difficulties in studying the synchronization of relaxation oscillations. These difficulties in this work were overcome by using the modified axiomatic Wiener-Rosebluth method in combination with the use of the means of uniform almost periodic functions. This method made it possible to use control signals of any form in the considered thalamus model. They can be sensory signals, signals from surrounding structures: from the cerebral cortex, hypothalamus, hippocampus. For example, the control signal can be a theta rhythm signal from the hippocampus.

Checking the relevance of the responses of the mathematical model and evaluating the relay selective amplifying properties were carried out by examining the responses to impulse input sensory signals. For this purpose, 10 small input impulse signals were fed to the input of a specific block of sensory nuclei, and activation was carried out with a small subthreshold signal of nuclei - neurons, specifically the fourth and sixth input signals. In this case, disinhibition of the corresponding channels of the "specific" part of the thalamus was assumed. The experimental results are as follows: 1) a relay selective hit of sensor signals into the required channels was observed; 2), a sharp explosive increase in the amplitude of the signal entering the planned channel was observed when the sum of small input signals and the control subthreshold signal exceeded the resulting threshold level. The results of the computational experiment are illustrated in Fig. 7.

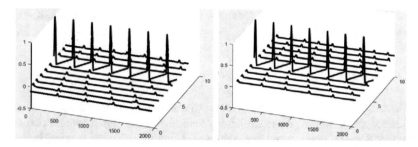

Fig. 7. Results of a computational experiment to test the relay, selective and amplifying properties of the mathematical model of the thalamus. Left - the passage and explosive amplification of the sensory signal in the fourth channel, on the right - the passage and explosive amplification of the sensory signal in the sixth channel

6 The Discussion of the Results

The basic physiological data on the structure and cognitive functions of the thalamus are presented. A mathematical model of the thalamus is proposed in the form of a network structure, including nonlinear dynamic structures, which are systems with self-organized criticality. In general, the structure of the thalamus can be represented by a selective neural network with selective neurons-nuclei and feedback from the neocortex.

Experimental studies of the proposed mathematical model in a computational experiment are unique and performed for the first time in the world. The uniqueness of the results is explained by the significant difficulties in studying the synchronization of relaxation oscillations. These difficulties in this work were overcome by using the modified axiomatic Wiener-Rosebluth method in combination with the use of the means of uniform almost periodic functions. This method made it possible to use control signals of any form in the considered thalamus model.

When testing the model, control signals of various shapes were used: 1) rectangular with different amplitudes; 2) sinusoidal; 3) linearly increasing and linearly decreasing. The results obtained are in qualitative agreement with the experimental results obtained by Ming-Li Zhao and Chun Fang Wu when the neuron of Drosophila was exposed to signals of various shapes: rectangular with different amplitudes; linearly increasing and linearly decreasing [17].

Checking the relevance of the responses of the mathematical model and evaluating the relay selective amplifying properties was carried out by examining the responses to pulse input sensory signals.

In this case, disinhibition of the corresponding channels of the "specific" part of the thalamus was assumed. The results of the experiment are as follows: 1) there was a relay selective hit of sensor signals into the required channels; 2), a sharp explosive increase in the amplitude of the signal entering the planned channel was observed when the sum of the small input signals and the control subthreshold signal exceeded the resulting threshold level. Qualitatively similar results take place in real physiological conditions. The general conclusion of studies of the mathematical model in a computational experiment showed the relevance of its properties to the properties of the thalamus in physiological conditions.

7 Conclusion

A mathematical model of the cognitive properties of the "specific" structure of thalamic nuclei is proposed, which allows explaining a number of its remarkable properties. The mathematical model is presented in the form of a nonlinear dynamic system in the form of a single-layer selective neural network with selective neurons-nuclei. The substantiation of the representation of the thalamus as a system with self-organized criticality is given on the example of studying the nature of isoclines by the Van der Pol-Fitzhu equation. The study of the relevance of the mathematical model of the thalamus using the modified axiomatic Wiener-Rosenbluth method and Kronecker inequalities from the theory of uniform almost periodic functions is carried out. It is shown that the results of the computational experiment qualitatively coincide with the experimental results obtained when the neuron of Drosophila is exposed to signals of various shapes: rectangular with different amplitudes; linearly increasing and linearly decreasing. The relevance of the mathematical model for the presence of relay, selective and amplifying properties for sensor signals is shown.

In the considered model of the thalamus, control signals of any form can be used. They can be sensory signals, signals from surrounding structures: from the cerebral cortex, hypothalamus, hippocampus. For example, the control signal can be a theta rhythm signal from the hippocampus. This signal provides more efficient storage of input sensory signals.

Further work suggests a more detailed modeling of cognitive processes in the thalamus for more specific cases of interaction of the thalamus with individual structures of the neocortex.

References

1. Jones, E.G.: The Thalamus, vol. 2, p. 915. Springer, New York (2012)
2. Iyanda, A.R., Ninan, O.D., Ajayi, A.O., Anyabolu, O.G.: Predicting student academic performance in computer science courses: a comparison of neural network models. IJMECS 10(6), 1–9 (2018). https://doi.org/10.5815/ijmecs.2018.06.01
3. Alkhathlan, A.A., Al-Daraiseh, A.A.: An analytical study of the use of social networks for collaborative learning in higher education. IJMECS 9(2), 1–13 (2017). https://doi.org/10.5815/ijmecs.2017.02.01
4. Moshref, M., Al-Sayyad, R.: Developing ontology approach using software tool to improve data visualization (case study: computer network). IJMECS 11(4), 32–39 (2019). https://doi.org/10.5815/ijmecs.2019.04.04
5. Cheah, C.S., Leong, L.: Investigating the redundancy effect in the learning of C++ computer programming using screencasting. IJMECS 11(6), 19–25 (2019). https://doi.org/10.5815/ijmecs.2019.06.03

6. Nurafifah, M.S., Abdul-Rahman, S., Mutalib, S., Hamid, N.H.A., Ab Malik, A.M.: Review on predicting students' graduation time using machine learning algorithms. IJMECS **11**(7), 1–13 (2019). https://doi.org/10.5815/ijmecs.2019.07.01
7. Alvarez-Dionisi, L.E., Balza, M.M.R.: Teaching artificial intelligence and robotics to undergraduate systems engineering students. IJMECS **11**(7), 54–63 (2019). https://doi.org/10.5815/ijmecs.2019.07.06
8. Adekunle, S.E., Adewale, O.S., Boyinbode, O.K.: Appraisal on perceived multimedia technologies as modern pedagogical tools for strategic improvement on teaching and learning. IJMECS **11**(8), 15–26 (2019). https://doi.org/10.5815/ijmecs.2019.08.02
9. Buck, P.: How Nature Works: The Theory of Self-organized Criticality, p. 276. Springer, New York (2013)
10. Kelso, J.A.S.: Dynamic Patterns: The Self-organization of Brain and Behavior. MIT Press, Cambridge (1995)
11. Belkin, M.K., Kravchenko, G.I., et al.: Superregenerators, p. 248. Radio and Communication Publishing House, Moscow (1983)
12. Mazurov, M.E.: Solution of diophantine inequalities in problems of synchronization of relaxation oscillations. ZhVM MF AN SSSR. **31**(11), 1619–1636 (1991)
13. Mazurov, M.E.: Synchronization of self-oscillating relaxation systems, synchronization in neural networks. Proc. Russ. Acad. Sci. Phys. Ser. **82**(1), 83–87 (2018)
14. Mazurov, M.E.: Non-linear dynamics, almost periodic summation, self-oscillating processes, information coding in selective pulsed neural networks. Izv. RAS. Phys. Ser. **82**(11), 1564–1570 (2018)
15. Mazurov, M.E.: Nonlinear dynamics and synchronization of neural ensembles in the formation of attention. Izv. RAN. Phys. Ser. **84**(3), 451–456 (2020)
16. Mazurov, M.E.: Mechanisms of invariant error-correcting coding in impulse neural networks. Izv. RAS Phys. **84**(1), 90–95 (2020)
17. Zhao, M.-L., Wu, C.-F.: Alterations in frequency coding and activity dependence of excitability in cultured neurons of drosophila memory mutants. J. Neurosci. **17**(6), 2187–2199 (1997)

Advances in Technological
and Educational Approaches

The Natural Intelligence of the Wind Castle Design with the World Natural Heritage of Jeju Island

Moon Ho Lee[1] and Jeong Su Kim[2(✉)]

[1] Chonbuk National University, Jeonju-Si, Korea
[2] Department of ICT Engineering, Korea Soongsil Cyber University, Seoul, Korea
kjs@mail.kcu.ac

Abstract. As a World Natural Heritage (WNH, listed in 2007), Jeju is a largest volcanic and elliptic type island which has 1,950 m Halla Mountain with much winds, stones and Oreums (Volcanic Cones). The Oreums are 368 small mountains and scattered around the Halla Mountain, and Batdam (farmland stone wall, world agricultural heritage, listed in 2014), which has 1.5 m tall and 22,000 km distance, encloses around the Jeju Island. The Jeju is 4.8 m/s wind blows during the year. In this paper, 3 Layer's wind castle of Hallasan-Oreum-Batdam that prevents wind is proposed in terms of natural intelligence and the height of Oreum is calculated in mathematics. The key point is $369 = 3$ $(1 + 2 + 3)$, and the base is 10^2, 10^1, 10^0. The reliability of Hallasan-Oreum-Batdam shows 99% when the real number mapping of it is designed of a parallel tree structure with correlation of $3,6,9 = 3(1,2,3)$. Previously, it was only studied in geological terms. However, this study differs from the existing geological measurement method, for the first time, the height of Oreum was mathematically calculated by the Bernoulli's theorem of fluid mechanics. The proposed experimental-theoretic result and the actual height of Oreum matched exactly, and the results were confirmed by simulation.

Keywords: Natural intelligence · Wind castle of hallasan-oreum-batdam · World natural heritage of Jeju Island

1 Instruction

For the Jeju people, the wind is a friend and Koendang (Professor Moon Ho Lee proved the source Koendang word) which lives in a neighbor's house. The wind overtakes 22,000 km of Batdam (Black Silk Road) and 368 Oreums (Volcanic Cones) and crosses the 1950 m height of Hallasan (Halla Mountain). Jeju has an average wind of 4.8 m/s during the season. It is not an exaggeration to say that it is the home of the wind. Typhoons over 30 m/s pass 4–6 times a year. In Jeju, 1.5 m of Batdam, 368 climbs, 1,950 m of Hallasan, wind speed and topography are all unique [1, 3, 4, 6–9]. The people who are absorbed about Halla Mountain are Gu Kim, who invented Batdam in 1234, Joo Myeong Seok of Mt. Halla butterfly in 1943, Joong Seop Lee, who painted the eyes of the wind in the eyes of a yellow bull in 1951, Jong Hyu Boo in

Z. Hu et al. (Eds.): AIMEE 2020, AISC 1315, pp. 237–247, 2021.
https://doi.org/10.1007/978-3-030-67133-4_22

1980 who found 330 plants and Mnjangguls in Halla Mountain, The climber's Jong Cheol Kim (loved Oreum so much that his dead body was sprinkled on WitsaeOreum) in 1994, Young Gap Kim of Duomoak Oreum photographer, Woon Cheol Baek of Tamra Mokseokwon in 1980, Si Jee Byeon who was the storm of Jeju wind painter, Professor of the National University of Jeju in 1990s, and Professor Moon Ho Lee of Oreum's Mathematics, etc. [1]. However, no research has been conducted on the linked wind castle between Oreum, Batdam and Hallasan. This study focused on this, and tried to solve the problem by taking inspiration from the will of Professor Moon ho Lee's father, Gab Boo Lee "Hanil Doi Seoksam (which words mean korean ancient 1-2-3)". The number of Jeju Oreum is usually 368. However, Hallasan is also a great Oreum. The author calculated the height of the Oreum by finding 'the Oreum in Oreum (folded Oreum). This paper focused on the number of 1 2 3 and analyzed the height of 368 Oreums, and found that the height of the neutral zone is 1/2; 'the Oreum in Oreum'. Based on this, the correlation of natural intelligence wind castle to Hallasan-Oreum-Batdam was proved through information experimental-theoretic analysis. Artificial intelligence is a combination of human brain and intelligence for big data, IoT, cloud computing, and 5G communication [11–17]. On the other hand, natural intelligence interacts with nature and climate, and modules and networks are inter-connected [2]. First, as the wind rises and falls along the 3Layer contour layer of Hallasan-Oreum-Batdam, the wind castle controls the fluid flow of the wind according to the entering angle and height. For example, in summer, the typhoon blowing in the South Pacific is blocked by Batdam below the 200 hills, Oreum from the 350–650 hills, and by Hallasan at the final 1,950 hills. Topographically, Jeju's 20–40 m/s typhoon storming from the east and west, departing from the top of Halla Mountain, climbs Korea's main Jirisan Mountain (1,900 m), and then heads north along Gyeongsang-do's east coast or west Jeolla-do coast. Because of natural environment intelligence. If Jeju Island was not of the oval shape and the high east (lower west: high east of the island, low west), the typhoon would penetrate the center of Korea. Second, when looking at Jeju's summer farming and winter farming, Batdam [3] is 1.5 m high. Depending on the wind influence, the wind passes over the top of the Batdam in summer, so tall barley is farmed, and in the winter, the wind passes to the bottom of the Batdam. Third, it is the height of the Oreum. There are 368 Oreums in the middle of Jeju. The reason is that the underground magma layer is thinly laid on the 200–500 highland in Jeju, and the volcanic eruption has the same atmospheric pressure and the underground magma space eruption pressure of 350–600 m. As a result, the height of Oreum was almost similar to 350–600 m, and the other was the Oreum in the Oreum because the surface became the neutral zone (social language is called 'Middle Mountain', Middle Mt. Village). For example, there is Namsongak (340 m) in Seog-wangni, and Seogwang green tea field is 170 m above sea level, which is twice the elevation. When climbing from Seongpanak to Halla Mountain, it can be arranged as $1.5 \times 200 \times (1 + 2.5 + 3) = 1,950$ m. The first protest 1.5 (triangle) is Batdam height, 300 m is steep slope from Hallasan Seongpanak (750 m) to Azalea field (1,500 m), 900 m is 450 m Mirror (height of the magma underground) from Jindalrae field to the top of Hallasan. Drunk and doubled. Oreum is about 60% around 400 m, 40% around 500–700 m, and is one Halla Mountain at 1,950 m. In this way, Oreum can be represented.

Donnaeko-BaeknokdamDongneung-Jindalraebat-Seongpanak-Gyorae- WitsaeOreum (1,700 m) is also known as the 'rain belt'. Since it is the neutral zone where −, +, which collides with the low-pressure air mass that feeds on the humidity of the North Pacific rising from the southeast and the high-pressure cold air of the peak of Halla Mountain, is zero, it rains a lot. The aim of this study is to help cultivate crops in Jeju, where typhoon is very strong and frequent (4–5 times/year), by analyzing and modeling the natural intelligence characteristics that can defend against the wind, and to reduce wind damage. The composition of this paper proposed theory and its real measured Oreum height calculation H/2 in Sect. 2, Sect. 3 analyzed model of Hallasan-Oreum-Batdam. Finally, Sect. 4 shows conclusion.

2 Proposed Theory and Its Real Measured Oreum Height Calculation H/2

Jeju is a world of uphill, Batdam and wind. Tens of thousands of times a day, the wind rises and falls on Hallasan Mountain, Oreum, and Batdam. That is why Oreum and Batdam breathe in and out. Looking at the shape of the uphill and thatched roof, the slope (15–20°) is similar. The slope of the rough climb is slightly steeper than that of the thatched roof. Jeju's wind is 4–20 m/s, and annual rainfall is 1,400 ∼ 1,600 mm, so it has a lot of wind and rain. Therefore, in order to overcome this, the thatch roof and the slope of Oreum are optimal conditions of 15–30°. Jeju Island is centered on Baeknokdam of Hallasan Mountain, which is 1,950 m long, and the slopes of the east and west sides have a gentle slope of 3 to 5°, and the slopes of the north and south sides are around 5°. Oreum is spread all over the center of Halla Mountain, but it is widely distributed in the east. The topography of Jeju Island is an elliptical east high-west low with high Seongsanpo and low Mossulpo. An Oreum colony formed when the magma belt was erupted along the fire-way at a height of about 300 m between Hallasan's middle mountain, where the underground structure was connected to the magma chamber [4]. The Bernoulli theorem was used to calculate the height of 369 Oreums. The Bernoulli Stack Effect theorem is the kinetic and potential energy plus pressure energy, and Torricelli theorem is a special case of Bernoulli's theorem [5], because the magma with high volcanic lava density in the underground enclosed space runs out into the space. Based on this, the height (m) from the neutral belt to the top of the shaft and the height (m) from the bottom to the neutral belt are obtained. The definition of terms is as follows. H: Height of the shaft (m), H_n: Height from the bottom to the neutral zone (m), A_b: Area of the lower opening of the neutral zone, A_a: Area of the upper opening of the neutral zone, T_s: Air temperature inside the shaft (K), T_o: Ambient temperature (K).

$$H_n = \frac{H}{1 + (T_s/T_o)(A_b/A_a)^2}. \tag{1}$$

where $H_n = \frac{H}{2}$, if $T_s = T_o$ $A_a = A_b$. For example, in the case of Seogwang-ri Osulloc Green Tea Field Namsongak Oreum, when the underground magma elutes to the ground in the beginning, the inside and outside temperature of the shaft are the same, and the opening area of the top and bottom is the same. It is $H_n = \frac{340}{2}$ m $= 170$ m and above the neutral zone and the ground above 170 m above sea level, and the total height is 340 m. Therefore, height of the neutral zone

$$H_n = \frac{H}{2}. \tag{2}$$

The Namsongak of Osulloc Green Tea Field, located above Seogwangseori (140 m above sea level) in Andeok-myeon, is 339 m above sea level. Underneath Namsongak Oreum is a height of 170 m above sea level, and if you multiply it by twice, the height of Oreum is proved as follows.

The addition of kinetic energy ($\frac{1}{2}mv^2$) and potential energy (mgz) to pressure energy (PV) can be written by Bernoulli's theorem as (3).

$$\frac{1}{2}mv_1{}^2 + mgz_1 + P_1V_1 = \frac{1}{2}mv_2{}^2 + mgz_2 + P_2V_2. \tag{3}$$

If (3) is divided by mg and ρg is r.

$$\frac{v_1^2}{2g} + z_1 + \frac{P_1V_1}{mg} = \frac{v_2^2}{2g} + z_2 + \frac{P_2V_2}{mg}, \ \frac{v_1^2}{2g} + z_1 + \frac{P_1}{r} = \frac{v_2^2}{2g} + z_2 + \frac{P_2}{r}. \tag{4}$$

Where $P_1 = P_2$, atmospheric pressure $v_1 = 0$, $\frac{v^2}{2g}$ is velocity head, z is position head, and $\frac{P}{r}$ is pressure head, (4) is $\frac{v_2^2}{2g} = z_1 - z_2 = h_2$. Therefore, Torricelli's theorem is given by (5).

$$v_2 = \sqrt{2gh_2}. \tag{5}$$

1) Air fluid velocity from inside to outside [m/s].

$$v_2 = \sqrt{2gh_2} = \sqrt{2g\frac{\Delta P}{r_i}} = \sqrt{2\frac{\Delta P}{\rho_i}}, \Delta P = (\rho_0 - \rho_i)gh_2 = \sqrt{2g\frac{(\rho_0 - \rho_i)}{\rho_i}h_2}. \tag{6}$$

On the other hand, if $z_1 = z_2$ in the Eq. (6), the *pressure energy* $\times \frac{1}{kinetic\ energy} =$ *constant* will satisfy the jacket property [10].

2) Air fluid velocity from outside to inside [m/s].

$$v_1 = \sqrt{2gH_n} = \sqrt{2g\frac{\Delta P}{r_0}} = \sqrt{2\frac{\Delta P}{\rho_0}}, \Delta P = (\rho_0 - \rho_i)gH_n = \sqrt{2g\frac{(\rho_0 - \rho_i)}{\rho_0}H_n}. \tag{7}$$

3) Discharge mass flow from the shaft [kg/s] is $\rho_i \times C \times A_a \times v_2$. Therefore,

$$Q_o = \rho_i \times C \times A_a \times \sqrt{2g\frac{(\rho_0 - \rho_i)}{\rho_i}h_2}\,[kg/s], Q_o = C \times A_a\sqrt{2g\rho_i(\rho_0 - \rho_i)h_2}\,[kg/s].$$

(8)

4) Inflow mass flow from the shaft [kg/s] as $\rho_0 \times C \times A_b \times v_1$.

$$Q_i = \rho_0 \times C \times A_b \times \sqrt{2g\frac{(\rho_0 - \rho_i)}{\rho_0}H_N}\,[kg/s], Q_i = C \times A_b\sqrt{2g\rho_0(\rho_0 - \rho_i)H_n}\,[kg/s].$$

(9)

5) Discharge mass flow equal Inflow mass flow. Then,

$$C \times A_a \times \sqrt{2g\rho_i(\rho_0 - \rho_i)h_2} = C \times A_b \times \sqrt{2g\rho_0(\rho_0 - \rho_i)H_n}.$$

(10)

Hence,

$$A_a^2 \times \rho_i \times h_2 = A_b^2 \times \rho_0 \times H_n, A_a^2 \times T_0 \times h_2 = A_b^2 \times T_s \times H_n$$

(11)

($\because$ density ρ is inversely proportional to temperature T).
From (11),

$$\frac{h_2}{H_n} = (\frac{A_b}{A_a})^2 \times \frac{T_s}{T_0}, \frac{H - H_n}{H_n} = (\frac{A_b}{A_a})^2 \times \frac{T_s}{T_0}, \text{ therefore, } \frac{H_n}{H} = \frac{1}{1 + (T_s/T_0)(A_b/A_a)^2}.$$

(12)

$$H_n = \frac{H}{1 + (T_s/T_0)(A_b/A_a)^2}.$$

(13)

If

$$T_s = T_0,$$

$$A_a = A_b$$

$$H_n = \frac{H}{2} = \frac{340}{2} = 170m.$$

(14)

In a similar way, Using the height of the Oreum from (6), atmospheric pressure (ΔP) 101,325 [N/m^2], product of density and gravity (ρg) 9,800 [N/m^3], volcanic lava characteristic value 0.0304 [kg/m^3]. The calculated heights of Namsongak Oreum and Hallasan are as follows, the heights of Namsongak Oreum:

$$H = \frac{\Delta P}{0.0304 \rho g} = \frac{101,325}{0.0304 \times 9,800} [\frac{N}{m^2}] \times [\frac{m^3}{N}] = 340 \, m. \tag{15}$$

Also, the heights of Hallasan:

$$H = \frac{101,325}{0.0053 \times 9,800} [\frac{N}{m^2}] \times [\frac{m^3}{N}] = 1,950 \, m. \tag{16}$$

Equations (15) and (16) are the same as the measured actual height values. In Jeju, 368 Oreums were caused by the simultaneous rise of Magma into the sky due to Simultaneous volcanic eruption between 25,000 and 100,000 years. The density of basalt is about 3.3 [gram/cm³], the density and temperature are inversely proportional as in (11), and the 1,950 m volcano of Hallasan has a density value of 0.0053 [kg/m³], which consists of high temperature multistage explosions over 1,500 °C.

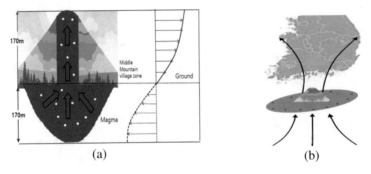

(a)   (b)

Fig. 1. Namsongak Oreum and the Wind Castle direction (a) A virtual diagram of magma eruption of Namsongak Oreum (339 m). (b) Jeju wind castle model diagram. The black line is flow of the wind. The wind castle model consists of 3Layers. 1. A big triangle with Hallasan (sea level 1,000 m) 2. Middle triangles with Oreum (sea level 300 m) 3. The small stone walls with Batdam (sea level 100 m), depend on the wind direction, vice versa as Jacket matrices [10].

When a typhoon is blowing, Hallasan and Oreum cry out when they defend the typhoon with their whole body, and Jeju people know the sound of the rising wind [6]. When a strong typhoon blows to Jeju, it hits the uphill of Hallasan. Then, the typhoon forces are slightly damped and displaced [7]. Without Hallasan and Oreum, a typhoon would head north on the Korean Peninsula, causing massive damage. The typhoon that winds up across the sea with Jeju's wind castle changes the course of the typhoon as it climbs on the rim of the Gimhae Plain east of Jiri Mountain (1,915 m) or the Gimje Mangyeong Plain west of Jiri Mountain. That is why Hallasan is called a guardian mountain.

3   Wind Castle Model of Hallasan-Oreum-Batdam

3.1   Natural Intelligence Wind Castle Parallel Modeling

There are 369 Oreum in Jeju including the large Hallasan (1,950 m). The base of $369 = 3(1 + 2 + 3)$ is $10^2, 10^1, 10^0$. Here, Mapping function is defined as follows.

$$3, 6, 9 = 3(1, 2, 3). \tag{17}$$

In formula (17), the right term 3 is 3 of Jeju Samdado (wind, stone, Oreum) and Layer 1 of the oval Jeju is about the left term 3.

$$1, 2, 3. \tag{18}$$

Layer 2 is 1, 2, 3 for the left term 6 of formula (17)

$$4, 5, 6. \tag{19}$$

Layer 3 is 1, 2, 3 for the left term 9 of formula (17)

$$7, 8, 9. \tag{20}$$

As shown in the figure, it can be represented as a tree, and the maximum values of (18), (19), and (20) expressions 3, 6, and 9 are parallel link points of three layers.

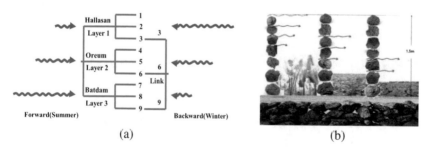

(a)   (b)

Fig. 2. Hallasan Wind Castle Tree and Batdam Crop (a) Hallasan-Oreum-Batdam Tree (b) Batdam crops and wind in summer, autumn and winter

The wind castle model is the wind direction along the four seasons. In summer, the south wind blows about 10 m/s from the Pacific Ocean. At this time, the wind is mainly defended in Batdam and Oreum. In winter, cold northern winds blow from Siberia at high altitude, assuming that they are defending on the side of Hallasan and Oreum, as shown in Fig. 2(a).

3.2 Simulation Analysis of the Wind Castle Model

In Oval Jeju, Batdam extends from the beach to Middle Mountain, and there are 368 Oreums on the upper 200–600 hills. The Hallasan mountain in the center of Jeju Island. This is like the number of forming a circular band. That is, the common similarity is that the spatial three-dimensional body blocks the fluid of wind. Here, the height of Batdam is $1.5 \times 22,000$ km $= (368$ Oreum $\times$ average height of Oreum 450 m + Hallasan height 1,950 m) $\times$ 196. If Batdam was built vertically in the space, the sum of the heights of Oreum and Hallasan is 196 times. Samdado Jeju is a place of stone wind rise, and the shape of Jeju is an oval, 3Layer beach, middle mountain, and Hallasan. Samdado Jeju is a town of stones, wind, and Oreum, and there are three layers of ovals, the seaside, the middle mountain, and the Hallasan. Hallasan is in the central center. Multiply 3 Layer times 3 of the Samdado to get 9. 9 is a real set {1 2 3 4 5 6 7 8 9}, covering each layer. Mapping the real set {1 2 3 4 5 6 7 8 9} to Hallasan (1 2 3), Oreum (4 5 6), and Batdam (7 8 9), respectively, according to their correlation to the shape resemblance and modified perfect number theorem [13], Hallasan becomes

$$1 + 2 + 3 = 6, 6 \text{ modular } 3 = 0, 1 \times 2 \times 3 = 6 \Rightarrow \text{ modular } 3 = 0 \qquad (21)$$

In case of Oreum

$$4 + 5 + 6 = 15, 15 \text{ modular } 3 = 0, 4 \times 5 \times 6 = 120 \Rightarrow \text{ modular } 3 = 0. \qquad (22)$$

If you decompose expression (22), it becomes like expression (23).

$$4 = 3 + 1, 5 = 3 + 2, 6 = 3 + 3. \qquad (23)$$

In case of Batdam

$$7 + 8 + 9 = 24, 24 \text{ modular } 3 = 0, 7 \times 8 \times 9 = 504 \Rightarrow \text{ modular } 3 = 0. \qquad (24)$$

Similarly, it becomes like recursive relation [10],

$$7 = 3 + 4 = 3 + 3 + 1, 8 = 3 + 5 = 3 + 3 + 2, 9 = 3 + 6 = 3 + 3 + 3. \qquad (25)$$

The modular of each result for sum and product is as shown in expression (26).

$$6 + 15 + 24 = 45, 45 \text{ modular } 3 = 0, 6 + 120 + 504 = 630, 630 \text{ modular } 3 = 0. \quad (26)$$

Just as the individual Eqs. (21), (22), and (24) are all zero in mod3, the Eq. (26), which combines Eqs. (21), (22), and (24), also becomes zero when mod3. It becomes 3 Layer field of Batdam-Oreum-Hallasan and circulates around 1, 2, 3. Therefore, the wind moves up and down in Jeju Island. Based on this equation, it can be drawn as shown in Fig. 3 like the periodic table of the contour elements [7]. Also, the rotation of Hallasan according to the wind direction was shown in the simulation result in Fig. 1. (b), 3 and 4.

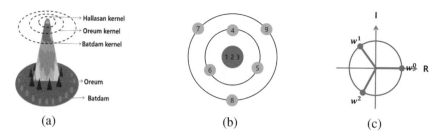

(a)   (b)   (c)

Fig. 3. The 3layer Characteristics of Hallasan-Oreum-Batdam (a) Hallasan-Oreum-Batdam (b) Hallasan-Oreum-Batdam like nuclear structure (c) Fourier Twiddle Factor

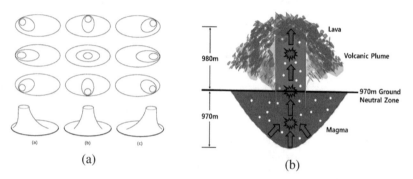

(a)   (b)

Fig. 4. Hallasan moving in the direction of the wind and Hallasan Multistage Volcanic Eruption (a) Hallasan moving in the direction of the wind (b) A conceptual diagram of a multistage explosion of Hallasan volcano. The base of Eoseungsaeng 970 m with ground peak 980 m; its sum height is 1,950 m. Underground 970 m and above ground 980 m; it is called symmetric 'the Oreum in Oreum'.

If the wind blows in the 0°, 120° and 240° directions simultaneously, the vector sum is zero. As shown in Fig. 3(c) [10],

$$w^0 + w^1 + w^2 = 0. \qquad (27)$$

Here, $w^0 = 1, w^1 = e^{j\frac{2\pi}{3}}, w^2 = \cos\frac{4}{3}\pi + j\sin\frac{4}{3}\pi$.

In the end, there is an Oreum and Batdam around the Kernel in the center of Hallasan to achieve the wind castle and defend the wind from high pressure and low pressure [8]. That is, the wind is circulating in space according to the periodic elements of the contour lines. All of these have an important key to Hallasan (1 2 3). There are 369 Jeju Oreum numbers, $369 = 3(1 + 2 + 3)$, $10^2, 10^1, 10^0$ is Base, 3 is 3 of Samdado, $1 + 2 + 3$ is Hallasan Kernel, In addition, 3,6,9 means that the last number of 1 2 3 is 3, the last number of 4 5 6 is 6, and the last number of 7 8 9 is 9, respectively. 3 6 9 is a link between Hallasan-Oreum and Batdam. $3 + 6 + 9 = 18$ is 18mod3 = 0, $3 \times 6 \times 9 = 162$ is 162mod3 = 0, It is 3, a medium of Samdado. In view of this, a group set of {1 2 3 4 5 6 7 8 9} was proposed.

3.3 Wind Castle Reliability

3.3.1 Reliability of Parallel System

The unreliability $F_p(t)$ is expressed as the product of the unreliability $F_i(t)$ of each part. $F_p(t) = \prod_{i=1}^{n} F_i(t)$. Therefore, the reliability of parallel system $R_p(t)$

$$R_p(t) = 1 - F_p(t) = 1 - \prod_{i=1}^{n} F_i(t) = 1 - \prod_{i=1}^{n} (1 - R_i(t)). \tag{28}$$

3.3.2 Reliability of Serial System

The reliability of serial system $R_s(t)$ is the product of the reliability of the components $R_i(t)$

$$R_s(t) = R_1(t) \cdot R_2(t) \cdots R_n(t) = \prod_{i=1}^{n} R_i(t). \tag{29}$$

Since Hallasan-Oreum-Batdam are in parallel, if Hallasan, Oreum, and Batdam are 80% components, $R_p(t) = (1 - (1 - 0.8)(1 - 0.8)(1 - 0.8)) = 99\%$ according to Eq. (28).

Reliability, so it acts as a castle against wind. In the case of serial, $R_s(t) = 0.8 \times 0.8 \times 0.8 = 51\%$ reliability is given by Eq. (29). From Sect. 3.2, the parallel reliability 99% based on the modified perfect number theorem. It is the perfectly the same. Therefore, God created Hallasan-Oreum-Batdam in parallel.

4 Conclusion

The wind castle for Hallasan-Oreum-Batdam on Jeju Island, a world natural heritage island, was defined, and the wind fluid flow was analyzed and proved as a concept of natural intelligence. As Pascal (1623–1662 French mathematician) said, 'The origin of all things is number'. All natural objects are combinations of natural numbers placed in space. For example, Oreum is a large stone of 350 m with 368 tumbling stones, and Batdam is 22,000 km with small stones of 1.5m in height. Hallasan is a very large 1,950 m rock body standing in the center. These three are independent but organic resemblance. Accordingly, the contour line periodic table of Hallasan-Oreum-Batdam was completed. Circulation around Batdam-Oreum-Hallasan 3Layer 1, 2, 3 satisfied the periodic Twiddle Factor function of Galois-Fourier. The height of Oreum is 200–600 m above sea level, and the Middle Mt. Village becomes the neutral zone, and there is Oreum in its surface. If we solve it virtually, adding the height of the Oreum in the ground and the Oreum of the ground, it is exactly the height of the measured Oreum. For example, Samdasoo, the largest natural mineral water in Korea, is located at 430 m above sea level in Gyoraeri, Jeju, and has Water Oreum, 840 m high, nearly twice the sea level of 430 m. The proposed theory was almost same with the measured values. Height of the neutral zone was found, and the height of was derived by considering the characteristics of the fluid mechanics of the volcanic magma eruption based on the

Bernoulli's theorem. The natural intelligence of the wind castle design was useful in Juju farmland, applied to people's life and wind control.

Acknowledgment. I would like to thank to my late father, Gab Boo Lee who given motivation of this paper.

References

1. Lee, M.H.: Wind Castle: Hallasan-Oreum-Batdam-Wind, Jeju Voice, 09 January 2020
2. Teng, X.: Discussion about artificial intelligence's advantages and disadvantages compete with natural intelligence. In: Journal of Physics: Conference (2019)
3. "Jeju Batdam", Jeju Research Institute, April 2017
4. Ko, G.-W., Park, J.-B., Moon, D.-C.: Geology and groundwater in Jeju Island, Volcanic Island. Hana Publishing (2017)
5. Kang, K.-W., Song, K.-C., Yoo, H.-J.: Special Theory of Firefighter Professional Engineer, Donghwa Engineer (2019)
6. Lee, M.H.: Batdam Kim Gu, Jeju Voice, 9.26-10.14 (2019)
7. Lee, M.H.: Halla Mountain simulation device functioning as typhoon Jinsan, Patent, registration number 10-1792745, Korea (2017)
8. Lee, M.H., Kim, J.-S.: Why won't the field wall collapse in the typhoon?: mathematical approach to non-orthogonal symmetric weighted Hadamard matrix. J. Inst. Internet Broadcast. Commun. (JIIBC) **19**(5), 211–217 (2019). https://doi.org/10.7236/JIIBC.2019.19.5.211
9. Lee, M.-H., Kim, J.-S.: Wind Castle: I, II, JIIBC **20**(3) (2020)
10. Lee, M.H., Matrices, J.: Construction and Its Application for Fast Cooperative Wireless Signal Processing, Germany LAMBERT (2012)
11. Lee, M.H., Hai, H., Lee, S.K., Petoukhov, S.V.: A mathematical proof of double helix DNA to reverse transcription RNA for bioinformatics. In: AIMEE 2017, pp. 53–67, 20 August 2017
12. Lee, M.H., Lee, S.K.: A life ecosystem management with base complementary DNA. In: AIMEE 2018, Moscow, 6–8 October 2018
13. Maor, E.: The Story of a Number. Princeton University Press, Princeton (1994)
14. Kajol, R., Akshay Kashyap, K., Keerthan Kumar, T.G.: Automated agricultural field analysis and monitoring system using IOT, no. 2, pp. 17–24. Published Online March 2018 in MECS (2018). https://doi.org/10.5815/ijieeb.2018.02.03
15. Bhagawati, K., Bhagawati, R., Jini, D.: Intelligence and its Application in Agriculture: Techniques to Deal with Variations and Uncertainties, no. 9, pp. 56–61. Published Online September 2016 in MECS (2016). https://doi.org/10.5815/ijisa.2016.09.07
16. Ahmed, W., Dasan, A., Babu Anto, P.: Developing an intelligent question answering system. I. J. Educ. Manag. Eng. (6), 50–61 (2017). https://doi.org/10.5815/ijeme.2017.06.06. Published Online November 2017 in MECS
17. Faisal, H., Usman, S., Zahid, S.M.: In What Ways Smart Cities Will Get Assistance from Internet of Things (IOT), no. 2, pp. 41–47. Published Online March 2018 in MECS (2018). https://doi.org/10.5815/ijeme.2018.02.05

The Impact of Interactive Visualization on Trade-off-Based Decision-Making Using Genetic Algorithm: A Case Study

Bui V. Phuong[1(✉)], Sergey S. Gavriushin[1,2], Dang H. Minh[3],
Phung V. Binh[4], Nguyen V. Duc[5], and Vu C. Thanh[6]

[1] Bauman Moscow State Technical University, 5c1 2nd Baumanskaya st,
105005 Moscow, Russian Federation
phuongbv1991@gmail.com, gss@bmstu.ru

[2] Mechanical Engineering Research Institute of the Russian Academy
of Sciences, 4, Malyi Kharitonievsky pereulok, 101990 Moscow, Russian
Federation

[3] Industrial University of Ho Chi Minh City, Ho Chi Minh City, Vietnam
danghoangminh@iuh.edu.vn

[4] Le Quy Don Technical University, Hanoi, Vietnam
phungvanbinh@lqdtu.edu.vn

[5] Thuyloi University, 175 Tay Son, Dong Da, Hanoi, Vietnam
ducnv@tlu.edu.vn

[6] Radar Institute, Academy of Military Science and Technology, Hanoi,
Vietnam
thanhvch@gmail.com

Abstract. This article presents a new visual interactive analysis method or VIAM that helps decision-maker (DM) to find the most favorable solution in the multi-criteria decision-making (MCDM) by evaluating the distribution of Pareto solutions (searched by genetic algorithm). The novelty of VIAM in comparison with others is that it is possible to re-orient search target. This might change the problem statement and consequently approach to the desired solution of DM. An application of VIAM for the multi-objective optimization of slider-crank mechanism (SCM) has been studied in order to demonstrate the efficiency of the method. The SCM design option obtained by VIAM has achieved the optimization of three criteria simultaneously such as power consumption, joint dynamic reaction, and mass. It is important to note that VIAM can be developed in the simple way, based on existing algorithms and it does not require the use of complex software.

Keywords: Data visualization · Interactive method · Trade-off · Strategic decision-making · Genetic algorithm · Multi-criteria decision making · Visual interactive analysis method-VIAM

Z. Hu et al. (Eds.): AIMEE 2020, AISC 1315, pp. 248–258, 2021.
https://doi.org/10.1007/978-3-030-67133-4_23

1 Introduction

Today, data visualization attracts the interest of many researchers in various fields. This appealing topic significantly increases users' intellect and confidence in data mining. Through data visualization and resultant conceptual descriptions (e.g. in the form of rules), qualitative overview of complex big data can be obtained [1–3]. However, the question is how to utilize this asset in the best way in order to improve human decision-making and problem-solving. Currently, the visual analysis and evaluation is one of the most commonly-used effective methods in the field of data analysis science.

Visual evaluation is a difficult issue and it often requires a lot of knowledge and auxiliary devices [4], even beyond the context of decision-making. In multi-objective optimization, the interactive multi-objective optimization (IMO) techniques play a key role in assisting decision-makers (DM) to find an optimal solution in the design space that complies with the objective functions, or Pareto optimal solution (POS). Normally, the multi-objective decision-making process has to take place on a large number of options. Hence, in most of practical cases, DM desires to have a small number of reasonable options, which are easy to understand and simply take the problem under control. To achieve this, the multi-criteria decision-making (MCDM) research society addresses two principal approaches: *i*) reducing the number of options based on the optimization process to yield the smaller optimal set of solutions, i.e. so-called Pareto frontier [5] and *ii*) visualizing the space of these solutions to help users select the best solution meeting their subjective criteria.

The decision-making process becomes more and more intricate, in case several targets or objective functions conflict with each other. In this contradictory relationship, one target could be detrimental, and another might be beneficial. In other words, there is no simultaneous optimal solution for all of them. Thus, visualization is used to assist users or DM in observation of the conflicting relationship in various formats. Figure 1 depicts the DM-Machine interaction system, in which there is an interaction between DM and the machine in IMO modes. On one hand, DM wishes to find the most favorable solution based on the selections, on the other hand the machine represents an algorithm and search methods, i.e. the machine is an optimization algorithm.

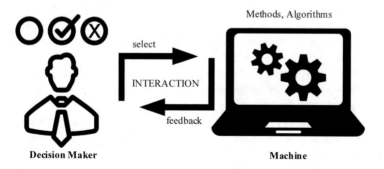

Fig. 1. DM – Machine interaction system

Visualization assessment auxiliary tools are really valuable, because they provide DM with a method for analyzing the POS space, in which a mutually-agreed solution among experts is determined. Indeed, in addition to acquiring POS, it is important to effectively visualize this space to make the decision-making process easier. However, it is not simple to demonstrate solutions so that DM can select a good points of POS. To solve this problem, apart from visual auxiliary tools such as self-organizing map [6] and the graphical visualization [7], at present one of common techniques for multi-dimensional images is a method of parallel coordinates (PC) [8]. PC has been proposed and studied by Alfred Inselberg since 1981 [9]. It has been widely used in the fields of robotics, statistics, computational geometry, etc. [10–12]. A multi-dimensional point representing the IMO's objective function is plotted in a two-dimensional graph. Each dimension of the original data is translated into x coordinates in this two-dimensional chart. There are two basic forms of PC such as linear and spherical, as illustrated in Fig. 2. Based on this, DM is able to observe the conflicting grade among target by looking for an "X" pattern (slash form) between axes.

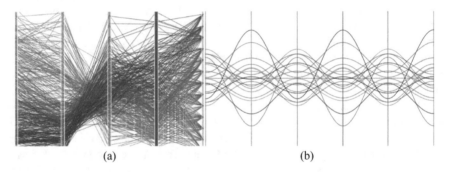

(a)   (b)

Fig. 2. Linear (a) and Spherical (b) parallel coordinates

Decision-making techniques, or simply decision-making tools assisting DM, are a growing area with fascinating and positive solutions in a priori methodology, progressive methodology and posteriori methodology [13–15]. Over the last decade, many interactive decision-making support systems have emerged, based on both standard and novel decision-making methods: a visual design method on the basis of Rasmussen's abstraction-aggregation hierarchy [16], spheres based on the even swap concept [17], PriEsT [18], AHP-GAIA [19], VIDEO [20].

Trade-off-based is one of the IMO methods based on comparing values of objective functions. In general, a trade-off means that taking into consideration a reference target, DM needs to quantify the beneficiary of some objective functions and to compensate for the detriment of other ones. Therefore, the trade-off can provide an accurate search instruction to find a better desirable solution. So far, there are several methods of IMO based on trade-off such as even swap method [21], interactive step exchange method based on slope GRIST [22], PROJECT method [23], and IMO method driven by evolutionary algorithm (EA) [24]. These methods often concentrate on data visualization in the fields of economics, management, etc., but the applications in the field of engineering design are still limited. In addition, the existing tools intend to visualize the Pareto frontiers of the solution, as well as plot only the solution distribution graphs. Yet, they are not really effective for the case when it needs to find better solutions or to deal with the situation that there are no solutions.

To enhance the flexibility of IMOs, EAs are used in combination with a search for POS. In this paper, a visual interaction analysis method (VIAM) assisting DM is proposed to find the most favorable solution in MCMD by evaluating the distribution range of the Pareto solutions, which were found by genetic algorithm (GA). The advantage of VIAM is an effective data visualization method for experts. In case that there is no solution, based on VIAM, the spatial domain, where the solution may exist, is shown that DM can reconsider and adjust the mathematical model. Besides, in some cases, DM may re-define the search target, which in turn the problem statement is altered, and approach to the most desirable solution. A case study will be discussed to evaluate the effectiveness of the proposed method.

2   Methodology

VIAM in combination with GA is illustrated in the form of a diagram in Fig. 3. The GA single-objective optimization is used to determine the upper bound (MAXΦ) and lower bound values (MINΦ) of the objective function Φ, which help engineers significantly to analyse and make a decision in the determination of solution [25]. Meanwhile, the GA multi-objective optimization is used to define the POS, which is the basis for decision making.

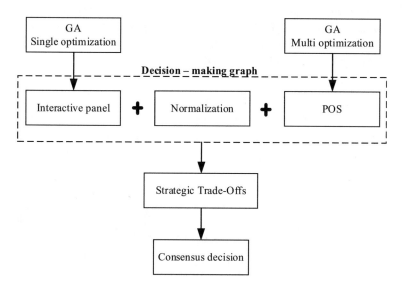

Fig. 3. Diagram of IMO in combination with VIAM

Data visualization of POS in VIAM is carried out by using the normalization of objective and PC visualization devices. The normalization of criteria is a parameter conversion technique in space, which is described in Ref. [26]. In order to avoid the difficulty of optimization algorithms, when the parameters $\underset{(N+Q)\times 1}{\mathbf{A}} = \left\{ \underset{N\times 1}{\boldsymbol{\alpha}}, \underset{Q\times 1}{\boldsymbol{\Phi}\mathbf{X}} \right\}$ are set in a limited value domain (with the spatial parameter $\boldsymbol{\alpha}$ (N-direction, limited), $\boldsymbol{\Phi}\mathbf{X}$ is function of α) into the spatial parameter $\mathbf{t}$ (unlimited). After spatial conversion, the parameter vector A is referred as $\underset{(N+Q)\times 1}{\mathbf{T}} = \left\{ \underset{N\times 1}{\mathbf{t}}, \underset{Q\times 1}{\boldsymbol{\Phi}\mathbf{T}} \right\}$, where $\mathbf{t} = \{t_1, t_2, \ldots, t_N\}$, $\boldsymbol{\Phi}\mathbf{T} = \left\{ \Phi T_1, \Phi T_2, \ldots, \Phi T_Q \right\}$. In the new space $\mathbf{T}$, the parameter is varied from $-\infty$ to $+\infty$, hence it is not limited as before in the space $\mathbf{A}$. This helps to improve the stability in all optimization algorithms used, due to the fact that there might be any invalid solution, which in turn the search process is ceased abruptly or error occurs. By this normalization, the value of objective function $\boldsymbol{\Phi}$ is replaced by the normalized value $\|\boldsymbol{\Phi}\|$ that appeared on PC.

Strategic trade-off of VIAM is used during search process of POS by auxiliary devices, in other words, this is a form of DM-Machine interaction while the algorithm is working. This means that DM can suspend the operation of the algorithm at will, and the algorithm will continue working right after DM issues new priorities to make a decision on a mutually-agreed solution in case it is required. Beside, the trade-off rule is made visually by mean of decision-making graph built on PC, as shown in Fig. 4. Assuming potential domain (PD) is the 2D space located below the POS. It can be seen that PD play a role as the domain representing the probability to find the better solutions than the existing ones in POS. However, for the complex multi-objective optimization problems (large number of objective functions, variables, and constraints), the search process takes a lot of time to achieve the desired POS.

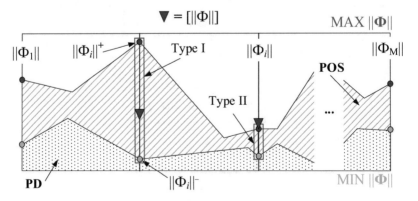

Fig. 4. Base of DM rules

At any moment during the solving process of multi-objective optimization problem, DM might decide to suspend the POS search based on the number and distribution of the solutions on the PC interactive tool, in order to immediately make a decision on the mutually-agreed solution. The threshold limits (denoted by ▼) to filter out the best and mutually-agreed solution in POS have values ranging from $\|\Phi^+\|$ to $\|\Phi^-\|$, where $\|\Phi^+\|$ and $\|\Phi^-\|$are the upper and lower bounds of the objective function $\|\Phi\|$ respectively in the POS space.

The rule of setting threshold is applied to various types of objective function as follows: Type I represents the one with a wide range of value and Type II is the one with a relatively narrow range of values in comparison with the previous (Fig. 4). It can be observed that, in the POS, the best values of Type I are often accompanied by the worst ones of Type II (Pareto principle). Determination of the solution with all the best values is very difficult, even impossible in some cases. At this point, experts use the method of setting the threshold according to the following principle: keep or accept to slightly loosen the threshold of the Type II, but to tighten the threshold of the Type I. The threshold is selected according to the following formula:

$$[\|\Phi_i\|] = \left[\begin{array}{l} \|\Phi_i\|^{\text{Type I}} = \dfrac{\|\Phi_i\|^+ + \|\Phi_i\|^-}{n}, \quad n \geq 2 \\ \|\Phi_i\|^{\text{Type II}} \approx \|\Phi_i\|^+ \end{array} \right. \tag{1}$$

After setting the thresholds, the filtering process of the best solution is automatically performed on the basis of checking p constraints of the objective functions:

$$\mathrm{f}_i = \|\Phi_i\| - [\|\Phi_i\|] \leq 0, \text{ where } i = 1...p \tag{2}$$

There exist three following possibilities:

(i)   If there is the solution, which is mutually-agreed by the experts, the final solution can be immediately concluded;

(ii) If there are several solutions, which are not mutually-agreed by the experts, it is necessary to tighten the threshold value $[\|\Phi\|]$ for the Type I by increasing the value of n. The filtering process continues until the final mutually-agreed one is achieved;

(iii) If there is no mutually-agreed solution after establishing all possible f_i constraints, this indicates that the POS does not possess the most desirable solution. Therefore, it is obligatory that users find a new POS domain in the PD area. At this point, VIAM allows users to reset the original multi-objective problem on the basis of adding new p constraints. For the Type I, these new constraints are p constraints in the latest decision-making process. For the Type II, it is required to loosen the threshold values as follows:

$$\|\Phi_i\|^{\text{Type II}} \geq \|\Phi_i\|^+ \tag{3}$$

DM continues to choose the moment to stop the search algorithms, when the new POS is eligible to make a decision on the most mutually-agreed solution. This process is repeated on the base of DM rules until the final mutually-agreed one is found. It can be seen that the search process of the mutually-agreed solutions is carried out on the feasible POS based on the visual evaluation and interaction of DMs. In the search process, VIAM allows users to stop searching and interacting directly with the algorithm only once, this helps to reduce the "burden" on DM but still ensure efficiency. of the search. Hence, the final solution is not only the POS, but also the compromission for decision making among experts.

3   A Case Study

Problem Statement: The proposed approach IMO-VIAM is used for solving multi-objective problem of the slider-crank mechanism (SCM) [27], which composes of 12 control variables α, 3 objective functions Φ, 16 constraints f:

$$
\begin{aligned}
&minimize : \Phi(\alpha) = \{\Phi_1(\alpha), \ldots, \Phi_3(\alpha)\} \\
&\alpha \in D(\alpha) \\
&subject\ to : D(\alpha) = \{\alpha | f(\alpha) = \{f_1(\alpha), \ldots, f_{16}(\alpha)\} \leq 0\} \subset \alpha = R^N \\
&a_i \leq \alpha_i \leq b_i,\ i = 1..12
\end{aligned}
$$

The requirement is to find a set of variables, which are detailed dimensions of SCM structure $\alpha = \{\alpha_1 \ldots \alpha_{12}\}$, so that at the same time minimize power consumption (Φ_1), SCM mass (Φ_2) and dynamic reaction at rotating joints (Φ_3) on the basis of meeting the constraints of structure, technology, and permissible geometry. The mathematical model is described in details in Ref. [28].

The search process of mutually-agreed solution is carried out as follows:

Step 1: DM's interaction during the search

When the POS space reaches 1260 possible options, the interaction process begins with checking the condition set by the DM: Whether there exist in this POS the solution better than the 1260 alternatives or not? By checking with DM's additional condition:

$$\|\Phi_{1,2,3}\| \leq 0.1$$

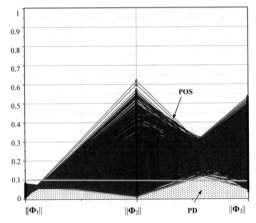

Result: There is no satisfactory solution, thus if the algorithm continues to look for POS, it may take longer to reach better solutions.

Step 2: Continue searching with additional conditions

Add three more constraints in Step 1 and run the POS search process again.

Result: Obtain the new POS including 31 solutions located entirely in the PD (below the threshold of 0.1).

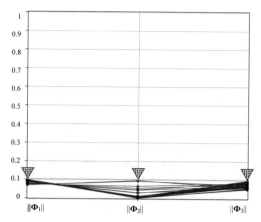

Step 3: Use filter base on DM rules

Comment: It needs to trade-off the "worse" value of Φ_1 to get the "better" value of Φ_2 and Φ_3 by setting the filter:

$$\begin{cases} ||\Phi_1|| \leq 0.1 \\ ||\Phi_{2,3}|| \leq 0.05 \end{cases}$$

Result: Obtain one valid solution. The decision-making process ends when DM agrees with this solution.

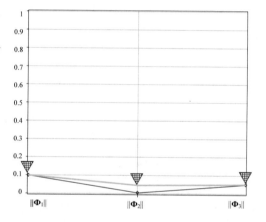

The obtained mutually-agreed solution is compared with the SCM design based on the traditional loop design process, which is presented in Ref. [29], the detailed outcome is described in the following Table 1:

Table 1. Comparative results

Solution	P_{max}, W	m, kg	RO_{max}, N
Traditional	40.022	2.497	120.825
VIAM	39.988	0.739	81.464
	(−0.09%)	(−70.42%)	(−32.58%)

It is important to note that, all optimization targets (objective functions) are achieved, in which the trade-off of power consumption (almost unchanged) brings a positive effect when the SCM mass and dynamic reaction are reduced with the proportions of 70.42% and 32.58% respectively. This indicates that the VIAM helps DM to interact and make an effective decision in the field of MCDM.

4 Conclusions

The visual interactive analysis method (VIAM) was proposed to help decision-maker (DM) to find the optimal and mutually-agreed solutions in the multi-criteria design process of complex mechanical systems. VIAM allows the DM to make a decision during the POS search by using the visual representation of solutions so that the DM can analyze and propose changes and finally select the best solutions based on the requirement of the experts. During the search, VIAM allows users to stop searching and interact with the algorithm only once, this helps to reduce the "burden" on DM while ensuring the effectiveness of search. An application of VIAM for the multi-objective optimization of SCM was presented to demonstrate the efficiency of the method. The solution obtained by VIAM has improved simultaneously three criteria such as power consumption, dynamic reaction and mass being 0.09%, 70.42% and 32.58% respectively. It was observed that the development of VIAM has been quite simple, convenient, based on existing algorithms and it did not require the intervention of complex software. Finally, VIAM could also be used for multi-criteria design of other mechanical systems.

References

1. Viktor, H.L., Paquet, E.: Visualization techniques for data mining. In: Data Warehousing and Mining: Concepts, Methodologies, Tools, and Applications, pp. 1623–1630. IGI Global (2008)
2. EL HAJI, E., Azmani, A.: Proposal of a digital ecosystem based on big data and artificial intelligence to support educational and vocational guidance. Int. J. Mod. Educ. Comput. Sci. (IJMECS) 12(4), 1–11 (2020). https://doi.org/10.5815/ijmecs.2020.04.01
3. Mahmoud, S.M., Habeeb, R.S.: Analysis of large set of images using MapReduce framework. Int. J. Mod. Educ. Comput. Sci. (IJMECS) 11(12), 47–52 (2019). https://doi.org/10.5815/ijmecs.2019.12.0
4. Inyang, U.G., Eyoh, I.J., Robinson, S.A., Udo, E.N.: Visual association analytics approach to predictive modelling of students' academic performance. Int. J. Mod. Educ. Comput. Sci. (IJMECS) 11(12), 1–13 (2019). https://doi.org/10.5815/ijmecs.2019.12.01
5. Ehrgott, M.: Multicriteria Optimization, vol. 491. Springer (2005). 248 p.
6. Obayashi, S., Jeong, S., Chiba, K.: Multi-objective design exploration for aerodynamic configurations. In: 35th AIAA Fluid Dynamics Conference and Exhibit, p. AIAA-2005-4666 (2005)
7. Lotov, A.V., Bushenkov, V.A., Kamenev, G.K.: Interactive decision maps: approximation and visualization of Pareto frontier, vol. 89. Springer (2013), 310 p.
8. Andrienko, N., Andrienko, G.: Informed spatial decisions through coordinated views. Inf. Vis. 2(4), 270–285 (2003). https://doi.org/10.1057/palgrave.ivs.9500058
9. Inselberg, A.: N-Dimensional Graphics Part I: Lines and Hyperplanes. IBM LA Science Center Report No. G320-2711. IBM Corporation, White Plains, NY (1981)
10. Cohan, S.M., Yang, D.C.H.: Mobility analysis of planarfour-bar mechanisms through the parallel coordinate system. J. Mech. Mach. Theory 21, 63–67 (1986)
11. Fiorini, P., Inselberg, A.: Configuration space representation in parallel coordinates. In: Proceedings of the IEEE Conference on Robotics and Automation, pp. 1215–1219. IEEE Press, Piscataway (1989)

12. Wegman, E.: Hyperdimensional data analysis using parallel coordinates. J. Am. Stat. Assoc. **411**(85), 664–665 (1990)
13. Chiandussi, G., Codegone, M., Ferrero, S., Varesio, F.E.: Comparison of multi-objective optimization methodologies for engineering applications. Comput. Math Appl. **63**, 912–942 (2012)
14. Branke, J., Deb, K., Miettinen, K., Slowinski, R.: Multiobjective Optimization: Interactive and Evolutionary Approaches. Springer, Heidelberg (2008). 481 p.
15. Cui, Y., Geng, Z., Zhu, Q., Han, Y.: Review: multi-objective optimization methods and application in energy saving. Energy **125**, 681–704 (2017). https://doi.org/10.1016/j.energy.2017.02.174
16. Rouse, W.B., et al.: Interactive visualizations for decision support: application of Rasmussen's abstraction-aggregation hierarchy. Appl. Ergon. **59**, 541–553 (2017). https://doi.org/10.1016/j.apergo.2016.03.006
17. Li, H.L., Ma, L.C.: Visualizing decision process on spheres based on the even swap concept. Decis. Support Syst. **45**(2), 354–367 (2008). https://doi.org/10.1016/j.dss.2008.01.004
18. Siraj, S., Mikhailov, L., Keane, J.A.: PriEsT: an interactive decision support tool to estimate priorities from pairwise comparison judgments. Int. Trans. Oper. Res. **22**(2), 217–235 (2015). https://doi.org/10.1111/itor.2015.22.issue-2
19. Ishizaka, A., Siraj, S., Nemery, P.: Which energy mix for the UK? An evolutive descriptive mapping with the integrated GAIA (graphical analysis for interactive aid)-AHP (analytic hierarchy process) visualization tool. Energy **95**, 602–611 (2016). https://doi.org/10.1016/j.energy.2015.12.009
20. Kollat, J.B., Reed, P.: A framework for visually interactive decision-making and design using evolutionary multi-objective optimization (VIDEO). Environ. Model. Softw. **22**(12), 1691–1704 (2007). https://doi.org/10.1016/j.envsoft.2007.02.001
21. Hammond, J.S., Keeney, R.L., Raiffa, H.: Even swaps: a rational method for making trade-offs. Harvard Bus. Rev. **76**, 137–150 (1998)
22. Yang, J., Li, D.: Normal vector identification and interactive tradeoff analysis using minimax formulation in multiobjective optimization. IEEE Trans. Syst. Man Cybern. Part A Syst. Hum. **32**(3), 305–319 (2002)
23. Luque, M., Yang, J., Wong, B.Y.H.: PROJECT method for multiobjective optimization based on gradient projection and reference points. IEEE Trans. Syst. Man Cybern. Part A Syst. Hum. **39**(4), 864–879 (2009)
24. Chen, L., Xin, B., Chen, J.: A trade-off based interactive multiobjective optimization method driven by evolutionary algorithms. J. Adv. Comput. Intell. Intell. Inform. **21**(2), 284–292 (2017)
25. Minh, D.H., Van Binh, P., Van Phuong, B., Duc, N.V.: Multi-criteria design of mechanical system by using visual interactive analysis tool. J. Eng. Sci. Technol. **14**(3), 1187–1199 (2019)
26. Dang, H.M., Phung, V.B., Nguyen, V.D.: Multi-objective design for a new type of frame saw machine. Int. J. Mech. Prod. Eng. Res. Dev. (IJMPERD) **9**(2), 449–466 (2019)
27. Nga, N.T.T., Minh, D.H., Hanh, N.T.M., Binh, P.V., Phuong, B.V., Duc, V.N.: Dynamic analysis and multi-objective optimization of slider-crank mechanism for an innovative fruit and vegetable washer. J. Mech. Eng. Res. Dev. (JMERD) **43**(2), 127–143 (2020)

28. Bui, V.P., Gavriushin, S.S., Phung, V.B., Dang, H.M., Dang, V.T.: A generalized mathematical model of slider-crank mechanism with spring used in an innovative fruit and vegetable washer. In: XXV International Symposium "Dynamic and Technological Problems of a Mechanics of Constructions and Continuous Mediums" dedicated to A. G. Gorshkov, Vyatichi, no. 1, pp. 75–77 (2020). (in Russian)
29. Bui, V.P., Gavriushin, S.S., Phung, V.B., Dang, H.M., Prokopov, V.S.: Dynamic and stress analysis of the main drive system in design process of an innovative fruit-vegetable washer. Eng. J. Sci. Innov. (4) (2020). https://doi.org/10.18698/2308-6033-2020-4-1970 (in Russian)

Advantages of Fuzzy Controllers in Control of Technological Objects with a Big Delay

Andrey Zatonsky[1] , Ruslan Bazhenov[2(✉)] , Andrey Dorofeev[3] , and Oksana Dudareva[3]

[1] Perm National Research Polytechnic University, Perm, Russian Federation
zxenon2000@yandex.ru
[2] Sholom-Aleichem Priamursky State University,
Birobidzhan, Russian Federation
r-i-bazhenov@yandex.ru
[3] Irkutsk National Research Technical University, Irkutsk, Russian Federation
dorbaik2007@mail.ru, odudareva@mail.ru

Abstract. This article comments on the issue that deals with a fuzzy logic control (FLC) of proportional–integral–derivative (PID)-type. It is used for managing complex chemical-technology objects with a big delay up to 25 s. A number of real chemical-technology devices used at the plants in the Perm Region (Russia) are considered as objects of study. The same devices are widely spread all over the world. An opportunity of a fuzzy logic control used for such objects is proved by comparison with traditional P-, PI- and PID-controllers. These controllers are tuned by Simulink PID-tuner; the fuzzy logical controls are designed and tuned by Matlab fuzzy logic toolbox. Transition processes are given by Simulink software. All necessary indicators of transition processes and quality characteristics of the controls are obtained, namely, mismatch maximum modulus and control time. PID-control indicators are used as a base for comparing to FLC. As a result, FLC indicators are significantly less than the same PID indicators are. The maximum module of mismatch decrease is about 27.2 times, while the time of transition process decrease is about 2.9 times if compared to the PID controller. So FLC usage leads to some definite improvement of technological processes ongoing in the devices. Consequently, an amount of non-conditional products could be decreased when some inlet disturbances come about.

Keywords: Fuzzy logic control · Technology processes · Delay · Modelling · Transition processes

1 Introduction

This article comments on the issue that deals with a fuzzy logic control (FLC) of proportional–integral–derivative (PID)-type. It is used for managing complex chemical-technology objects with a big delay up to 25 s. A number of real chemical-technology devices used at the plants in the Perm Region (Russia) are considered as objects of study. The same devices are widely spread all over the world. An opportunity of a fuzzy logic control used for such objects is proved by comparison with

© The Author(s), under exclusive license to Springer Nature Switzerland AG 2021
Z. Hu et al. (Eds.): AIMEE 2020, AISC 1315, pp. 259–269, 2021.
https://doi.org/10.1007/978-3-030-67133-4_24

traditional P-, PI- and PID-controllers. These controllers are tuned by Simulink PID-tuner; the fuzzy logical controls are designed and tuned by Matlab fuzzy logic toolbox. A number of unique chemical, mining and chemical and chemical-metallurgical industries are located within the territory of Berezniki-Solikamsk industrial hub in the Perm Region (Russia). Their manufactured products obtained by extraction enrichment and processing of commercial minerals of various types are in high demand in the world market. The product sales income significantly contributes to the budgets of different level. The management of the complicated production facilities in order to improve the qualitative, quantitative and financial performance indices of such factories is a currently applied research objective that does not have a single solution at present [1, 2]. In addition, the processes that proceed at such plants are often characterized by transport delays, i.e. their control response or perturbance effect happens with time delay. The quality of methods like a standard control in such conditions can be unsatisfactory. It results in a decline in the efficiency of processing mineral resources into products-completed and impairs the ability of the industry to compete.

2 Literature Review

As stated previously, there is no single approach to the management of complicated technical objects at present and, in particular, the objects with a delay. One possible method is to develop special adjustment procedures for standard proportional–integral–derivative (PID) controllers, for example, through the conditional optimization method [3], fractional controllers [4], the regression model of parameters [5], etc. Additional complexity is a necessity to carry out an active experiment in the conditions of the existing products in this case. The classical theory of automatic control involves standard signals used to control object identification. The generation of these signals makes the performance of continuous production worse, so it is difficult to apply it at the enterprise. The current methods of controlling object identification based on random changes in factor values are worse conditioned, and they require a long search for such changes in the technological data trends.

Some authors suggest an adaptive approach based on the choice of PID controller parameters that are the most suitable for a given specific state of the control object from several sets of them [6]. This approach is widely used in practice, but it is complicated for two reasons. First, it is necessary to determine the controller parameters for each special state of the control object. These parameters could be determined either by designing complex models of the control object, or experimentally. The latter method leads to the difficulties described above. Second, it is necessary to decide on the transitions of technological equipment from one special state to another. It is a complex task that should be solved separately but with no one solution, though.

Besides, according to [7], a PID controller with two degrees of freedom can be used to compensate for dead time. Predictive control of objects with a delay is widespread [8, 9], including Smith predictor [10]. However, such predictive systems are characterized by low steadiness and robustness. It is necessary to obtain a lot of data on the control object to design a predictive controller that is why they are used at the enterprises rarely.

So there is no valid method for PID controller adjustment procedure nowadays, and optimization of its settings according to different quality criteria gives different solutions. It is quite natural that the configuration of its modifications listed above is still a more difficult task. Therefore, the fuzzy logic control based is gaining an increasingly strong foothold.

Fuzzy logic controls (FLC) can both simulate the operation of a conventional PID controller [11, 12] and actualize special laws of control [13]. Furthermore, the issue of the opportunity to use them for managing complicated chemical and technological objects has not been discovered well enough yet.

The problems of synthesizing PID-like FLCs of various types, their settings following the selected control objects at the enterprises in Berezniki-Solikamsk industrial hub, and also assessing the quality of the corresponding fuzzy control laws are observed within this survey.

The authors used the following transition functions (TF) obtained from different scientific researches devoted to typical objects with the delay:

- The steam temperature in a boiler depending on burning natural gas flow

$$W(s) = \frac{0,35}{40s^2 + 12s + 1} \exp(-1.5s);$$

- The exhaust air temperature for the fluidized bed furnace, depending on gas

$$\text{flow } W(s) = \frac{0,2}{205s^2 + 30s + 1} \exp(-5s);$$

- The flue gas temperature in the furnace of the drying drum, depending on gas

$$\text{flow } W(s) = \frac{0,32}{18s^2 + 9s + 1} \exp(-6s);$$

- The temperature in the middle of the brick kiln, depending on gas flow

$$W(s) = \frac{0,03}{5775s^2 + 127s + 1} \exp(-10s);$$

- The pH of an ammonium nitrate solution in a neutralizing nitric acid with ammonia apparatus depending on ammonia line flow

$$W(s) = \frac{6}{6974s^2 + 152s + 1} \exp(-25s).$$

Different scientists investigated the objects earlier. Their TFs are reviewed in [16] among the others.

3 Materials and Methods

At present, the FLC theory has not been developed properly yet. There is neither classification nor a common algorithm for constructing them. We can cite a wide range of scientific works where the authors conclude the synthesis of such a controller as an art more than a craft in each case. This is especially distinct when the control objects have significant delays or other features that made their transmission functions significantly non-linear.

According to the review, the following FLC factors can be distinguished, allowing for their conditional classification to be developed.

1. The number of input parameters. Each parameter is a control error; it is derivative of the first or second-order or the result of its integration according to the control time. Most scientists consider PID-like FLCs with the number of input parameters from one to three (error signal/discrepancy $\Delta(t)$, it is integral $\int_0^t \Delta(t)dt$ and time derivative $\frac{d\Delta(t)}{dt}$. Probably, this approach originates from the classical automatic control theory based on P-, I- and D-components. Other approaches are also supposed to be. For example, DD^2-components ($PIDD^2$-regulators) are used both for controlling objects with a big delay and those with a rapidly changing output such as aircraft stabilization systems. That is why the limitation of choice only in error, derivative and integral is not final and undeniable.

2. The number of the regulator components. A one-component (simple) FLC converts input parameters to an output one from a single rule base [14]. The disadvantage of this approach is a fast increase in the number of rules in the database with the increase in the number of input parameters of the controller and their term-values. This results in a more complicated development and implementation of the regulator, and the regulation itself since the number of operations necessary to produce control activities is growing. A multi-component (complicated) FLC consists of two or more one-component ones. Each of them converts input parameters to output ones based on its base of rules. This is a typical application of a classical system approach, when a complex object is replaced by a system of simple elements. Aggregation of simple solutions for each element allows creating a common solution for an object. Such controllers can also implement hybrid/mixed control algorithms combining fuzzy and classic types of them [15]. The resulting output signal is generated as the sum of the output signals of separate components following the principles of system analysis mentioned above.

According to that classification, 14 types of PID-like FLCs can be synthesized: 3 simple ones with one input (P-, I- and D-like FLC); 3 simple ones with two inputs (PI-, PD- and ID-like FLC); 1 simple one with three inputs (PID-like FLC); 3 complicated two-component ones with two inputs (PI-, PD- and ID-like FLC); 3 complicated two-component ones with three inputs (PID-like FLC with components PI + D, PD + I and ID + P); 1 complicated three-component one with three inputs (PID-like FLC).

Integrally differentiating (ID-) controllers are used in practice limited due to the weak conditionality of the transition from the regulation error itself to its derivative or integral. Therefore, in the above list, they are added rather for a generality. They are not considered further in the article.

The authors carry out the design, implementation, and configuration of some types of PID-like FLCs in graphical simulation environment Simulink of Matlab software.

The researchers choose the Heaviside function / a unit step function $1(t) = \begin{cases} 0, t < 0; \\ 1, t \geq 0 \end{cases}$ as a disturbance.

The scholars choose process units as control objects as a component of production lines at various enterprises in Berezniki-Solikamsk industrial hub.

Since FLC synthesis procedure is rather complicated, and the number of PID-like FLCs is great, on the purposes of this study, the authors perform the synthesis of the following different types within the classification given above: a simple P-like FLC; a simple PI-like FLC; a three-component PID-like FLC.

The number and composition of input parameters of PID-like fuzzy controllers are defined by their control algorithms (Table 1). The only output parameter of FLC reviewed is the position change of the regulating unit, i.e. the control activity generated to compensate for the input disturbance is U_i, $i = 1,2,3$ and for a three-component PID-like FLC, the output parameter U_3 is composed of output parameters of P-component U_{31}, I-component U_{32}, and D-component U_{33}.

In accordance with [13], the number of terms is usually the same for all linguistic variables. Their number is also limited by the capabilities of FIS processor Matlab, which allows one to specify from one to nine values of each index of the controller. As mentioned above, too many terms complicate the logical rules base and the process of generating the control activity, and very few of them do not allow achieving a proper control quality.

Table 1. Input parameters of PID-like FLC presented

Type of regulator	Input parameters			Output parameter
	P-component	I-component	D-component	
Simple U-like FLC	$X_{11} = \Delta(t)$	–	–	U_1
Simple PI-like FLC	$X_{21} = \Delta(t)$	$X_{22} = \int_0^t \Delta(t)dt$	–	U_2
Three-component PID-like FLC	$X_{31} = \Delta(t)$	$X_{32} = \int_0^t \Delta(t)dt$	$X_{33} = \frac{\Delta(t)}{dt}$	$U_3 = U_{31} + U_{32} + U_{33}$

For the research purpose, the authors use the sets of seven terms of the following forms: {big negative (BN), medium negative (MN), small negative (SN), zero (Z), small positive (SP), medium positive (MP), big positive (BP)}, as well as from five terms of the form {big negative (BN), small negative (SN), zero (Z), small positive (SP), big positive (BP)} (Table 2).

Table 2. Number and composition of terms for PID-like FLC indices

Regulator Indices	The number and composition of terms						
	1	2	3	4	5	6	7
X_{11}, X_{31}, X_{32}, X_{33}, U_1, U_{31}, U_{32}, U_{33}	BN	MN	SN	Z	SP	MP	BP
X_{21}, X_{22}, U_2	BN	SN	Z	SP	BP	–	–

Matlab FIS processor supports eleven different kinds of membership functions (MFs), including triangular and trapezoidal ones, allowing to describe linguistic variables shown in Figs. 1a–1b. Similar membership functions are often used in cases where the definition of *approximately equal* is a characteristic of a fuzzy variable, and there is no specific reason to choose another type of function. Using S-shaped, Z-shaped, Z-sigmoidal, etc. membership functions can be based on the proper characteristics of an automation object. However, for measurable properties, signs, and attributes, such as speed, time, temperature, pressure, it is still more often used exact triangular functions.

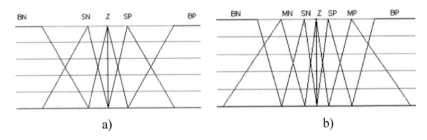

a)   b)

Fig. 1. Description of the linguistic variable in triangular and trapezoidal MF (a) the number of terms n = 5; b) the number of terms n = 7)

The general initial periods for the supporters of input and output indices of FLC MF (Table 3) are identified by examining the behaviour of controlled objects as follows.

1. For each object, using Matlab Simulink software, the authors define the indices of the classical PID controller, which neutralizes a unit step function best of all, and they construct dynamic series graph charts for the error values (Error), its Error derivative and Error integral, as well as generated by Controller output. This is done through the PID tuner component. Here Error integral means an integral of the Error is calculated from the beginning of the transition process to the end of them.
2. As an initial interval for X_{k1}, $k = 1,2,3$, an interval of the form symmetric concerning zero is chosen as $[-\max_t|\text{Error}|; +\max_t|\text{Error}|]$, when Error mismatches value is obtained at the previous stage.
3. As an initial interval for X_{k2}, $k = 2,3$, an interval of the form symmetric concerning zero is chosen as $[-\max_t|\text{Error integral}|; +\max_t|\text{Error integral}|]$.

Table 3. Initial intervals of MF supporters of FLC indices

No	Input indices			Output index
	P- component	I- component	D- component	
1	[–0.157; 0.157]	[–3.206; 3.206]	[–0.018; 0.018]	[0.934; 1.066]
2	[–0.062; 0.062]	[–2.558; 2.558]	[–0.004; 0.004]	[0.945; 1.055]
3	[–0.228; 0.228]	[–4.763; 4.763]	[–0.026; 0.026]	[0.934; 1.066]
4	[–0.090; 0.090]	[–1.701; 1.701]	[–0.0001; 0.0001]	[0.929; 1.071]
5	[–1.205; 1.205]	[–372.3; 372.3]	[–0.020; 0.020]	[0.863; 1.137]

4. As an initial interval for $X_{33} = \Delta(t)$, an interval of the form symmetric concerning zero is chosen as $[-\max_t|\text{Error derivative}|; +\max_t|\text{Error derivative}|]$.

5. Since the interval for the generated control activity is symmetric concerning the disturbance value, when a unit step function is applied to the input of the controlled object, this interval is symmetric concerning to unity. Then, as the initial interval for U_1 and U_2, an interval of the form $[2-\max_t|\text{Controller output}|; \max_t|\text{Controller output}|]$ is chosen. The numerical value of the Controller output is observed on the Scope blocks of model. This parameter sometimes changes because the Controller output depends on the final results of modeling. That is why it changes at every iteration of model tuning, here as well as below.

6. For a three-component PID-like FLC, the interval U_3 of the form $[2-\max_t|\text{Controller output}|; \max_t|\text{Controller output}|]$ was divided for P-, I-, and D-components in relation to 2: 1: 1, respectively. Then, as an initial interval for U_{31}, a symmetric interval concerning 0.5 is chosen as $[(2-\max_t|\text{Controller output}|)/2; \max_t|\text{Controller output}|/2]$, and for U_{32} and U_{33} is an interval of the form symmetric concerning 0.25 $[(2-\max_t|\text{Controller output}|)/4; \max_t|\text{Controller output}|/4]$.

According to the review, the authors can conclude the following rules for generating a matrix of PID-like FLCs knowledge.

1. A bigger absolute value of the input index FLC corresponds to a bigger absolute value of the output index and vice versa.
2. The signs of the input and output index values of FLC are the same.
3. If FLC has more than one input, the values of the opposite signs for the input indices leads to a decrease in the value of the output index in absolute value.

From the structure of the knowledge matrix, the first two rules mean that each line corresponding to the output index contains a subsequence of non-decreasing values. The third rule means that the two-dimensional knowledge matrix corresponding to a simple FLC with two inputs has a secondary diagonal packed with a zero-point (Z) control value and is antisymmetric with respect to it. Incompleteness and inconsistency of the knowledge matrices considered by FLC, constructed following the formulated rules, selected by the number and composition of terms (Tables 4 and 5), are excluded from the table view.

In Matlab FIS processor, the Mamdani algorithm is used automatically as the fuzzy inference procedure. It uses the logic minimum operation to specify the truth degree of prerequisites of each rule, and the logical maximum operation [13] to generate the membership function of the resulting output variable. A possible replacement of this algorithm with one of the other ones (Tsukamoto, Sugeno, Larsen, etc.) does not lead to significant changes in the results of fuzzy inference for triangular membership functions.

Table 4. Knowledge matrices for P-like, and PID-like FLC component

Input index (X_{11}, X_{31}, X_{32}, X_{33})	BN	MN	SN	Z	SP	MP	BP
Output index (U_1, U_{31}, U_{32}, U_{33})	BN	MN	SN	Z	SP	MP	BP

Table 5. Knowledge matrix for PI-like FLC

X_{21}	X_{22}				
	BN	SN	Z	SP	BP
BN	$U_2 = $ BN	$U_2 = $ BN	$U_2 = $ MN	$U_2 = $ MN	$U_2 = $ Z
MN	$U_2 = $ BN	$U_2 = $ MN	$U_2 = $ MN	$U_2 = $ Z	$U_2 = $ MP
Z	$U_2 = $ MN	$U_2 = $ MN	$U_2 = $ Z	$U_2 = $ MP	$U_2 = $ MP
MP	$U_2 = $ MN	$U_2 = $ Z	$U_2 = $ MP	$U_2 = $ MP	$U_2 = $ BP
BP	$U_2 = $ Z	$U_2 = $ MP	$U_2 = $ MP	$U_2 = $ BP	$U_2 = $ BP

FLC by Simulink setting is performed experimentally by changing the lengths of MF supporter intervals of the input and output indices of the controllers. It is done to achieve the resulting transfer characteristics of acceptable quality level compared to the corresponding processes which are obtained by controlling by a PID controller automatically configured by Simulink. The authors can preliminarily note the following:

1. The transition process for FLC-controllers is better than one for PID-controller from all traditional indicators, such as time of transient process or a maximum error.
2. Moreover, the trend does not exceed the limits of the error value of 2%, which is usually recognized as reaching the settings of the PID-controller.
3. The difference between the transient processes for FLC-controllers is much smaller than between FLC- and PID-controllers.

4   Experiment Results and Discussion

The evaluation of the control quality of technological facilities with the delay using FLC is made by comparing regulator performances by the maximum level of disagreement absolute magnitude $\max_t |\Delta_i^j(t)|$ where $j = 1,2,3,4,5$ is the number of controlled facility, $i = 0,1,2,3$ is the number of the controller, and $i = 0$ corresponds to the classic PID controller, and the response time T_i^j, $i = 0,1,2,3$; $j = 1,2,3,4,5$ defined as the moment when the value of the step response is set within the limits of interval $\pm\,0.01\cdot\max_t\left|\Delta_0^j(t)\right|$. To compare results, the authors set a single interval to all four step-function responses of each controlled facility, the smaller the values of both indicators, the faster and more accurate the regulator neutralization of the input disturbance.

According to the review of absolute and relative values, $\max_t\left|\Delta_i^j(t)\right|$, $i = 0,1,2,3$; $j = 1,2,3,4,5$ (Fig. 2) it can be argued that all designed FLCs provide a much better value of this indicator compared to a conventional PID controller. The rate drops for $\max_t\left|\Delta_1^1(t)\right|$ by 7.85 (by 87.34%) times in the worst-case scenario, and it drops by 65.56 times (98.47%) for $\max_t\left|\Delta_3^4(t)\right|$ in the best case. Moreover, the least value for different objects is provided by different controllers. The best control is achieved by using PID-like FLC for the first and fourth objects. It is due to PI-like FLC for the third and fifth ones, and it is due to P-like FLC for the second one.

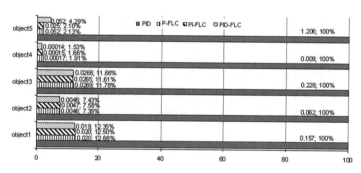

Fig. 2. Relative and absolute values $\max_t\left|\Delta_i^j(t)\right|$, $i = 0,\ 1,\ 2,\ 3;\ j = 1,\ 2,\ 3,\ 4,\ 5.$

It is to be noted that it is a basic result, emphasizing the advantages of FLC-controllers over PID-controllers. The considered FLC-controllers of proper technological control objects are designed manually, obviously, influenced by voluntarism of the authors regarding the choice of their parameters and tuning. Nevertheless, they provide the highest quality control regardless of the type of FLC-controller. As a result of using P-FLC controller, the authors can see no static error, which always occurs when a regular P-controller is used. That is the main reason why it is necessary to complicate P-controllers to PI- and PID-, and therefore, calculate more settings for each specific state of the technological equipment for FLC-controller. FLC-controller does not have this disadvantage. Only the configuration of a single logical unit is required to achieve high-quality control.

Response time, T_i^j, $i = 1,2,3$; $j = 1,\ 2,\ 3,\ 4,\ 5$ when using FLC is also reduced if compared to the corresponding index when using a PI controller T_0^j (Fig. 3). The index decreases in the worst case for T_3^1 by 7.81% and in the best case for T_2^5 by 83.95%. However, the least value for different objects is also provided by different controllers: the best control is achieved by using P-like FLC for the first and third objects, it is due to PI-like FLC for the second and fifth ones, and it is due to PID-like FLC for the fourth one.

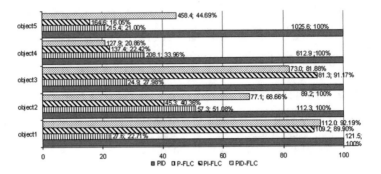

Fig. 3. Relative and absolute values T_i^j, $i = 0, 1, 2, 3$; $j = 1, 2, 3, 4, 5$.

It is worth pointing out as $\max_t \left| \Delta_i^j(t) \right|$ and T_i^j, $j = 1, 2, 3, 4, 5$; $i = 1, 2, 3$ are inversely related, so in theory, this is possible to reduce the amount of regulation time if maximizing the absolute magnitude of discrepancy.

5   Conclusion

In conclusion, according to the study, it can be argued that using PID-like FLCs allows one to control chemical-technical objects of various types better in comparison with regular controllers. In terms of production, this means loss reduction of ultimate products as manufacturing defects, and, consequently, resource conservation when producing them. It enhances the efficiency of the enterprises in Berezniki-Solikamsk industrial hub and other similar territories in the world. Of course, the same approach can be applied at any similar enterprise, for example, at chemical-technological ones with transport delays in devices. There is no reason to believe that if there is no delay, the use of FLC-controller tuned in the same way will give a worse result than using a PID-controller. The authors find out no obvious limitation in applying this method as it can function with the described objects, and they are quite representative.

Nevertheless, the authors do not succeed in putting a single type of FLC that would provide better control of all the objects observed. However, the difference in using different types of FLC-regulators is not significant. Moreover, FLC setting is carried out experimentally, i.e. by expanding and narrowing the intervals of supporters of initial triangular and trapezoidal MF regulator parameters to achieve a proper quality of the step response characteristic. Although, neither a complete sorting out of such intervals with a particular circle is performed, nor changes of the relative position of MF terms on it. The MF alternative forms are not considered, either. It means that the authors have not used FLC setting capabilities completely. Their experimentally selected parameters may not be the best ones and can be improved. This not only highlights the objectives of the follow-up research but also provides a great opportunity to promote control performance of complicated chemical-engineering facilities.

References

1. Ji, S., Xue, X.: The stochastic maximum principle in singular optimal control with recursive utilities. J. Math. Anal. Appl. **471**, 378–391 (2019)
2. Zatonskiy, A.V., Varlamova, S.A.: Use of reflection flare spots for automatic recognition of froth parameters in potassium ores flotation. Obogashchenie Rud **2**, 49–56 (2016)
3. Grandi, G.L., Trierweiler, J.O.: Tuning of fractional order PID controllers based on the frequency response approximation method. IFAC-PapersOnLine **52**, 982–987 (2019)
4. Hamamcia, S.E., Koksal, M.: Calculation of all stabilizing fractional-order PD controllers for integrating time delay systems. Comput. Math. Appl. **59**, 1621–1629 (2010)
5. RamaKoteswara, R.A., Lekyasri, N., Rajani, K.: PID control design for second order systems. Int. J. Eng. Manuf. (IJEM) **4**, 45–56 (2019)
6. Keyvan-Ekbatania, M., Papageorgioub, M., Knoop, V.L.: Controller design for gating traffic control in presence of time-delay in urban road networks. Transp. Res. Procedia **7**, 651–668 (2015)
7. Beevi, P.F., Kumar, T.K.S., Jacob, J.: Two degree of freedom controller design by AGTM \AGMP matching method for time delay systems. Procedia Technol. **25**, 20–27 (2016)
8. Morteza, S., Zahra, R., Behrooz, R.: Fuzzy predictive control of step-down DC-DC converter based on hybrid system approach. Int. J. Intell. Syst. Appl. (IJISA) **2**, 1–13 (2014)
9. Trimboli, S., Di Cairano, S., Bemporad, A., Kolmanovsky, I.V.: Model predictive control for automotive time-delay processes: an application to air-to-fuel ratio control. In: 8th IFAC Workshop on Time-Delay Systems, vol. 42, pp. 90–95 (2009)
10. Gamal, M., Sadek, N., Rizk, M.R.M., Abou-elSaoud, A.K.: Delay compensation using Smith predictor for wireless network control system. Alexandria Eng. J. **55**, 1421–1428 (2016)
11. Tamilselvan, G.M., Aarthy, P.: Online tuning of fuzzy logic controller using Kalman algorithm for conical tank system. J. Appl. Res. Technol. **15**, 492–503 (2017)
12. Christian, A.A., Olaniyi, O.M., Dogo, E.M., Aliyu, S., Arulogun, O.T.: Nature-inspired optimal tuning of scaling factors of Mamdani fuzzy model for intelligent feed dispensing system. Int. J. Intell. Syst. Appl. (IJISA) **9**, 57–65 (2018)
13. Gostev, V.I.: Fuzzy-system of automatic control of the transmitter power of the adaptive channel of radio communication. J. Autom. Inf. Sci. **42**, 65–77 (2010)
14. Burakov, M.V., Kurbanov, V.G.: Neuro-pid control for nonlinear plants with variable parameters. ARPN J. Eng. Appl. Sci. **12**, 1226–1230 (2017)
15. Kumar, A., Kumar, V., Gaidhane, P.J.: Optimal design of fuzzy fractional order $PI\lambda D\mu$ Controller for redundant robot. Procedia Comput. Sci. **125**, 442–448 (2018)
16. Zatonskii, A.V.: Programmnye sredstva global'noi optimizatsii nastroek sistem avtomaticheskogo regulirovaniya [Software of global optimization of systems of automatic control: monograph], p. 136. Izdatel'skii Tsentr RIOR Publ, Moscow (2013). (in Russion)

Synthesis of *l*-coordinate Parallel Mechanism Without Singularities

Gagik Rashoyan[1(✉)], Narek Maloyan[2], Anton Antonov[1],
and Andrey Romanov[1]

[1] Blagonravov Mechanical Engineering Research Institute of the Russian
Academy of Sciences, 101000 Moscow, Russia
gagik_r@bk.ru, ant.ant.rk@gmail.com, Dru.ny@mail.ru
[2] Lomonosov Moscow State University, 119991 Moscow, Russia
maloyan.narek@gmail.com

Abstract. This article considers relevant topics related to the research and development of the new types of parallel manipulators. These devices move by linear drives located outside their working space. On the example of a novel parallel *l*-coordinate manipulation, the paper demonstrated some aspects of its singularity analysis and geometrical synthesis. The proposed method uses grapho-analytical techniques based on the direct kinematics solution, which has a closed-form solution for the presented mechanism. The approach studies the singularities concerning the spatial geometrical figures formed by the mechanism links. The analytical part of the method allows finding such geometrical parameters of the mechanist so that it will not have singularities in its working area. A numerical example demonstrates the implementation of the proposed approach.

Keywords: Kinematic analysis · Direct kinematics · Geometric synthesis · Singular position · *l*-coordinate mechanism · Parallel manipulator

1 Introduction

Parallel mechanisms are the ones that have the output link connected to the base by several kinematic chains [16]. The spatial parallel manipulators are widely used for performing technological operations [3, 7], in 3D printers [24, 25], metal-cutting machines [23], medical robots [5], devices for rehabilitation [19], and others [6, 12, 18]. One of the first mechanisms of this type is the well-known Stewart-Gough platform [9].

Among the wide variety of spatial parallel mechanisms, *l*-coordinate ones have a special place [11]. The structure similar to a frame allows them to take large external loads: the kinematic chains of the mechanisms operate under compression and tension forces. *l*-coordinate mechanisms differ by high rates of rigidity and speed. In terms of these characteristics, they are superior to robots with open kinematic chains.

However, along with the great functionality, these mechanisms have a significant drawback due to the singularities in the working area. This circumstance significantly affects the performance of manipulators and robots, based on the *l*-coordinate spatial

Z. Hu et al. (Eds.): AIMEE 2020, AISC 1315, pp. 270–282, 2021.
https://doi.org/10.1007/978-3-030-67133-4_25

mechanisms. In these positions, the manipulator loses controllability, and uncontrolled mobility appears [4]. The presence of singularities reduces the useful working area of the manipulator, and it acquires variable rigidity in the positions close to the singular ones.

Many works study the singularity analysis of such parallel mechanisms. There are approaches based on screw theory [1], Grassmann geometry [2, 15], Jacobian analysis [8, 14]. Some authors proposed to measure closeness to singular positions using different indices [10, 17, 22]. Though all these and other works suggest different approaches to singularity analysis, they lack information on how to synthesize a mechanism without singular configurations.

In this paper, on the example of a novel *l*-coordinate manipulator, we present an algorithm for its kinematic analysis that can be used to synthesize mechanisms without singular positions inside their workspaces. First, we introduce the structure of the novel manipulator, obtained from a basic *l*-coordinate mechanism. The direct kinematics problem of this manipulator has a closed-form solution. Next, we show how to use this solution to detect a singularity within the working area using both graphical and numerical methods. After that, we consider a numerical example in which we show how to select the mechanism's parameters to eliminate singular positions from its workspace. The conclusion recapitulates the results of the research.

2 Mechanism Structure

A classical spatial *l*-coordinate mechanism with a parallel structure is considered as a prototype. The device consists of a base and an output link connected by six SPS kinematic chains. Each of them contains a driven linear kinematic pair-P and two non-driven spherical pairs-S (Fig. 1a).

Figure 1b shows a new modified mechanism developed from the basic one (Fig. 1a). It is obtained by transferring the output link under its base. In this case, three "piercing" rods connect the output link with the base. Note that the modified mechanism retains the structure of the basic one. For example, in the novel mechanism, three kinematic chains intersect at a common point, two other chains – at another one, and the last one is alone (Fig. 1b).

Singularity determination deals with the position and velocity analysis and can be studied using the direct kinematics. Only a small number of *l*-coordinate mechanisms have an explicit solution for the direct kinematics problem [21]. For the most structures of *l*-coordinate manipulators, it is solved by obtaining a high-degree polynomial equation or using iterative algorithms [13, 16]. The multiple solutions of the direct kinematics problem determine the theoretically possible assembly modes of the mechanism. In this case, the transition from one assembly to another occurs through the singular position. In the next section, we consider the direct kinematics for the modified mechanism.

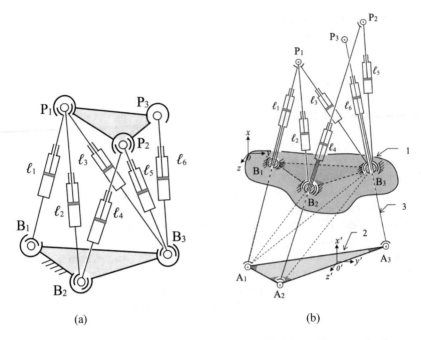

Fig. 1. a) Basic *l*-coordinate mechanism; b) Modified *l*-coordinate mechanism

3   Direct Kinematics

The given mechanism (Fig. 1b) has the output link 2 (points A_1, A_2, A_3) connected to the base 1 (points B_1, B_2, B_3) by three "piercing" rods with the known lengths l_{P1A1}, l_{P2A2}, l_{P3A3}. Coordinates of the points $B_1(X_{B1}, Y_{B1}, Z_{B1})$, $B_2(X_{B2}, Y_{B2}, Z_{B2})$, $B_3(X_{B3}, Y_{B3}, Z_{B3})$ in a fixed coordinate system $OXYZ$ are known. Coordinates of the points A_1, A_2, A_3 in $O'X'Y'Z'$ are given.

Six *l*-coordinates are the generalized ones and represent the distance between the base points B_1, B_2, and B_3, and the points P_1, P_2, and P_3 of the rods' ends (Fig. 1b). The values of the generalized coordinates vary between l_{imin} and l_{imax}, $i = 1, \ldots, 6$, which are usually the same for all the driven pairs.

First of all, it is necessary to find position of the points $A_1(X_{A1}, Y_{A1}, Z_{A1})$, $A_2(X_{A2}, Y_{A2}, Z_{A2})$, $A_3(X_{A3}, Y_{A3}, Z_{A3})$ in the fixed coordinate system $OXYZ$. These coordinates can be found from the following equations:

$$\begin{cases} (x_{B1} - x_{A1})^2 + (y_{B1} - y_{A1})^2 + (z_{B1} - z_{A1})^2 = l_{A1B1}^2 \\ (x_{B2} - x_{A1})^2 + (y_{B2} - y_{A1})^2 + (z_{B2} - z_{A1})^2 = l_{A1B2}^2 \\ (x_{B3} - x_{A1})^2 + (y_{B3} - y_{A1})^2 + (z_{B3} - z_{A1})^2 = l_{A1B3}^2 \end{cases} \quad (1)$$

$$\begin{cases} (x_{A1} - x_{A2})^2 + (y_{A1} - y_{A2})^2 + (z_{A1} - z_{A2})^2 = l_{A1A2}^2 \\ (x_{B2} - x_{A2})^2 + (y_{B2} - y_{A2})^2 + (z_{B2} - z_{A2})^2 = l_{B2A2}^2 \\ (x_{B3} - x_{A2})^2 + (y_{B3} - y_{A2})^2 + (z_{B3} - z_{A2})^2 = l_{B3A2}^2 \end{cases} \quad (2)$$

$$\begin{cases} (x_{A1} - x_{A3})^2 + (y_{A1} - y_{A3})^2 + (z_{A1} - z_{A3})^2 = l_{A1A3}^2 \\ (x_{A2} - x_{A3})^2 + (y_{A2} - y_{A3})^2 + (z_{A2} - z_{A3})^2 = l_{A2A3}^2 \\ (x_{B3} - x_{A3})^2 + (y_{B3} - y_{A3})^2 + (z_{B3} - z_{A3})^2 = l_{A3B3}^2 \end{cases} \tag{3}$$

The above systems of equations have an explicit solution with the algorithm presented in [20]. To solve the systems of Eqs. (1)–(3), we should find the unknown additional coordinates – the lengths of the segments on the right sides of these systems. To find the lengths of the rods' segments, we can use the following expressions:

$$l_{A1B1} = l_{P1A1} - l_1, l_{A2B2} = l_{P2A2} - l_4, l_{A3B3} = l_{P3A3} - l_6.$$

Let's consider the following steps for determining additional coordinates l_{A1B2}, l_{A1B3}, and l_{A2B3}. To find the length of the segment l_{A1B2}, consider $OA_1B_2P_1$ and $OB_1B_2P_1$ (Fig. 2). From $\Delta B_1B_2P_1$, we find

$$\cos \beta = \frac{l_1^2 + l_2^2 - l_{B1B2}^2}{2 \cdot l_1 \cdot l_2},$$

then we find l_{A1B2} from $\Delta A_1B_2P_1$:

$$l_{A1B2} = \sqrt{l_{A1P1}^2 + l_2^2 - l_{A1P1} \cdot \frac{(l_1^2 + l_2^2 - l_{B1B2}^2)}{l_1}} \tag{4}$$

Similarly, considering $\Delta A_1B_3P_1$ and $\Delta B_1B_3P_1$, we find the length l_{A1B3} (Fig. 2):

$$l_{A1B3} = \sqrt{l_{A1P1}^2 + l_3^2 - l_{A1P1} \cdot \frac{(l_1^2 + l_3^2 - l_{B1B3}^2)}{l_1}} \tag{5}$$

Further, to solve system (2), similarly to the previous one, from the triangles $\Delta A_2B_3P_2$ and $\Delta B_2B_3P_2$, we find the value of the additional coordinate l_{A2B3}:

$$l_{A2B3} = \sqrt{l_{A2P2}^2 + l_5^2 - l_{A2P2} \cdot \frac{(l_4^2 + l_5^2 - l_{B2B3}^2)}{l_4}} \tag{6}$$

Given the values of the generalized and additional coordinates, it is possible to find the position of the output link of the mechanism. Therefore, the direct kinematics problem is solved.

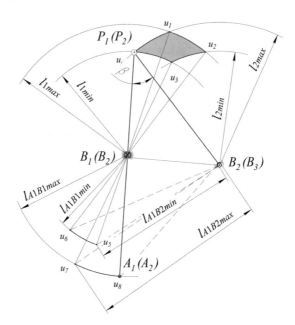

Fig. 2. Determination of the additional *l*-coordinates of the mechanism

4   Grapho-Analytical Method

The solution of the systems (1)–(3) has a geometric interpretation and means the determination of the vertices coordinates of three tetrahedrons (Fig. 1b):

- tetrahedron with the base $B_1B_2B_3$ and a vertex at the point A_1;
- tetrahedron with the base $A_1B_2B_3$ and a vertex at the point A_2;
- tetrahedron with the base $A_1A_2B_3$ and a vertex at the point A_3.

The analytical solution analysis of the direct kinematics problem shows that in singular positions these tetrahedrons shrink into their base planes and become flat figures. It means that instead of a spatial geometrical figure, the mechanism has a "flattened" tetrahedron. In this case, there can be from one to three tetrahedron at the same time. Works [6, 15, 21] present graphical and analytical methods to analyze singular positions and to synthesize the basic *l*-coordinate mechanism without singularities in its workspace.

To analyze the singular positions of the new mechanism using graphical constructions, it is necessary at first to find the limiting values of the of the additional coordinates – the lengths of the segments $l_{A1B2min}$ and $l_{A1B2max}$, $l_{A1B3min}$ and $l_{A1B3max}$, $l_{A2B3min}$ and $l_{A2B3max}$.

Consider the segment B_1B_2. We plot the concentric circles with the radii l_{1min}, l_{1max} u l_{2min}, l_{2max} centered at the points B_1 and B_2 (Fig. 2). As a result of circles intersection, we obtain the area of existence domain for the point P_1. That area is formed by the arcs of the circles that intersect at the points u_1, u_2, u_3, and u_4.

Then, we consider the problem of describing the existence domain of the point A_1 located on the other end of the rod with the length l_{P1A1}. Circular arcs with the radii $l_{A1B1min}$ and $l_{A1B1max}$ bound this region, and the points u_5, u_6, u_7, and u_8 place on these arcs. The lengths of the arcs are: $l_{A3B3max} = l_{A3P3} - l_{6min}$, $l_{A3B3min} = l_{A3P3} - l_{6max}$.

The exact position of these points can be found by connecting the points u_1, u_2, u_3, and u_4 with a straight line passing through the point B_2 until the intersection with the circles of those radii.

The limiting values $l_{A1B2min}$ and $l_{A1B2max}$ will correspond to the minimum and maximum distances measured between the points u_5, u_6, u_7, and u_8 and the point B_2.

To find the lengths of the additional coordinates $l_{A1B3min}$ and $l_{A1B3max}$, we will use a similar procedure. Let us find the existence domain of the point P_1. It depends on the length change of the *l*-coordinates l_{1min}, l_{1max}, l_{3min}, and l_{3max}, as well as the given lengths l_{B1B3} and l_{P1A1}. To find the lengths $l_{A1B2min}$ and $l_{A1B2max}$, the existence domain of P_1 should be connected by a straight line, passing through the point B_1, with the intersection of the circles with the radii $l_{A1B1min}$ and $l_{A1B1max}$. Next, we need to find the minimum and maximum distances from the point B_3 to the existence domain of the point A_1 (this region can be represented by the points u_5, u_6, u_7, and u_8).

The segments $l_{A2B3min}$ and $l_{A2B3max}$ can be found similarly (Fig. 2). In this case, the variation domain of the point P_2 depends on the length change of the *l*-coordinates l_{4min}, l_{4max}, l_{5min}, and l_{5max}, and the specified lengths l_{B2B3} and l_{P2A2}. To find the lengths $l_{A2B3min}$ and $l_{A2B3max}$, we need to find the existence domain of the point A_2 which is located between the circles of the radii:

$$l_{A2B2max} = l_{A1P2} - l_{4min}, l_{A2B2min} = l_{A1P2} - l_{4max}.$$

Next, we find the lengths $l_{A2B3min}$ and $l_{A2B3max}$ by measuring the distance from the point B_3 to the ones characterizing the A_2 existence domain.

5   Graphic Approach to the Analysis

To analyze the singularities, consider a graphical interpretation of the tetrahedron with the base $B_1B_2B_3A_1$ and the edges l_{A1B1}, l_{A1B2}, and l_{A1B3} (Fig. 3). The singular configuration is equivalent to the "flattened" tetrahedron $B_1B_2B_3P_1$ (on the plane of the base $\triangle B_1B_2B_3$) since the points P_1 and A_1 belong to the same straight line and these tetrahedrons have a common base. The intersection of the circles with the radii l_{2min}, l_{2max}, l_{3min}, and l_{3max}, centered at B_2 and B_3, results in two existence domains for the point P_1. These regions are formed on the tetrahedron base plane and designated as U_L and U_R, respectively (Fig. 3).

Next, we plot two circles with the radii l_{1min} and l_{1max} centered at B_1. If they intersect the U_L and U_R regions, then this tetrahedron becomes a flat figure. It indicates the presence of a singularity.

Measuring the lengths of the segments from the point B_1 to the points u_1– u_8, we find the limit values of the coordinates l_{1min} and l_{1max} for each area. If the given variation domain of l_1 does not coincide with the obtained measured lengths l_{B3uj} ($j = 1,2,3,4$ and $j = 5,6,7,8$) or does not intersect with them, then there is no singular position. If the intersection exists, then the length l_1 should be corrected to avoid singularities for the first tetrahedron.

We can apply a similar graphical analysis to the tetrahedron $A_1A_2A_3B_3$ with the edges l_{A1A3}, l_{A1A3}, and l_{A3B3} (Fig. 4). To determine graphically whether the given tetrahedron becomes a flat figure, it is necessary:

– to draw concentric circles of the radii $l_{A1B3min}$ and $l_{A1B3max}$, $l_{A2B3min}$ and $l_{A2B3max}$ centered at the points A_1 and A_2 respectively;
– to draw concentric circles of radii $l_{A3B3min}$ and $l_{A3B3max}$ centered at the point B_3, where: $l_{A3B3max} = l_{A3P3} - l_{6min}$ and $l_{A3B3min} = l_{A3P3} - l_{6max}$.

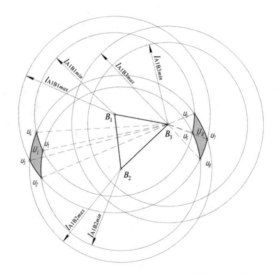

Fig. 3. Graphics for the tetrahedron with the base $B_1B_2B_3$ and a vertex at A_1

If there is no intersection between the circles with the radii $l_{A3B3min}$ and $l_{A3B3max}$ with the B_3 variation area (which is bounded by the circles with the radii $l_{A1B3min}$, $l_{A1B3max}$, $l_{A2B3min}$, $l_{A2B3max}$), then there is no singularity.

Next, let's consider a graphical analysis of the tetrahedron $A_1B_2B_3A_2$ with the edges l_{A1A2}, l_{A2B2}, and l_{A2B3} (Fig. 5) to find out whether it becomes a flat figure. We can plot the following concentric circles:

– with the radii $l_{A1B2min}$ and $l_{A1B2max}$, $l_{A2B2min}$ and $l_{A2B2max}$ centered at B_2;
– with the radii $l_{A1B3min}$ and $l_{A1B3max}$, $l_{A2B3min}$, and $l_{A2B3max}$ centered at B_3.

The intersections of these circles result in two existence domains for the points B_2 and B_3 (D_1 and D_2 in Fig. 5). These regions are on the plane of the possible "flattening" of the considered tetrahedron.

Next, one should check whether the segment A_1A_2 is on the flattening plane in such a way that the point A_1 is in D_1, and the point A_2 is in D_2. If such an arrangement is not possible, then there is no singularity. Otherwise, a singular position can be excluded by varying the length of the segment A_1A_2 or the tetrahedron edges, depending on the given problem.

6 Numerical-Analytical Approach

This approach allows us to determine the presence of singularities and to synthesize a mechanism without singular positions in its workspace. Let's consider the problem in more detail. To find additional coordinates, we will find the minimum and maximum values of the functions (4)–(6) by variation of variables with the given limits (Fig. 5).

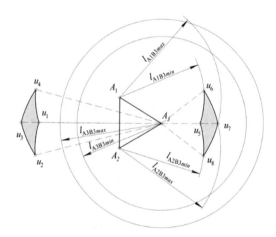

Fig.4. Graphics for the tetrahedron with the base $A_1A_2A_3$ and a vertex at B_3

Let's perform calculations for the tetrahedron $B_1B_2B_3A_1$ (Fig. 3). We will plot the concentric circles with the radii $l_{A1B2min}$ and $l_{A1B2max}$ centered at B_2, and concentric circles with the radii $l_{A1B3min}$ and $l_{A1B3max}$ centered at B_3.

In a singular position (on the base plane $\triangle B_1B_2B_3$), we find the coordinates of the points u_j, $j = 1,...,8$, as the intersection of the circles l_2 and l_3 in their extreme values. We can compose four systems of equations:

$$\begin{cases} (x_{B2} - x_{u1})^2 + (y_{B2} - y_{u1})^2 = l_{2min}^2 \\ (x_{B3} - x_{u1})^2 + (y_{B3} - y_{u1})^2 = l_{3min}^2 \\ (x_{B2} - x_{u3})^2 + (y_{B2} - y_{u3})^2 = l_{2max}^2 \\ (x_{B3} - x_{u3})^2 + (y_{B3} - y_{u3})^2 = l_{3max}^2 \end{cases} \quad \begin{cases} (x_{B2} - x_{u2})^2 + (y_{B2} - y_{u2})^2 = l_{2min}^2 \\ (x_{B3} - x_{u2})^2 + (y_{B3} - y_{u2})^2 = l_{3max}^2 \\ (x_{B2} - x_{u4})^2 + (y_{B2} - y_{u4})^2 = l_{2max}^2 \\ (x_{B3} - x_{u4})^2 + (y_{B3} - y_{u4})^2 = l_{3min}^2 \end{cases} \quad (7)$$

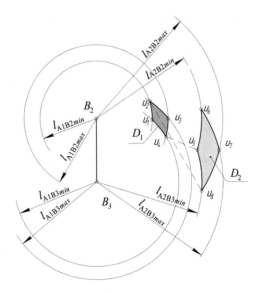

Fig.5. Graphics for the tetrahedron with the base $A_1B_2B_3$ and a vertex at A_2.

Solving the systems of Eqs. (7), we find the coordinates from the intersection of the circles with the radii of l_2 and l_3. Two existence domains are obtained, designated as U_L and U_R, respectively. Next, we find the lengths of eight segments representing the distances between the points B_1 and u_j, $j = 1, ..., 8$:

$$l_{uj} = \sqrt{\left(x_{B1} - x_{uj}\right)^2 + \left(y_{B1} - y_{uj}\right)^2}.$$

Further, we determine the maximum and minimum values of the obtained lengths: $\min\limits_{j=1,...,4} l_{uj}$ for the U_L region and $\min\limits_{j=5,...,8} l_{uj}$ for U_R. To eliminate singular positions, the obtained results should not coincide (neither partially nor completely) with the variation range of the coordinates l_{1min} and l_{1max}.

Calculations for the tetrahedron with the base $A_1A_2A_3$ and a vertex at the point B_3 are similar. In this case, the limiting values of the coordinates l_{6min} and l_{6max} are used to check for a singular position.

Consider the tetrahedron with the base $A_1A_2B_3$ and a vertex at the point A_3. Similar to the systems of Eqs. (7), we compose and solve the systems of equations for the intersection of circles:

－ with the radii $l_{A1B2min}$, $l_{A1B2max}$, $l_{A1B3min}$, and $l_{A1B3max}$ to find the coordinates of the points $u_j, j =1,...,4$, for the region D_1;
－ with the radii $l_{A2B2min}$, $l_{A2B2max}$, $l_{A3B3min}$, and $l_{A3B3max}$ to find the coordinates of the points $u_j, j =5,...,8$, for the region D_2.

Next, we find the maximum and minimum distances between the points $u_j, j=1,...,4$, of the region D_1 and the points $u_j, j = 5,...,8$, of the region D_2 from the set of distances:

$$d = \left\{ \sqrt{\left(x_{ui} - x_{uj}\right)^2 + \left(y_{ui} - y_{uj}\right)^2} \,\middle|\, i = 1, \ldots, 4, j = 5, \ldots, 8 \right\}.$$

We compare $min\ d$ and $max\ d$ with l_{A1A2}. If $l_{A1A2} > max\ d$, then the tetrahedron will not become a flattened one, therefore, there is no singularity. If $min\ d < l_{A1A2} < max\ d$, then there are singular positions. Finally, if $l_{A1A2} < min\ d$, then we cannot design a mechanism with the given parameters.

7   Numerical Experiment

Consider a numerical experiment for the l-coordinate mechanism shown in Fig. 1b with the following parameters (in dimensionless quantities):

$$l_{A1A2} = 20, l_{A2A3} = 25, l_{A1A3} = 35, l_{B1B2} = l_{B2B3} = l_{B1B3} = 20, l_{A1P1} = l_{A2P2} = l_{A3P3} = 70,$$
$$l_{imin} = 38, l_{imax} = 45, i = 1, \ldots 6.$$

The problem can be solved both graphically and analytically with the subsequent correction of the geometric parameters to exclude possible singularities in the working space.

The results of the experiment, which are given in Table 1, showed that:

– the tetrahedron $B_1B_2B_3P_1$ has no singular positions;
– the tetrahedron $A_1A_2A_3B_3$ has a singularity with the value $l_{A1B3} = 29.82$, since this value is in the range $25 < l_{A3B3} < 32$, where the limiting values of l_{A3B3} are found as:

$$l_{A3B3max} = l_{A3P3} - l_{6min} = 70 - 38 = 32$$
$$l_{A3B3min} = l_{A3P3} - l_{6max} = 70 - 45 = 25.$$

Table 1. Analysis of $B_1B_2B_3P_1$ and $A_1A_2A_3B_3$

Point	$B_1B_2B_3P_1$		$A_1A_2A_3B_3$	
	Distance from u_i to B_1	Within $[l_{1min}, l_{2max}]$	Distance from u_i to A_3	Within $[l_{A3B3min}, l_{A3B3max}]$
u_1	19.34	No	17.68	No
u_2	25.05	No	10.92	No
u_3	26.55	No	22.12	No
u_4	25.05	No	**29.82**	**Yes**
u_5	53.98	No	60.26	No
u_6	56.93	No	59.73	No
u_7	61.19	No	66.95	No
u_8	56.93	No	65.86	No

Table 2. Distances between u_1–u_4 and u_5–u_8 for $A_1B_2B_3A_2$

	u_5	u_6	u_7	u_8
u_1	18.22	33.06	18.66	**5.99**
u_2	11.16	23.99	7.06	8.06
u_3	22.58	37.06	20.90	10.64
u_4	30.39	**45.10**	31.55	18.75

To eliminate singular positions, let us put the limit for the minimum length of the link $l_{A3B3min} > l_{A3P3} - l_{6max} > 29.82$, from where we find

$$l_{A3P3} > l_{A3B3min} + l_{6max},$$
$$l_{A3P3} > 29.82 + 45 > 74.82$$

For example, we can take it equal to $l_{A3B3} = 75$. Then, the limiting values of l_{A3B3} will be the following:

$$l_{A3P3max} = l_{A3P3} + l_{6min} = 75 - 38 = 37,$$
$$l_{A3P3min} = l_{A3P3} + l_{6max} = 75 - 45 = 30.$$

Analysis of the experimental results for the tetrahedron $A_1B_2B_3A_2$ (Table 2) showed that for $5.99 < l_{A1A2} < 45.10$ the tetrahedron can become a planar one. To exclude that, one needs to increase the length of the segment $A_1A_2 > 45.10$. After that, the geometry of the triangle $A_1A_2A_3$ will change, so it will be necessary to check the conditions for the tetrahedron $A_1A_2A_3B_3$ again. Analysis of the second iteration showed the absence of singularity positions for this tetrahedron.

The results of the numerical experiment showed that our proposed methods help to find singular positions. If we detect them, then it is possible to obtain such geometric parameters of the mechanism that it will not have singularities in its working area.

8 Conclusion

This paper proposed the grapho-analytical method, which allows us to solve the problem of synthesizing a new l-coordinate mechanism without singular positions in its working space. The graphical part has an intuitively clear sense. It considers the analysis of the spatial tetrahedrons, formed by the mechanism's links, and their degeneracy into planes in singular configurations. The analytical part uses the fact that the direct kinematics solution for the considered manipulator has a closed-form solution.

In this paper, we only considered the constraints on the l-coordinates. The future research should also concern the constraints on the joint angles and the link interference. These additional constraints also affect the working area of the manipulator and should be included in the analysis to get more accurate results.

The proposed method allows us to find the geometrical parameters of the mechanism, which will not have singularities in its working area, and can be used on the design stage. The suggested approach can also be applied to other basic *l*-coordinate mechanisms where a direct kinematics problem has a closed- form solution.

Acknowledgement. This work was supported by the Russian Science Foundation, the agreement number 19-19-00692.

References

1. Aleshin, A., Glazunov, V., Rashoyan, G., Shai, O.: Analysis of kinematic screws that determine the topology of singular zones of parallel-structure robots. J. Mach. Manuf. Reliab. **45**(4), 291–296 (2016)
2. Ben-Horin, P., Shoham, M.: Application of grassmann—cayley algebra to geometrical interpretation of parallel robot singularities. Int. J. Rob. Res. **28**(1), 127–141 (2009)
3. Chunikhin, A.Y., Glazunov, V.: Developing the mechanisms of parallel structure with five degrees of freedom designed for technological robots. J. Mach. Manuf. Reliab. **46**(4), 313–321 (2017)
4. Conconi, M., Carricato, M.: A new assessment of singularities of parallel kinematic chains. IEEE Trans. Rob. **25**(4), 757–770 (2009)
5. Fernandes, J., Arockia, S.: Kinematic and dynamic analysis of 3puu parallel manip- ulator for medical applications. Procedia Comput. Sci. **133**, 604–611 (2018)
6. Glazunov, V.: Parallel structure mechanisms and their application: robotic, tech- nological, medical, and training systems. Institute of Computer Science, Izhevsk (2018). (in Russian)
7. Glazunov, V., Rashoyan, G., Aleshin, A., Shalyukhin, K., Skvortsov, S.: Structural synthesis of spatial *l*-coordinate mechanisms with additional links for technological robots. In: Hu, Z., Petoukhov, S., He, M. (eds.) Advances in Artificial Systems for Medicine and Education II, pp. 683–691. Springer, Cham (2020)
8. Gosselin, C., Angeles, J.: Singularity analysis of closed-loop kinematic chains. IEEE Trans. Rob. Autom. **6**(3), 281–290 (1990)
9. Gough, V.: Contribution to discussion of papers on research in automobile stability, control and tyre performance. Proc. Autom. Div. Inst. Mech. Eng. **171**, 392–394 (1957)
10. Joumah, A.A., Albitar, C.: Design optimization of 6-rus parallel manipulator using hybrid algorithm. Int. J. Inf. Technol. Comput. Sci. **10**(2), 83–95 (2018)
11. Koliskor, A.: The *l*-coordinate approach to the industrial robots design. IFAC Proc. vol. **19** (2), 225–232 (1986)
12. Krishan, G., Singh, V.R.: Motion control of five bar linkage manipulator using conventional controllers under uncertain conditions. Int. J. Intell. Syst. Appl. **8**(5), 34–40 (2016)
13. Kumar, V., Sen, S., Roy, S.S., Shome, S.N.: Inverse kinematics of redundant manipulator using interval newton method. Int. J. Eng. Manuf. **5**(2), 19–29 (2015)
14. Li, H., Gosselin, C., Richard, M., St-Onge, B.: Analytic form of the six-dimensional singularity locus of the general gough-stewart platform. J. Mech. Des. **128**(1), 279–287 (2005)
15. Merlet, J.P.: Singular configurations of parallel manipulators and grassmann geometry. Int. J. Rob. Res. **8**(5), 45–56 (1989)
16. Merlet, J.P.: Parallel Robots. Springer, Cham (2006)
17. Nawratil, G.: New performance indices for 6-dof ups and 3-dof rpr parallel manipulators. Mech. Mach. Theory **44**(1), 209–221 (2009)

18. Patel, Y., George, P.: Parallel manipulators applications – a survey. Mod. Mech. Eng. **2**(3), 57–64 (2012)
19. Rashoyan, G., Shalyukhin, K., Antonov, A., Aleshin, A., Skvortsov, S.: Analysis of the structure and workspace of the isoglide-type robot for rehabilitation tasks. In: Hu, Z., Petoukhov, S., He, M. (eds.) Advances in Artificial Systems for Medicine and Education III, pp. 186–194. Springer, Cham (2020)
20. Rashoyan, G., Aleshin, A., Antonov, A., Gavrilina, L., Glazunov, V., Skvortsov, S., Shalyukhin, K.: Analysis and synthesis of parallel structure mechanism without singularities. J. Phys. Conf. Ser. **1260**(11), 112023 (2019)
21. Rashoyan, G., Demidov, S., Aleshin, A., Antonov, A., Skvortsov, S., Shalyukhin, K.: The direct position problem for *l*-coordinate mechanisms of various types. J. Mach. Manuf. Reliab. **48**(5), 392–400 (2019)
22. Voglewede, P., Ebert-Uphoff, I.: Measuring "closeness"to singularities for parallel manipulators. In: IEEE International Conference on Robotics and Automation, 2004. Proceedings. ICRA 2004, vol. 5, pp. 186–194 (2004)
23. Weck, M., Staimer, D.: Parallel kinematic machine tools – current state and future potentials. CIRP Ann. **51**(2), 671–683 (2002)
24. Ye, W., Fang, Y., Guo, S.: Design and analysis of a reconfigurable parallel mechanism for multidirectional additive manufacturing. Mech. Mach. Theory **112**, 307–326 (2017)
25. Zi, B., Wang, N., Qian, S., Bao, K.: Design, stiffness analysis and experimental study of a cable-driven parallel 3D printer. Mech. Mach. Theory **132**, 201–222 (2019)

SWOT Analysis of Computer Distance Learning Systems

N. Yu. Mutovkina[✉] [ID]

Tver State Technical University, Tver, Russia
`letter-boxNM@yandex.ru`

Abstract. The article considers the problem of development and implementation of distance technologies in the educational process. Over the past six months, this problem has become more urgent in connection with the transfer of all educational institutions to remote operation due to the spread of new coronavirus infection. In these conditions, computer-based distance learning is the only option for implementing educational programs. However, despite the advantages of computer-based distance learning, it is not without drawbacks and some risks. To better understand the advantages and disadvantages of computer-based distance learning, as well as optimize this system, it is proposing to perform a SWOT analysis of the distance learning system. In this paper, we consider a method of extended SWOT analysis, which assumes its periodic implementation in each educational organization and region. The methodology is base on the complex application of various tools for analyzing and evaluating the effectiveness of distance learning. It contains such elements as standard SWOT analysis matrix, weighted scoring SWOT analysis (ranking), statistical SWOT analysis, McKinsey matrix model, the model of weighted potentials, and implementation levels. The proposed method will allow for a more complete and diverse analysis of all significant aspects of the computer distance learning system.

Keywords: Distance learning · SWOT analysis · Strengths and weaknesses · Opportunities · Threats · Risks · Ranking

1 Introduction

At present, computer-based distance learning (CBDL) is becoming more and more widespread. Distance technologies in education are a natural response to the challenges of time and a consequence of scientific and technological progress. In scientific works, distance education technologies (DET) are considering as one of the main types of modern educational technology [1], a means of organizing the activities of students with disabilities [2], a trend in the modern electronic educational environment of the university [3], a means of network interaction of educational institutions [4] and in other areas that emphasize the importance and effectiveness of CBDL. DET is used both in training and in monitoring the development of educational material.

The advantages of distance learning include its flexibility, modularity, parallelism, long-range operation, asynchrony, coverage, and profitability. Now, in the context of the pandemic, DET has become the only possible means of providing the learning process.

However, it is impossible not to notice problems in the implementation of DET. Thus, the disadvantages of distance learning are the lack of direct communication between students and teachers, which makes it difficult to create a creative atmosphere in a group of students; the lack of a well-established system of legal regulation of the implementation of DET, in particular, the preservation and protection of copyright for e-courses; the high cost and complexity of building a distance learning system; the problem of student authentication when checking knowledge [5, 6]. Distance learning does not yet meet the requirements of either student, teachers, or organizers in full. It is because of its low quality that the very need to use DET in training causes a lot of controversies [7].

This ambiguity in the perception of the CBDL caused the need to perform an analysis of the strengths and weaknesses of this educational system from the side of opportunities and threats of the external environment, that is, SWOT analysis. It is advisable to conduct a step-by-step study of all the most significant external and internal factors that affect the development of CBDL and its implementation in the educational process. To do this, you need to use statistical and expert analysis methods.

The author draws attention to the fact that the article deals specifically with distance learning, not distance education. There is a difference in terms of "learning" and "education". Education is a process and result of mastering and consolidating a set of general and specific professional competencies. The base way of getting an education is learning in educational institutions. It is a purposefully organized, systematic, and systematic acquisition of knowledge, skills, and abilities under the guidance of teachers. Thus, education is the result of personal learning (gaining professional competence), and learning is the process of acquiring knowledge, skills, and abilities [8].

The purpose of the research is to create analytical tools that allow you to develop strategies for the development of the information environment and computer technologies for distance learning.

The developed method of SWOT analysis can be using by specialists engaged in the promotion and implementation of remote technologies in the educational process for strategic purposes.

2   Literature Review

SWOT analysis is a brief analysis of marketing information, the results of which determine the direction of development of the research object, distribute resources by its constituent elements [9]. The result of this work is the development of a strategy for the development and implementation of CBDL. Upon completion of the analysis, the advantages and disadvantages of the CBDL, it's potential, and possible external threats are identifying. Their assessment of the demand for DET among students and leading teachers is also give.

To conduct the most objective, grounded SWOT analysis, you need to adhere to the following rules:

- precisely define the area of analysis;
- correctly determine what the factor in question relates to (strengths or weaknesses, opportunities or threats);
- the analysis should demonstrate the actual position of the research object in its environment;
- it is necessary to make specific statements to better understand the impact of this factor on the segments of the research object.

Classic SWOT analysis is one of the most common methods used in strategic business planning [9, 10]. Its peculiarity is that it can be to separate what depends on performers on the ground from what does not depend on them. At the first stage, data on the current state of the research object is collecting. The next step summarizes the strengths and weaknesses of the internal structure of the research object, opportunities, and threats from the external environment. The third stage is to formulate individual strategies to minimize foreign menaces and internal weaknesses of the research object.

When identifying the strengths and weaknesses of the computer-based distance learning system, its resources are assessing, and competitiveness is considering. Traditional SWOT analysis has its advantages and disadvantages. The worths of a SWOT analysis are:

- It has an extensive scope of application;
- It provides a free choice of analyzed elements depending on the goals set;
- It can be performing for operational and strategic evaluation;
- It does not require special knowledge.

Cons of SWOT analysis:

- It shows common factors;
- The results are presenting in the form of a qualitative description, and there are no quantitative parameters;
- This type of analysis is subjective, depending on the performances of the person who conducts his.

However, these disadvantages can be minimizing by using an extended SWOT analysis methodology, which provides for a transition from qualitative to quantitative assessments and a procedure of agreement expert assessments.

Using the terminology of the mathematical theory of optimal processes [11–13], it can be argued that the SWOT analysis forms the positional management of the CBDL based on the situation {S, W, O, T}. Since these four components characterize a volatile situation, the SWOT analysis should be performing every time a previously made management decision needs to be corrected. Dynamic SWOT analysis will be formalizing in this article in this is the context. It can be called monitoring the implementation of the planned scenario of CBDL functioning. Monitoring involves adjusting management decisions if necessary.

3 Advanced SWOT Analysis Methodology

An advanced SWOT analysis should be performing using the following methodology.

1. At time t_0, factors affecting the object of research are identifying. Factors are dividing into internal and external. Internal factors are the strengths and weaknesses that characterize the research object and are under the control of the object owner. External factors are opportunities and threats that describe the environment external to the object and are beyond the control of the object owner. The identified strengths, weaknesses, opportunities, and threats are summarizing in a four-field matrix, which is using in a classical SWOT analysis (Fig. 1).
2. Expert assessment of the significance and the level of implementation of the selected elements of the CBDL system. The calculation of the weighted grades and ranks according to the formulas shown in Fig. 2. In the columns "Importance" credits are awarded from 3 to 5 (3 is the least significant element, 5 is the most weighty element), and in the "Score" provides expert assessment of the feasibility of these elements in a specific region at time t_0 on a scale from 1 (if the answer is negative) to 5 (definitely with an affirmative reply).
3. Statistical evaluation of indicators of SWOT analysis of the CBDL system, obtained in the form of ranks. In this case, the SWOT analysis matrix is basing on the principle of multiplying ranks, for example:

$$k_{ij}^{\{SO\}} = R_i^{\{S\}} \cdot R_j^{\{O\}}. \tag{1}$$

To calculate the level of fulfillment (F) of opportunities and threats, and potential (P) of strengths and weaknesses of the CBDL system, the following formulas are used:

$$
\begin{cases}
P_i^{\{S\}} = \sum k_{ij}^{\{SO\}} - R_i^{\{S\}} \cdot \sum k_{ij}^{\{ST\}} \\
P_i^{\{W\}} = \sum k_{ij}^{\{WO\}} - R_i^{\{W\}} \cdot \sum k_{ij}^{\{WT\}} \\
F_i^{\{O\}} = \sum k_{ji}^{\{SO\}} - R_i^{\{O\}} \cdot \sum k_{ji}^{\{WO\}} \\
F_i^{\{T\}} = \sum k_{ji}^{\{ST\}} - R_i^{\{T\}} \cdot \sum k_{ji}^{\{WT\}}
\end{cases} \tag{2}
$$

An example of the implementation of this stage of the method shown in Fig. 3.

4. Analysis of the success of the implementation and development of the CBDL system using the McKinsey model. The choice of this model due to the need to determine the level of implementation possibilities combined with the positive factors of the internal environment of the system of CBDL in the process of impact of threats and negative factors of the internal environment of the system under consideration (Fig. 4). An adapted model of McKinsey is formed based on the data of the previous stage and represented in Fig. 5.
5. Determining acceptable levels of potentials $P_{opt}^{\{S\}}$ and $P_{opt}^{\{W\}}$, as well as borderline levels fulfillment of opportunities ($F_{opt}^{\{O\}}$) and threats ($F_{opt}^{\{T\}}$). The calculation of the

appropriate potential of strengths and the boundary level of fulfillment of opportunities carried out using the following formulas:

$$P_{opt}^{\{S\}} = \lambda \cdot \min P_i^{\{S\}} + (1 - \lambda) \cdot \max P_i^{\{S\}} \tag{3}$$

$$F_{opt}^{\{O\}} = \lambda \cdot \min F_i^{\{O\}} + (1 - \lambda) \cdot \max F_i^{\{O\}} \tag{4}$$

where $\min P_i^{\{S\}}$, $\min F_i^{\{O\}}$, $\max P_i^{\{S\}}$, $\max F_i^{\{O\}}$ are the minimum and maximum values of the potential of strengths and the level of fulfillment of opportunities (Fig. 3), λ is the coefficient of pessimism, $(1 - \lambda)$ is the coefficient of optimism. The value of λ varies from zero to one, depending on the opinion of experts.

Based on the results of calculations, first of all, you should pay attention to those strengths and opportunities for which the values of $P_i^{\{S\}}$ and $F_i^{\{O\}}$ were lower than $P_{opt}^{\{S\}}$ and $F_{opt}^{\{O\}}$. It is for these elements of SWOT analysis that development strategies must be developing in conjunction with other indicators.

The calculation of the acceptable potential of weakness and the permissible level of threat implementation is performing using the following formulas:

$$P_{opt}^{\{W\}} = \lambda \cdot \max P_i^{\{W\}} + (1 - \lambda) \cdot \min P_i^{\{W\}} \tag{5}$$

$$F_{opt}^{\{T\}} = \lambda \cdot \max F_i^{\{T\}} + (1 - \lambda) \cdot \min F_i^{\{T\}} \tag{6}$$

where $\max P_i^{\{W\}}$ and $\min P_i^{\{W\}}$ are the maximum and minimum values of the potential of weaknesses, $\max F_i^{\{T\}}$ and $\min F_i^{\{T\}}$ are the maximum and minimum values of the threat (Fig. 3), λ is the risk coefficient, which varies from zero to one, depending on the opinion of experts.

All weights of SWOT analysis threat indicators that exceed the calculated limit of acceptable risk are expected risks with a higher probability. Those weights that do not exceed $F_{opt}^{\{T\}}$ are unexpected risks, the manifestation of which can lead to minor losses. The same can be said about the weaknesses of the object of analysis. In this case, when elaborating strategies for the development of the CBDL system, exceptional attention should be paid to those weaknesses and threats for which the values of $P_i^{\{W\}}$ and $F_i^{\{T\}}$ were higher than $P_{opt}^{\{W\}}$ and $F_{opt}^{\{T\}}$, respectively.

6. The formulated strategies can be dividing into implementation periods. In the short term, the strategies of "adaptation" {ST} and "survival" {WT} should be using. In the medium term, it is advisable to apply the "defense" strategy {WO}. And, accordingly, in the long term, a strategy of "offensive" {SO} is necessary [14].
7. The repeating steps 1 through 6 after a certain period. The recommended period is one academic semester (trimester).

4 Results and Discussion

According to the above methodology, an extended SWOT analysis of the computer distance learning system in the Tver region (Russia) is carrying out.

Figure 1 shows the most significant strengths and weaknesses of the CBDL system, as well as opportunities and threats to its implementation and development.

Strengths {S}		Weakness {W}	
S_1	is the expanding of the scope of training	W_1	there is a lack of precise methods for implementing a computer-based
S_2	there is an openness of educational resources of the University, accessibility of training, including for people with disabilities	W_2	there is a lack of legal mechanisms for protecting e-learning courses, obtaining copyrights
S_3	there is an individual approach to students	W_3	there is a discrepancy between the complexity of developing e-learning courses and the remuneration of teachers
S_4	there is an increase in the profitability of the learning process	W_4	there is insufficient development of information and communication
S_5	there is the use of various possibilities of computer technology and modern information technologies in the delivery of educational information	W_5	there is a high cost of building a distance learning system
S_6	there is an increase in the interest of students in applying new technologies using the Internet	W_6	Remote programs are not suitable for mastering professions that require a lot of practice
Opportunities {O}		Threats {T}	
O_1	there is a possibility of the fullest realization of the needs of society in training, the constitutional right of every citizen to education	T_1	there is a threat of a large number of non-accredited and non-certified distance courses of low quality due to an imperfect legal framework
O_2	there is a possibility of improving the demographic situation in the regions, reducing the outflow of applicants from regions to large cities	T_2	there is a threat of high competition with foreign educational institutions that provide distance learning
O_3	there is a possibility of increasing the economic indicators of the country by educating the population without complete interruption from work	T_3	there is a threat to reduce the solvent demand for new personnel, innovative developments, and products
O_4	there is an opportunity to develop and improve the legal framework, organizational and legal conditions for creating and implementing elements of a computer-based distance learning system	T_4	there is an unpredictability of the development of the computer distance learning system, of its effectiveness for the country's economy
O_5	there is the possibility of integrating economic sectors, expanding the number and availability of interdisciplinary research and developments	T_5	there is any uncertainty of social effect as a result of the widespread introduction and development of computer-based distance learning

Fig. 1. Standard SWOT table

Figure 2 shows the results of the second stage of the methodology for the Tver region in the Russian Federation. According to the agreed expert estimates, the weighted score (rank) of the SWOT analysis elements presented in Fig. 1 is calculating.

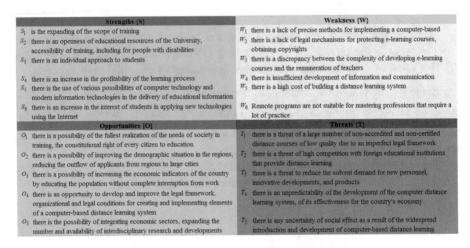

{S}	Importance $(Z_i^{\{S\}})$	Score $(A_i^{\{S\}})$	Weighted score in points $D_i^{\{S\}}=Z_i^{\{S\}}\cdot A_i^{\{S\}}$	Rank $R_i^{\{S\}}=\dfrac{D_i^{\{S\}}}{\sum D_i^{\{S\}}}$	{W}	Importance $(Z_i^{\{W\}})$	Score $(A_i^{\{W\}})$	Weighted score in points $D_i^{\{W\}}=Z_i^{\{W\}}\cdot A_i^{\{W\}}$	Rank $R_i^{\{W\}}=\dfrac{D_i^{\{W\}}}{\sum D_i^{\{W\}}}$
S_1	4	4	16	0,16	W_1	3	3	9	0,09
S_2	5	4	20	0,20	W_2	5	3	15	0,14
S_3	4	5	20	0,20	W_3	5	4	20	0,19
S_4	3	4	12	0,12	W_4	5	4	20	0,19
S_5	5	3	15	0,15	W_5	4	4	16	0,15
S_6	4	4	16	0,16	W_6	5	5	25	0,24
Sum:			99	1	Sum:			105	1
{O}	Importance $(Z_i^{\{O\}})$	Score $(A_i^{\{O\}})$	Weighted score in points $D_i^{\{O\}}=Z_i^{\{O\}}\cdot A_i^{\{O\}}$	Rank $R_i^{\{O\}}=\dfrac{D_i^{\{O\}}}{\sum D_i^{\{O\}}}$	{T}	Importance $(Z_i^{\{T\}})$	Score $(A_i^{\{T\}})$	Weighted score in points $D_i^{\{T\}}=Z_i^{\{T\}}\cdot A_i^{\{T\}}$	Rank $R_i^{\{T\}}=\dfrac{D_i^{\{T\}}}{\sum D_i^{\{T\}}}$
O_1	5	5	25	0,40	T_1	4	4	16	0,21
O_2	4	4	16	0,26	T_2	4	2	8	0,10
O_3	4	3	12	0,19	T_3	3	4	12	0,16
O_4	3	1	3	0,05	T_4	4	4	16	0,21
O_5	3	2	6	0,10	T_5	5	5	25	0,32
Sum:			62	1	Sum:			77	1

Fig. 2. Matrix of SWOT analysis of weighted score estimate (the ranking)

At the same time, expert estimates can be coordinated by any of the known methods, for example, by calculating the concordance coefficient or using the method proposed in [15].

Also, ranking elements of SWOT analysis allows you to determine the weight of each component and select the most significant ones.

Then we calculate the potential (P) of strengths and weaknesses, as well as the level of fulfillment indicators (F) of opportunities and threats (Fig. 3).

	Rank	{O}					{T}					Potential
		$R_1^{\{O\}}$	$R_2^{\{O\}}$	$R_3^{\{O\}}$	$R_4^{\{O\}}$	$R_5^{\{O\}}$	$R_1^{\{T\}}$	$R_2^{\{T\}}$	$R_3^{\{T\}}$	$R_4^{\{T\}}$	$R_5^{\{T\}}$	P
{S}	$R_1^{\{S\}}$	0,0652	0,0417	0,0313	0,0078	0,0156	0,0336	0,0168	0,0252	0,0336	0,0525	0,14
	$R_2^{\{S\}}$	0,0815	0,0521	0,0391	0,0098	0,0196	0,042	0,021	0,0315	0,042	0,0656	0,16
	$R_3^{\{S\}}$	0,0815	0,0521	0,0391	0,0098	0,0196	0,042	0,021	0,0315	0,042	0,0656	0,16
	$R_4^{\{S\}}$	0,0489	0,0313	0,0235	0,0059	0,0117	0,0252	0,0126	0,0189	0,0252	0,0394	0,11
	$R_5^{\{S\}}$	0,0611	0,0391	0,0293	0,0073	0,0147	0,0315	0,0157	0,0236	0,0315	0,0492	0,13
	$R_6^{\{S\}}$	0,0652	0,0417	0,0313	0,0078	0,0156	0,0336	0,0168	0,0252	0,0336	0,0525	0,14
{W}	$R_1^{\{W\}}$	0,0346	0,0221	0,0166	0,0041	0,0083	0,0178	0,0089	0,0134	0,0178	0,0278	0,08
	$R_2^{\{W\}}$	0,0576	0,0369	0,0276	0,0069	0,0138	0,0297	0,0148	0,0223	0,0297	0,0464	0,12
	$R_3^{\{W\}}$	0,0768	0,0492	0,0369	0,0092	0,0184	0,0396	0,0198	0,0297	0,0396	0,0618	0,15
	$R_4^{\{W\}}$	0,0768	0,0492	0,0369	0,0092	0,0184	0,0396	0,0198	0,0297	0,0396	0,0618	0,15
	$R_5^{\{W\}}$	0,0614	0,0393	0,0295	0,0074	0,0147	0,0317	0,0158	0,0237	0,0317	0,0495	0,13
	$R_6^{\{W\}}$	0,096	0,0614	0,0461	0,0115	0,023	0,0495	0,0247	0,0371	0,0495	0,0773	0,18
Fulfillment	F	0,24	0,19	0,16	0,05	0,09	0,16	0,09	0,13	0,16	0,22	

Fig. 3. Statistical evaluation of indicators of SWOT analysis of the CBDL system

Thus, the opportunity of O_1 and the threat of T_5 have the highest probability of fulfillment. The strengths of S_2 and S_3 have the maximum potential, while the weakest side is W_6. Based on the data in Fig. 3, the SWOT analysis indicators are grouping into negative and positive ones (Fig. 4).

	Level	
Indicator	*Negative factors*	*Positive factors*
Opportunities $R^{\{O\}} = \sum F^{\{O\}}$		0,72
Strengths $R^{\{S\}} = \sum P^{\{S\}}$		0,83
Weakness $R^{\{W\}} = \sum P^{\{W\}}$	0,82	
Threats $R^{\{T\}} = \sum F^{\{T\}}$	0,77	
Sum:	1,59	1,55

Fig. 4. Aggregated level of opportunities, threats, and internal environment factors

Figure 5 shows the McKinsey model, adapted to the specifics of the object of analysis. The estimation of the level of fulfillment of the opportunities of the CBDL system is performing on a scale from 0 to 4. The value of 0 indicates the complete lack of implementation of the capabilities of the CBDL system and the four points on their full realization.

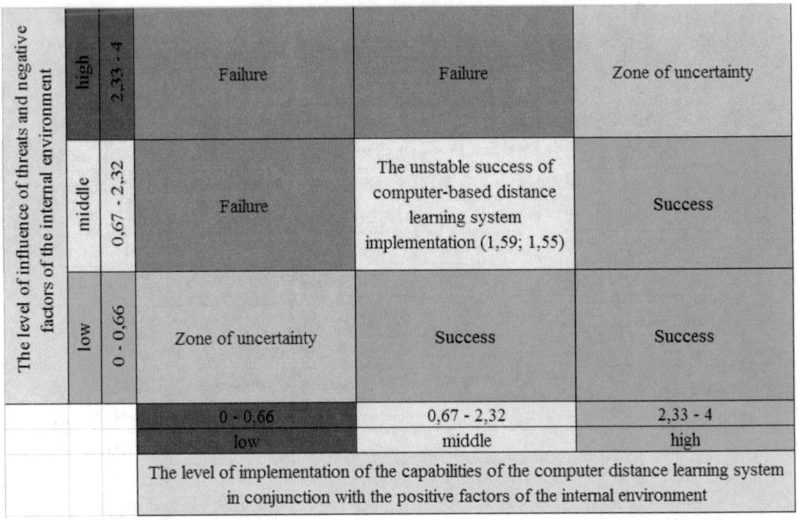

| | | | 0 - 0,66 | 0,67 - 2,32 | 2,33 - 4 |

Table (McKinsey model):

The level of influence of threats and negative factors of the internal environment	high 2,33 - 4	Failure	Failure	Zone of uncertainty
	middle 0,67 - 2,32	Failure	The unstable success of computer-based distance learning system implementation (1,59; 1,55)	Success
	low 0 - 0,66	Zone of uncertainty	Success	Success
		0 - 0,66 low	0,67 - 2,32 middle	2,33 - 4 high
		The level of implementation of the capabilities of the computer distance learning system in conjunction with the positive factors of the internal environment		

Fig. 5. McKinsey model in determining the success of implementing the capabilities of the CBDL system

As can be seen from Fig. 5, the implementation of the capabilities of the CBDL system under the influence of threats and internal environment factors has unstable success. This means that educational organizations and of the government of the Tver region need to pay more attention to creating conditions suitable for the introduction and development of the CBDL system.

Calculation using formulas (3)–(6) for $\lambda = 0,5$ gives the following results: $P_{opt}^{\{S\}} = 0,135$, $F_{opt}^{\{O\}} = 0,145$, $P_{opt}^{\{W\}} = 0,13$, and $F_{opt}^{\{T\}} = 0,155$. These results confirm the conclusion made above: at present, the Tver region cannot be considered fully ready for the comprehensive implementation of the CBDL system.

5   Conclusions

SWOT analysis is a model for forming an information base and developing on its basis the most effective option for managing a computer-based distance learning system.

As shown by the extended SWOT analysis of the modern Russian CBDL system, the technologies that provide this training require significant improvement. Currently, there are no mechanisms for protecting e-courses, obtaining authorship, the low payment for developing e-courses, so teachers are poorly motivated to create new and improve existing courses.

The results of the SWOT analysis showed that to improve the quality of the CBDL system in Russia as a whole, and in the Tver region, in particular, it is necessary to permanently monitor and implement such measures as:

- increase in funding and clarity of control over its distribution;
- development of information infrastructure;
- improving the quality of the legal framework;
- application of modern distance learning technologies;
- professional development of teachers;
- increasing the number of digitized textbooks by Russian authors;
- improving the quality of electronic educational and methodical complexes of disciplines.

The results of SWOT analysis can be used not only to develop optimizing measures but also to build predictive models of the development of a distance learning system.

References

1. Belenkova, I.V.: Distance education technologies as one of the main types of modern educational technology. In: The Collection: Innovations in Professional and Professional-Pedagogical Education Materials of the 20th All-Russian Scientific and Practical Conference, pp. 215–218 (2015)
2. Fialko, A.I., Senan, A.M.: Distance educational technologies as a means of organizing the activities of students with disabilities in the educational field "Technology". Tech. Econ. Educ. **12**, 44–46 (2019)
3. Kozilova, L.V.: Distance education technologies as a trend of the modern electronic educational environment of the University: from the experience of the Department. In: The Collection: Results of Modern Research and Development. Collection of Articles of the IX All-Russian Scientific and Practical Conference, pp. 105–107 (2020)
4. Sidnenko, T.I.: Distance education technologies as a means of network interaction of educational institutions (analysis of scientific approaches). Prob. Econ. Manage. Trade Ind. **2** (2), 114–120 (2013)
5. Illarionov, S.V., Illarionova, L.P.: The Pros and cons of training with the application of distance educational technologies. Center Innov. Technol. Soc. Expert. **2**(2), 27–32 (2015)
6. Popova, V.V., Tyukavina, I.A.: Problems of the introduction of remote technologies and their use in the educational process of higher education institutions. In: The Collection: Socio-economic, Political, and Historical Aspects of the Development of the Northern and Arctic Regions of Russia. Proceedings of the All-Russian Scientific Conference (with International Participation), pp. 167 – 170 (2018)
7. Orlova, E.R., Koshkina, E.N.: Problems of distance learning development in Russia. Natl. Interests Priorities Secur. **9**(23), 12–20 (2013)
8. Koshkina, E.N.: SWOT analysis of the use of distance learning in Russia. Bull. Int. Instit. Econ. Law **4**(13), 27–31 (2013)
9. Tong, J., Wen, H., Fan, X., Kummer, S.: Designing and decision making of transport chains between china and Germany. Int. J. Intell. Syst. Appl. **2**, 1–9 (2010). https://doi.org/10.5815/ijisa.2010.02.01

10. Singh, J.: Study on challenges, opportunities and predictions in cloud computing. Int. J. Modern Educ. Comput. Sci. (IJMECS) **9**(3), 17–27 (2017). https://doi.org/10.5815/ijmecs.2017.03.03

11. Barichello, L.B., Garcia, D.M., Siewert, C.E., Westbrook, R.: SWOT analysis: it's time for a product recall. Long Range Plann. **30**(1), 46–52 (1997). https://doi.org/10.1016/S0024-6301 (96)00095-7

12. Kotler, F., Roland, B., Bickhoff, N.: Strategic Management by Kotler. Best Techniques and Methods, p. 143. Alpina publisher, Moscow (2012)

13. Pontryagin, L.S., Boltyansky, V.G., Gamkrelidze, R.V., Mishchenko, E.F.: Mathematical Theory of Optimal Processes, p. 391. Nauka, Moscow (1961)

14. Baklushina, I.V., Koreshkova, P.S.: Application of SWOT analysis to develop strategies for the development of distance education technologies in the conditions of self-isolation. Trends Develop. Sci. Educ. **61–6**, 43–46 (2020)

15. Mutovkina, N.Yu.: Methods of Coordinated Optimization of Technical Re-equipment of Industrial Enterprises, p. 219. TvSTU, Tver (2009)

Research on Student Achievement Prediction Based on BP Neural Network Method

Deng Pan[1]([⊠]), Shuanghong Wang[1], Congyuan Jin[2,3], Han Yu[2,3], Chunbo Hu[2,3], and ChunZhi Wang[1]

[1] School of Computer Science, Hubei University of Technology, Wuhan 430068, China
475388483@qq.com, 1394756525@qq.com, chunzhiwang@vip.163.com
[2] Wuhan Fiberhome Technical Services Co., Ltd., Wuhan, China
[3] Fiberhome Telecommunication Technologies Co., Ltd., Wuhan, China

Abstract. Accurate analysis and prediction of student achievement is of great significance for improving the quality of teaching. By taking the student's scores as the experimental data, this paper proposes a BP neural network based student learning performance prediction method, which adopts four gradients of SGD, Momentum, AdaGrad and Adam. By comparing with the traditional machine learning algorithm of random forest, the effectiveness of proposed method is verified, and the prediction of students' learning performance and the early warning of abnormal situation of the students' learning performances are finally implemented, which helps teachers to make more efforts to improve worse situation of learning performance.

Keywords: Performance forecast · BP neural network · Random forest · Early waning of failure

1 Introduction

The goal of higher education is to improve the quality of teaching and create comprehensive talents who are benefit for the society. Student performance is one of important indicators for evaluating teaching quality [1]. How to effectively use students' scores data and analyze its valuable information are problems which are worth to attention. Due to many factors, the learning effect of students will be quite different [2]. If we can use the existing student achievement data to predict the student's future achievement based on predicted results, it is believed to help teachers make appropriate teaching strategies and provide rational suggestions for student development. Teaching quality is therefore improved significantly. Initially, performance prediction was mainly based on manual prediction methods. Students' scores as test data collected from daily teaching and examine results, the scores are estimated by teachers and researchers based on experience. This process is not easy and time-consuming somehow. There are available traditional achievement prediction methods based on machine learning [3], such as random forest [4], support vector machine [5], etc., but fitting effects of these method are not satisfied, it is hard to accurately predict learning

Z. Hu et al. (Eds.): AIMEE 2020, AISC 1315, pp. 293–302, 2021.
https://doi.org/10.1007/978-3-030-67133-4_27

achievement, thus, neural network-featured prediction method [6] with enhanced nonlinear modeling capability are rapidly applied. BP (Back Propagation) neural network [7] is a multi-layer feed forward network based on error back propagation, it is one of the most widely used neural networks so far [8].

A BP neural network model is proposed for nonlinear performance prediction in this paper, compared with a variety of gradient descent functions, the best prediction accuracy is achieved. In order to prove outstanding features of proposed model, traditional random forest method is compared in the article, the performance of proposed model has been proven to be the best in aspects of accuracy and efficiency.

2   Pre-processing of Student Achievement Data

To achieve research results close to real productive cases, 20 feature attributes are represented by Gs2, Xxds, Gll, Dxyyysj, Dxyydxy1, Dxyydxy2, Dxyw, Zgjdsgy, My Mg, Sx, Szlj, Dxwl_1, Dxwl_2, Wlsy_1, Wlsy_2, C_plus, Sjjg, Jsjwl, Sjk. The details are shown in Table 1.

Table 1. Course information

Course title	Attribute representation	Number of participating students
Higher Mathematics (1)-2	Gs2	411
Linear Algebra	Xxds	409
Probability Theory and Mathematical Statistics (1)	Gll	409
…		…
Computer Network	Jsjwl	411
Database Principles and the Application	Sjk	411

After data processing, a sample data of 401 are obtained. The relevant data of 401 samples are named as 15Grade_All data set. The specific form of the data set is shown in Table 2.

Table 2. Course information

Serial number	Gs2	Xxds	…	Sjk
1	38	23	…	77
2	78	63	…	87
3	65	81	…	67
4	69	70	…	56
5	82	89	…	87
6	56	60	…	90

According to the demand analysis of both BP neural network algorithm and random forest algorithm, the student performance is divided into five levels as shown in Table 3. The course name is expressed in letters and combined with the grade level, such as A-Excellent, which is represented in Table 4.

Table 3. Classification of student grade

Score segment	Grade >= 90	90 > Grade >= 80	80 > Grade >= 70	70 > Grade >= 600	Grade < 60
Level	Excellent	Good	Medium	Risk	Warn

Table 4. The letter for the Course Name

Course	Higher Mathematics (1)-2	Linear Algebra	...	Database Principles and the Application
Letter representation	A	B	...	F

3 Method of BP Neural Network

3.1 BP Neural Network

BP neural network is an information processing method with capability of independent learning to simulate the process of transmitting, storing and processing information by human neurons, it is a kind of multi-layered forward-feed neural networks trained by reverse propagation update weights according to the output error [9]. BP neural networks include input layers, hidden layers, and output layers. Its structure is shown in Fig. 1, the set $[x_1, x_2, ..., x_n]$ indicates the input data, the set $[W_{i1}, W_{i2}, ..., W_{in}]$ indicates the neural network connection weight, W_{i0} is the threshold, $f(e)$ is the activation function, and y_i is the output of the neural network.

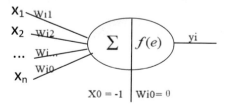

Fig. 1. Neuron model

3.2 BP Neural Network Model and Learning Steps

When a set of training samples are input into the network from the input layer, the activation value of the neuron begins to pass from input layer to hidden layer and then to the output layer. Each neuron gets the input response in the output layer [10]. As shown in Fig. 2, next, starting from the output layer, each connection weight of the network is adjusted in the direction of error reduction according to certain rules. With progress of back propagation step by step, the correct rate of network responding to input mode keeps increasing [11].

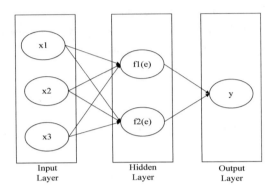

Fig. 2. BP neural network model

The diagram of the BP neural network training steps is shown in Fig. 3.

Step 1: Provide input data $x_1, x_2, \ldots, x_n$ to the network;

Step 2: Calculate the weighted sum product $e = \sum_{j=0}^{n} W_{ij}X_i$;

Step 3: Activate function output $Y_1 = f_1\left(W_{(x_1)_1 x_1} + W_{(x_2)_1 x_2} + W_{(x_3)_1 x_3}\right)$;

Step 4: Calculate the output $\theta = W_{(f_2)_1 \theta}$;

Step 5: Calculate the error between output value and true value. If the error is within ideal threshold, the training is ended, otherwise, the training continues to adjust the weight in the reverse direction;

Step 6: Adjust the weights $\theta = Z - y$;

until the output error reaches ideal error threshold;

Step 7: Re-select input and target functions from training set and continue the third step until the global error is less than the preset threshold, i.e. convergence, until the training for all training is over;

Step 8: Stop training.

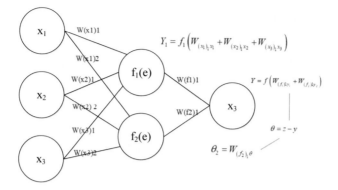

Fig. 3. BP neural network model

4   Method of Decision Tree and Random Forest

4.1   Decision Tree

Decision tree is the basic unit of random forest. Before introducing the random forest algorithm, a brief introduction is given to decision tree.

Decision tree algorithm is a classic classification algorithm. It first processes the original data and generates readable rules by induction. The rules are generally embodied in a tree structure, so it is called a decision tree. When it is necessary to classify new data, classification results are obtained by using the decision tree to analyze the new data, which has been widely used in big data mining algorithms.

The decision tree is a kind of directed acyclic trees. Each non-leaf node of the tree corresponds to an attribute in the training sample set, the branch of the non-leaf node corresponds to a numerical division of the attribute. Each leaf node represents a class, the path from the root node to the leaf is called a classification rule.

4.2   Random Forest

The random forest is proposed by Breiman et al., it consists of several decision trees. For classification questions, a test sample is sent to each decision tree to make predictions and then voting is conducted with the class of the most votes being the final classification result. The predictive output for the regression problem random forest is the mean of all decision tree outputs. For example, there are 10 decision trees in random forests, the prediction result of 8 lesson trees is Class 1, the prediction result of 1 decision tree is Class 2, and the prediction result of 2 decision trees is Class 3, the sample to category 1 will therefore be determined.

Random Forest is an integrated learning algorithm that integrates multiple decision trees to make predictions. For the classification problem, the prediction result is the vote of all the decision tree prediction results; for the regression problem, it is an average value of all the decision tree prediction results. During training process, Bootstrap sampling is used to form the training set of each decision tree. When training each node of a decision tree, the selected features are also part of the features extracted from the entire feature vector. By integrating multiple decision trees, training all decision trees with both obtained samples and feature components each time, the variance of the model can be effectively reduced. The algorithm flow is shown in Fig. 4.

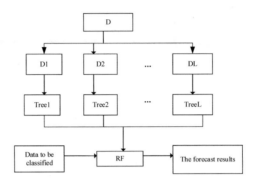

Fig. 4. BP neural network model

5   Experimental Results

5.1   Results Display

The purpose of BP neural network training continuously adjusts the weights to minimize the objective function value E, as shown in formula (1). y_r is the predicted score level (e.g. A-Excelent), and y is the actual score level. In this paper, four gradient descent methods are used to compare them, and finally the gradient descent method with the highest accuracy is determined.

The four gradient descent methods are shown below:

(1) The SGD gradient descent function with the maximum accuracy of 85.5%, as shown in Fig. 5 and Fig. 6.

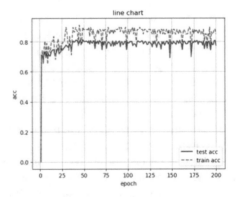

Fig. 5. SGD accuracy

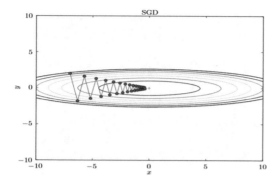

Fig. 6. The most optimized update path for SGD

(2) The Momentum gradient descent function with the maximum accuracy of 85.5%, as shown in Fig. 7 and Fig. 8.

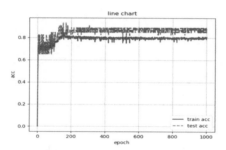

Fig. 7. Momentum accuracy

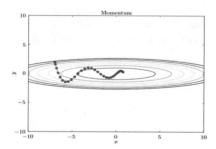

Fig. 8. The most optimized update path for Momentum

(3) The AdaGrad gradient descent function with the maximum accuracy of 85.5%, as shown in Figs. 9 and 10.

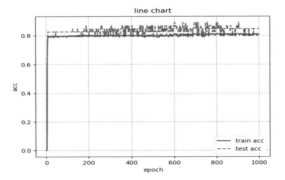

Fig. 9. AdaGrad accuracy

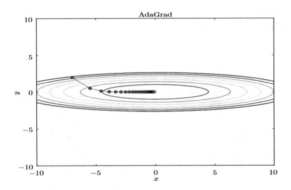

Fig. 10. The most optimized update path for AdaGrad

In order to verify the effectiveness of the method in this paper, the BP neural network is compared with the classical machine learning algorithm random forest. The random forest accuracy rate change chart is shown in Fig. 11, the highest prediction accuracy rate as a comparison result is shown in Table 5.

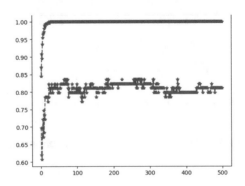

Fig. 11. Random forest accuracy

Table 5. Course information

	Accuracy rate
BP Neural Network	85.5%
Random Forest	83.4%

5.2 Analysis of Experimental Results

More than 400 records of the students of grade 15 in Department of Computer Science of Hubei University of Technology in the past four years have been divided into five categories, a study on multiple category classification has been introduced in this paper.

Different gradient descent function and BP neural network have been used to research the influence of different gradient search path in the results. Finally, SGD is the best method for analyzing experimental data, which has achieved the highest accuracy of 85.5%. At the same time, SGD is compared with classical machine learning algorithm of random forest, comparison result reveals the accuracy of random forest is 83.4%. In the specific experimental process, the actual students' scores have been calculated by the model, the difference value between forecast results and real results is less than absolute value of 5 points in general, which suggest the experimental results are satisfied.

6 Conclusions

As an innovation, regression model has been proposed to predict leaning scores. By using BP neural network, a variety of gradient descent methods are utilized to integrate with BP neural networks, the best prediction results are achieved finally. The experiment proves that the difference value between forecast results and real results is less than absolute value of 5 points, which suggests satisfied prediction result has been achieved. Comparing with traditional classification model, it can achieves more accurate prediction result of interval prediction, which helps make appropriate teaching strategies and therefore improve the quality of teaching in higher education. On the other hand, it makes possible to evaluate teaching quality promptly and to analyze from students' learning performance reflected by learning scores.

Although experimental results have achieved more accurate prediction result, there are still some further improvement work to carry on with:

1. For BP neural networks, the initialization of data has a great impact on the final results. The method of normalizing data for initialization was used in this study, different initialization methods could be utilized for trial to achieve better results with higher accuracy.

2. For the limitation of data amount for this experiment, some methods of data distribution have been tried to compare their distribution results, the best data distribution method is determined afterwards. A large volume of experimental data could be created by automatic data generating software which is believed more effective to determine the internal relationship among data, it can help achieve better results fast and effectively.

Acknowledgements. This work was supported by the National Natural Science Foundation of China (61772180); the Hubei Provincial Technology Innovation Project (2019AAA047).

References

1. Rifat, M.R.I., Al Imran, A., Badrudduza, A.S.M.: Educational performance analytics of undergraduate business students. Int. J. Modern Educ. Comput. Sci. (IJMECS) **11**(7), 44–53 (2019). https://doi.org/10.5815/ijmecs.2019.07.05
2. Yan, S.: Analysis of the causes of alienation of modern information technology and the possibility of its elimination. Modern Educ. Technol. **1**, 8–11 (2009)
3. Vijayalakshmi, V., Venkatachalapathy, K.: Comparison of predicting student's performance using machine learning algorithms. Int. J. Intell. Syst. Appl. (IJISA) **11**(12), 34–45 (2019). https://doi.org/10.5815/ijisa.2019.12.04
4. Song, Y.: Study on evaluation of students' achievement based on random forest. J. Qiqihar Univ. **11**, 1–4 (2017)
5. Djulovic, A., Li, D.: Towards freshman retention prediction: a comparative study . Int. J. Inf. Educ. Technol. **5**, 494–500 (2013)
6. Rumelhart, D., Hinton, G., Williams, R.: Learning representations by back-propagating errors. Nature **323**, 533–536 (1986). https://doi.org/10.1038/323533a0
7. Kotsiantis, S., Pierrakeas, C., Pintelas, P.: Predicting students' performance in distance learning using machine learning techniques . Appl. Artif. Intell. **5**, 411–426 (2004)
8. Mahedy, A.S., Abdelsalam, A.A., Mohamed, R.H., El-Nahry, I.F.: Utilizing neural networks for stocks prices prediction in stocks markets. Int. J. Inf. Technol. Comput. Sci. (IJITCS), **12** (3), 1–7 (2020). https://doi.org/10.5815/ijitcs.2020.03.01
9. Romero, C., Espejo, P.G., Zafra, A., et al.: Web usage mining for predicting final marks of students that use moodle courses . Comput. Appl. Eng. Educ. **1**, 135–146 (2013)
10. Ben-Hur, A., Siegelmann, H.T., et al.: Support vector clustering . J. Mach. Learn. Res. **2**(2), 125–137 (2002)
11. Noble, W.S.: What is a support vector machine. Nature Biotechnol. **24**(12), 1565–1567 (2006)

Post-quantum Digital Signature Scheme for Personal Data Security in Communication Network Systems

Maksim Iavich[1], Giorgi Iashvili[1], Razvan Bocu[2], and Sergiy Gnatyuk[3(✉)]

[1] Caucasus University, Tbilisi, Georgia
{miavich,giiashvili}@cu.edu.ge
[2] Transilvania University of Brasov, Brasov, Romania
razvan@bocu.ro
[3] National Aviation University, Kyiv, Ukraine
s.gnatyuk@nau.edu.ua

Abstract. The reliable collection, processing and storage of the sensitive personal data constitute an important research topic in the European context, especially after the General Data Protection Regulation started to produce its legal effects. The data encryption has been the traditional way of ensuring the sensitive data is legitimately accessed. Nevertheless, traditional encryption systems have become vulnerable to quantum computer-based attacks. This fostered the research efforts that look for encryption schemes that are immune to quantum computers-based attacks. This paper describes one of the few digital signature schemes, which is essentially immune to quantum computers-based attacks. The optimized version of the scheme is offered in the article. This scheme is much more efficient than the original one and has the stronger security guarantees. Furthermore, it may run efficiently on traditional computer systems. The suitability of the proposed digital signature scheme is assessed using real-world data, which demonstrates that it can be used to secure large amounts of personal sensitive data.

Keywords: Communication network systems · Digital signatures · Post-quantum · Cryptography · Quantum computers based attack · Hash · One-way functions

1 Introduction

Encryption of personal data is widely regarded as a privacy preserving technology which could potentially play a key role for the compliance of innovative IT. The big data stores, which are dedicated for the long-term maintenance of the sensitive personal data, cannot use traditional encryption schemes for the implementation of the data privacy considering two reasons: they are susceptible to be broken by quantum computers-based attacks, and they are not always run efficiently on various hardware architectures. This paper presents a novel encryption model that addresses all the mentioned requirements. Google Corporation, NASA and the Universities Space

© The Author(s), under exclusive license to Springer Nature Switzerland AG 2021
Z. Hu et al. (Eds.): AIMEE 2020, AISC 1315, pp. 303–314, 2021.
https://doi.org/10.1007/978-3-030-67133-4_28

Research Association (USRA) have teamed up with D-Wave, the global manufacturer of the quantum processors. D-Wave 2X is a quantum processor, which is based on a fabric of more than 1000 qubits and more the 3000 of couplers. The ability to attain.

this level of processing power is based on a CPU assembly that contains more than 128,000 Josephson Junctions, which are considered to be the most complex super-conductor incorporated circuits ever created. Current classic cryptographic schemes, which are massively used in practice, are vulnerable to the attacks of quantum computers. In 1994, Shor described a factorization algorithm, which would be able to break "RSA" and "Diffie-Hellman" by using quantum computers [1]. Nevertheless, all the sensitive information should stay safe in the age of quantum computers. This information should be encrypted in such a way that it will withstand the quantum computers-based attacks.

The scientific field of classical cryptography relates to various problems, such as encryption, the distribution of keys, the management of digital signatures, the pseu-dorandom number generators, and the one-way functions. For example, the RSA algorithm is safe only when factorization is deemed as a hard computational problem for classical computers. The assumptions regarding the computational tasks that are potentially hard to resolve are regarded as hypotheses. Therefore, finding such algo-rithms, which resist at the possible attack of a quantum computer represents a very difficult task, which involves a significant amount of research effort. Furthermore, traditional digital signature systems, which are used in a real world processes, are not safe against the attacks of quantum computers [2–4]. The level of security that they provide depends on the hardness of the discrete logarithmic calculations, and the related large numbers factorization. Scientists are working at the development of RSA alternatives, which might be protected from attacks that are run through quantum computers. Thus, the class of hash-based digital signature schemes is a relevant cate-gory. They use a cryptographically secure hash function. The level of security, which the respective digital signature systems provide, depends on the collision resistance of the hash features that they use.

The research aim of the paper is to offer the improved digital signature system, which will be more efficient than the previous offers and will be secure against attacks of quantum computer.

2   Literature Review

In [1] is offered the algorithms for quantum computation. The authors of [2–4] offer different types of quantum attacks. They adjust the algebraic structure of the cryp-tosystems, considering a quantum attacker limited only to classical queries and various quantum computations, which he can perform offline. The constructions of hash-based digital signatures is offered in [5, 6], the authors analyze the security and the efficiency of the schemes. In [7] the disadvantages of one time signature schemes are analyzed and the scheme of Merkle is offered. The author of [8] offers the technical report about the certified digital scheme and analyzes its efficiency. The authors of [9] analyze SHA2 hash functions and offer preimage attacks on up to 43-step of this functions. In [10] the fast quantum mechanical algorithm for database search is proposed. The

authors of [11–13] offer the cryptanalysis of hash functions and different hash algorithms. The authors of [14–16] analyze the hash function BLAKE, they.
analyze its efficiency and security, and they also offer its implementation. In [17,
18] the authors offer and analyze BLAKE 2 hash function. In 19 the authors offer the
optimized implementation of BLAKE using the CPU instructions, such as AVX,
AVX2, and XOP, which makes the hash function much faster. In [20] BLAKE3 is
offered, its security and efficiency is analyzed. The authors of [21–23] offer the different
efficient models of one-time functions and pseudorandom number generators. In [24–
26] the authors offer different improvements if post-quantum digital signature schemes
and analyze their security.

3   Hash-Based Digital Signature Schemes

3.1   The Lamport–Diffie One-Time Signature Scheme

The Lamport-Diffie one-time signature scheme is a hash-based digital signature scheme
[5]. The signature key X of the signature consists of 2n random lines, and each of them
has the length n.

$$X = (x_{n-1}[0], x_{n-1}[1], x_{n-2}[0], x_{n-2}[1], \ldots, x_0[0], x_0[1]) \in \{0, 1\}^{n,2n}. \tag{1}$$

The verification key Y of the signature consists of $2n$ lines, and each of them has
the length n.

$$Y = (y_{n-1}[0], y_{n-1}[1], y_{n-2}[0], y_{n-2}[1], \ldots, y_0[0], y_0[1]) \in \{0, 1\}^{n,2n} \tag{2}$$

The verification key is obtained by applying the one-way function to the corresponding signature key. Thus, the following relations stands:

$$y_i[j] = f_o(x_i[j]), 0 == n - 1, j = 0, 1,$$

while f_o is the one-way function $f_o : \{0, 1\}^n \rightarrow \{0, 1\}^n$.
It can be observed that the function must be applied 2n times in order to obtain the
verification key. Thus, the signing of a certain message m involves that it is transformed into the specific size n through the usage of the following hash function: h
(m) = hashed = (hashedn-1, hashedn-2, ..., hashed0), where h is the cryptographic
hash function that is defined as h: $\{0, 1\}^* \rightarrow \{0, 1\}^n$.
The message is signed considering the following function:

$$signature = (x_{n-1}[hashed_{n-1}], x_{n-2}[hashed_{n-2}], \ldots, x_0[hashed_0]) \in \{0, 1\}^{n,n} \tag{3}$$

The line with the rank j in the signature equals to xj[0], if the corresponding bit in
the signature equals to zero. The line equals to xj[1], if the corresponding bit in the
signature is equal to one. It can be noticed that the length of the signature is $n2$, and the

one-way function is not used during the computation of the signature. Furthermore, the signature verification phase involves that the message is hashed as **hashed** $=$ ($hashed_{n-1}, hashed_{n-2}, \dots, hashed_0$).

Moreover, the following equality must be verified:

$$(f_o(signature_{n-1}), f_o(signature_{n-2}), \dots, f(signature_0))$$
$$= (y_{n-1}[hashed_{n-1}], y_{n-1}[hashed_{n-2}], \dots, y_0[hashed_0]).$$

Considering that the equation is successfully verified, then the signature is deemed as valid. The one-way function must be used n times during the verification phase.

3.2   The Winternitz One-Time Signature Scheme

The Winternitz one-time signature scheme was proposed in order to decrease the size of the signature [6]. The signature size optimization is based on the principle to sign several bits of the hashed message simultaneously by using one line of the key. Consequently, this approach is suitable in order to decrease the size of the signature.

The signature key X of the signature consists of $2n$ random lines of the length n. The Winternitz parameter must be selected as $w \geq = 2$, and it must be equal to the number of bits, which must be signed simultaneously. Furthermore, two values must be calculated as $v_1 = n/w$ and $v_2 = (log_2 v_1 + 1 + w)/w$, with $v = v_1 + v_2$ The key X of the signature consists of v random lines, which have the length n, and it is defined by the following equation:

$$X = (x_{v-1}[0], x_{v-1}[1], x_{v-2}[0], x_{v-2}[1], \dots, x_0[0], x_0[1]) \in \{0, 1\}^{v,2n}.$$

The verification key Y of the signature is defined by $2n$ lines of the length v through the following equation:

$$Y = (y_{v-1}[0], y_{v-1}[1], y_{v-2}[0], y_{v-2}[1], \dots, y_0[0], y_0[1]) \in \{0, 1\}^{v,2n},$$

where $y_i = f_o^{2^w-1}(x_i)$, and $0 < = i < = v - 1$. It can be noticed that the length of the signature and the verification keys equals to np bits. The one-way function must be applied $v(2^w - 1)$ times in order to get the verification key. The message is signed and hashed using the function $h(m)$. The parameter w must divide the hashed value. Therefore, the needed number of zeros must be prepended to this value. During the ensuing process, it must be divided into $v1$ parts of length w. Considering that the function $h(m) = p_{v-1}, \dots, p_{v-v1}$, the checksum is calculated using the following relation: $c = \sum_{i=v-v1}^{v-1}(2^w - p_i)$. Considering that $c < = v_1 2^w$, the length of the binary representation is equal to $log_2 v_1 2^w + 1$. The minimum number of zeros must be prepended to the binary representation in order to get the length of the representation, which is divided by w. Consequently, it is divided into $p2$ parts of length w. The binary representation should be divided, so that the needed number of zeroes is prepended to it. Furthermore, it is divided into $v2$ pieces of the length w. Let us consider that $c = p_{v2-1}, \dots, v_0$. Thus, the message is signed considering the following equation:

$$signature = (f_o \wedge p_{v-1}(x_{v-1}), \ldots, f_o \wedge p_0(x_0)). \qquad (4)$$

It can be deduced that the one-way function is used $v(2^w - 1)$ times in the worst case. The signature size is vn. The calculation of the signature is conducted considering the relation $signature = (signature_{n-1}, \ldots, signature_0)p_{v-1}, \ldots, p_0$(the meaning of these red variables should be clearer must be calculated. Furthermore, the following equality must be verified:

$$(f_o \wedge (2^w - 1 - p_{v-1})) (signature_{n-1}), \ldots, (f_o \wedge (2^w - 1 - p_0))(signature_0) = y_{n-1}, \ldots y_0$$

Consequently, the one-way function must be used $v(2^w - 1)$ times during the verification stage.

3.3 Merkle Signature Scheme

The one-time signature cannot be used in practice, considering that for every signature the new key pair is needed. In 1979, Ralph Merkle described the Merkle signature scheme [7]. The principle of the scheme is to use a binary tree of hashes in order to reduce the validity of some specific numbers of the onetime verification keys to the validity of only one public key, which is the root of this tree. The size of the tree, which is specific to this scheme, has to be H > = 2. Thus, by means of one public key, 2^H documents can be signed securely, and 2^H pairs of signatures and verification keys must be generated, and X_i, Y_i, $0 < = i < = 2H$. The leaves of the tree are obtained through the consideration of the hashed keys, using the secure hash function:

$$h : \{0, 1\} * \rightarrow \{0, 1\}^n$$

The parent node is obtained through the concatenation and hashing of the previous pair of nodes. The public key is the root of the tree.

In Fig. 1, the binary tree with $H = 3$ is illustrated, and $a[i,j]$ are the nodes of the tree. The signing process of any message involves that it is transformed to the size n using the

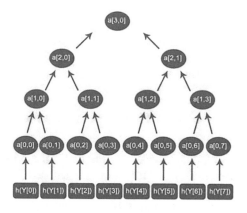

Fig. 1. The Merkle tree with $H = 3$

secure hash function. Thus, signing the message requires that one of the signature keys is chosen, so that the message is signed using one-time signatures keys. The final signature is a combination between the one-time signature, the one-time verification key, the index of the chosen key, and all the fraternal nodes according to the chosen key. The signature is verified by using the chosen one-time signature. If it is verified, the root of the tree must be calculated. The signature is verified completely, if the public key and the root coincide. It can be observed that the Merkle signature scheme uses the one-time hash-based digital signature scheme, and the hash function in Merkle is used $2^{H+1}-1$ times [8]. We offer to use the hash function, as the one-way function in one-time hash-based digital signature scheme, and to use the same hash function in Merkle.

4 Methodology

4.1 SHA-2 Hash Functions

SHA-2 is a family of cryptographic hash functions created by United States National Security Agency – NSA in 2001 [9]. These functions are constructed using Merkle–Damgård structure and the one-way compression function created by means of Davies–Meyer structure from the special block cipher. This family includes the following hash functions: SHA-224, SHA-256, SHA-384, SHA-512, SHA-512/224, SHA-512/256. These functions use the different amount of shifts and additive constants, but they have the identical structure, which differs in the number of rounds.

In the class of quantum algorithms, the Grover's algorithm can be used to perform the unstructured quantum search. This algorithm was developed by the American mathematician Lov Grover in 1996 [10]. It uses the property of quantum interference in order to solve the extremely important task of finding the value of a parameter on which a given function produces a certain result. This algorithm does not show an exponential gain for the task compared to the classical computational model, but it ensures that the process runs according to a quadratic speedup. Considering large values, this gain is significant. However, this is a general algorithm for solving a rather generalized problem, and it has been proved that it is impossible to achieve a better result in the framework of the quantum computing model. Therefore, the Grover's algorithm is a quantum algorithm, which can find out the concrete output value with a rather high probability to the input value hat is passed to the black box function by using the $O(\sqrt{N})$ evaluation of the function, where N is the function domain's size. It can be noticed that by using the Grover's algorithm, SHA-256 can be broken in 264 iterations, while SHA-512 can be broken in 2128 iterations. Therefore, SHA-256 is not secure against the attacks of quantum computers, and SHA-512 is secure against these types of attacks [11–13].

4.2 The Integration of the SHA-2 Hash Function into the Merkle Scheme

It must be emphasized that efficiency is a very important aspect in cryptography. Considering that the encryption scheme uses the hash function many times, we analyzed the efficiency of Merkle's signature using the integrated SHA-512 hash function, and we compared it to the implementation that is based on the usage of the SHA-256 hash function. The scheme is defined by the steps, which are enumerated in the following pseudo code.

4.3 Pseudocode of the Scheme

1. Importing necessary libs
2. Defining the function
3. Defining the string
4. Defining " alt_hashes(hashes) "
5. Set list " arr "
6. If hashes == " ", raise Exception
7. Generating the signature keys
8. Foreach loop
   8.1. Sorting hashes and appending into arr
9. Length_of_block == length of arr
10. While loop, if length is odd, copy last element in list
   10.1. Append it into arr list
11. Set list " another_arr "
12. Foreach loop
   12.1. For loop with range from 0 to length of " arr " and iteration by 2
      12.1.1. Define variable with " sha512()/sha256() " value
      12.1.2. Hash elements that are in " arr " list
      12.1.3. Append them into new " another_arr " list
      12.1.4. Return this list in hex
13. Create message and save it in " st " variable
14. Convert " st " value in binary
15. Generate " one-time signature "
   15.1. If st == 0
      15.1.1. Choose " First_signature_key " bits
   15.2. Else
      15.2.1. Choose "Second _signature_key " bits
16. First_pub_key = SHA2(hash_arr[0]) (Using sha512()/sha256())
17. Second_pub_key = SHA2 (hash_arr[1]) (Using sha512()/sha256())
18. Encryption
   18.1. Concatenate " one-time signature " with message's hash
19. Verification of " one-time signature "
   19.1. If bit of hashed " one-time signature " == 0
      19.1.1. Compare with " First_ key " bits
   19.2. Else
      19.2.1. Compare with "Second _ key " bits
20. Verification of " signature "
   20.1. Concatenate siblings with each other
   20.2. If this equals to public key
      20.2.1. Sign is correct
   20.3. Else
      20.3.1. Sign is not correct

The performance tests were performed on a machine that is powered by the processor i7-8565U CPU. Thus, the implementation that considers the SHA-256 hashing model produced a key generation time of 0.2522974 s, a signature time of 0.000409400

0000000042 s, and a verification time of 0.0025690000000000435 s. Furthermore, the implementation that considers the SHA-512 hashing model produced a key generation time of 0.3956368 s, a signature time of 0.0032782999999999562 s, and a verification time of 0.0103498999999999912 s. It can be noticed that the implementation, which considers the SHA-512 model is significantly less efficient than the implementation that is based on SHA-256. The efficiency factor is very important in cryptography. Nevertheless, considering that SHA-256 is not secure against the attacks of quantum computers, we propose a model that improves the Merkle scheme by considering a more efficient and secure hash function.

5 The BLAKE Hash Function

The BLAKE function is a cryptographic hash function, which is based on Dan Bernstein's ChaCha stream cipher. Thus, its input block is permuted, it is XORed with the round constants and added before every ChaCha round. There exist different variants in this general category of hashing functions: BLAKE-256, BLAKE-224, BLAKE-512 and BLAKE-384 [14–16]. In 2012, BLAKE2 was designed, which is based on the original version of the BLAKE function.

5.1 The BLAKE2 Hash Function

The BLAKE2 is a family of cryptographic hash functions designed in 2012 by Jean-Philippe Aumasson, Samuel Neves, Christian Winnerlein and Zooko Wilcox-O'Hearn. This cryptographic scheme is considered to be rather efficient, as it is faster than MD5, SHA-1, SHA-2, and SHA-3. It is also considered rather secure, as its security is based on the immunity to length extension, and the undifferentiability from a random oracle [17]. The BLAKE2 family includes two main functions: BLAKE2b and BLAKE2s. Thus, BLAKE2s is well optimized for the platforms of 8-bit and 32-bit, and it outputs hash values between 1 and 32 bytes long. Furthermore, BLAKE2b is well optimized for 64-bit platforms, which include NEON-enabled ARMs. It outputs the hash values of the size between 1 and 64 bytes. It is significant to note that both variants can work on any CPU, but when they are optimized, they produce up to 100% speed up [18, 19]. The BLAKE2b makes 12 rounds, and BLAKE2s makes 10 rounds, while in the original BLAKE 16 and 14 rounds were made, respectively. The work that is reported in this paper proves that the reduction in the number of the rounds does not affect the security of the respective cryptographic scheme. This reduction produces a 25% and 29% result optimization on the big inputs, respectively. Furthermore, the BLAKE-512 performs the rotations of four 64-bit words of 32, 25, 16, and 11 bits, while BLAKE2b replaces the number 25 with 24, and the number 11 with 63. Consequently, using the 24-bit rotation it allows SSSE3-capable processors to perform 2 rotations with only one SIMD instruction through an in-place bytes shifting operation that is called packed shuffle bytes (pshufb). In the case of new processors, the arithmetic cost of the involved processing is reduced by 12%. Additionally, the 63-bit rotation can also be translated through the addition and shift operations, which are followed by the application of the logical OR operations. This approach generates a speed up on the platforms where the addition and shift operations can be implemented in parallel. In 2020, the BLAKE3 hash function was designed as an evolution of the BLAKE2 scheme [20].

5.2 The BLAKE3 Hash Function

The BLAKE3 scheme is significantly more efficient than SHA-1, SHA-2, SHA-3, and BLAKE2. In the context of BLAKE3, the number of rounds is reduced from 10 to 7. Thus, most block ciphers specify a round, which includes the building blocks that are assembled together in order to design a cryptographic function, which runs multiple times. The BLAKE3 scheme considers a binary tree structure, in order to enable the use of parallelism, which is provided by SIMD instructions that are included in most of the modern central processing units. The BLAKE3 is a 128-bit security mechanism that can be used to prevent preimage, collision, or differentiability attacks. The 128-bit security level is enough for modern cryptography, but it cannot be used in the post-quantum epoch. Considering the analysis that has been performed, it can be stated that BLAKE2b is the optimal hash function. Consequently, we propose to use BLAKE2b as the hash function of the Merkle signature scheme, and as the one-way function in one-time signatures schemes. Accelerating one-way functions and pseudo random number generators [21–23] greatly improve the efficiency of the scheme.

6 The Improved Digital Signature

The authentication of 2^H documents involves that the signing entity selects the height of the tree, with $H \in N$ and $H \geq 2$, and generates 2^H pairs of one-time keys. The calculation of the leaves involves that the signing entity hashes the concatenation of the left and right children through the usage of the BLAKE2b hash function. The message is transformed into one of the size n using the BLAKE2b hash function. Thus, signing the message requires that one of the signature keys is chosen, so that the message is signed using one-time signatures keys, which use BLAKE2b as the one-way function. The final signature is a combination between the one-time signature, the one-time verification key, the index of the chosen key, and all the fraternal nodes according to the chosen key. The signature is verified by using the chosen one-time signature. If it is verified, the root of the tree must be calculated. The signature is verified completely, if the public key and the root coincide.

7 Experiment Results and Discussion

The signature scheme was implemented using Python programming language according to the pseudo code that is described in Sect. 3.1, and BLAKE2b was integrated into the implementation. The scheme performance was assessed on the same machine with an i7-8565U CPU. Thus, the key generation time is 0.1879717 s, the signature time is 0.0002919000000000006 s, and the verification time is 0.001286700000000008 s. The results show that in the case of the integration of BLAKE2b into the scheme, the system works even faster than in the case of SHA-256.

The speed result of the scheme are rather good so the scheme can be used on traditional computer systems, even before the massive release of quantum computers. It must be emphasized that the signature of the scheme is still much bigger than in the case of RSA.

7.1   Remarks Regarding the Security of the Proposed Scheme

The inherent security that is provided by the usage of the Merkle scheme is valued by the cryptographic model that is proposed in this paper. Moreover, its efficiency is increased as the improved implementation of the BLAKE2b model is integrated into it, and also into the one-time signature scheme. The scheme of Merkle is almost the same, but the improved implementation of BLAKE2b is integrated into it and into one-time signature scheme. The proposed model has the advantage that it preserves the advantages of the Merkle scheme, while enhancing the security and efficiency by integrating the BLAKE2b encryption core, with a key size of 256-bit. It is demonstrated that BLAKE2b is safe against the attacks of quantum computers. Thus, the proposed model fully preserves the safety of the respective data, while the computational and runtime efficiency is not affected in the case of real world applications. Consequently, the improved signature scheme is more efficient, and it is safe against the attacks of quantum computers [24–26].

8   Conclusions and Future Plans

This paper demonstrates that a secure, quantum-resistant encryption model can be designed and implemented. It offers the elegant integration BLAKE2b hash function into Merkle signature scheme. Furthermore the paper offers the scientific model and the practical implementation of the improved digital signature scheme, which is proved to be secure against the attacks of quantum computers. Furthermore, the proposed model runs faster than most of the existing similar encryption models. The practical performance assessment results completely support these assertions. In future it would be very interesting to enhance the security of BLAKE3 to 256-bit and integrate it into the proposed model. Furthermore, the cryptographic routines will be optimized, so that the data is processed efficiently.

Acknowledgement. The work was conducted as a part of PHDF-19–519 financed by Shota Rustaveli National Science Foundation of Georgia and the grant financed by Caucasus University.

References

1. Shor, P.W.: Algorithms for quantum computation: discrete logarithms and factoring. In: Proceedings 35th Annual Symposium on Foundations of Computer Science, pp. 124–134 (1994)

2. Ambainis, A., Rosmanis, A., Unruh, D.: Quantum attacks on classical proof systems: the hardness of quantum rewinding. In: 2014 IEEE 55th Annual Symposium on Foundations of Computer Science, pp. 474–483 (2014)
3. Bonnetain, X., Hosoyamada, A., Naya-Plasencia, M., Sasaki, Y., Schrottenloher, A.: Quantum attacks without superposition queries: the offline simon's algorithm. In: Galbraith, S., Moriai, S. (eds.) Advances in Cryptology ASIACRYPT 2019. Lecture Notes in Computer Science, vol 11921. Springer, Cham (2019)
4. Gagnidze, A., Iavich, M., Iashvili, G.: Analysis of post quantum cryptography use in practice. Bull. Georgian Natl. Acad. Sci. 11(2), 29–36 (2017)
5. Ajtai, M.: Generating hard instances of lattice problems. In: Complexity of computations and proofs, volume 13 of Quad. Mat., pp. 1–32. Dept. Math., Seconda Univ. Napoli, Caserta (2004). Preliminary version in STOC 1996. 8. Babai, L.: On Lovász lattice reduction and the nearest lattice point problem. Combinatorica, 6:1*13 (1986)
6. Buchmann, J., Dahmen, E., Ereth, S., Hülsing, A., Rückert, M.: On the security of the winternitz one-time signature scheme. In: Nitaj, A., Pointcheval, D. (eds.) Progress in Cryptology – AFRICACRYPT 2011. Lecture Notes in Computer Science, vol 6737. Springer, Heidelberg (2011)
7. Buchmann, J., Dahmen, E., Klintsevich, E., Okeya, K., Vuillaume, C.: Merkle signatures with virtually unlimited signature capacity. In: Katz, J., Yung, M. (eds.) Applied Cryptography and Network Security. ACNS 2007. Lecture Notes in Computer Science, vol 4521. Springer, Heidelberg (2007)
8. Merkle, R.: Secrecy, authentication and public key systems/A certified digital signature Ph. D. dissertation. Department of Electrical Engineering, Stanford University (1979)
9. Aoki, K., Guo, J., Matusiewicz, K., Sasaki, Y., Wang, L.: Preimages for step-reduced SHA-2. In: Matsui, M. (eds.) Advances in Cryptology – ASIACRYPT 2009. ASIACRYPT 2009. Lecture Notes in Computer Science, vol 5912. Springer, Heidelberg (2009)
10. Grover, K.: A fast quantum mechanical algorithm for database search. In: Proceedings of the twenty-eighth annual ACM symposium on Theory of Computing, pp. 212–219 (1996)
11. Fowler, A.G., Devitt, S.J., Jones, C.: Surface code implementation of block code state distillation. Scientific Reports 3, 1939 EP (2013). https://doi.org/10.1038/srep.01939
12. Kim, P., Han, D., Jeong, K.C.: Time–space complexity of quantum search algorithms in symmetric cryptanalysis: applying to AES and SHA-2. Quant. Inf. Process. 17, 339 (2018). https://doi.org/10.1007/s11128-018-2107-3
13. Debnath, S., Chattopadhyay, A., Dutta, S.: Brief review on journey of secured hash algorithms. In: 2017 4th International Conference on Opto-Electronics and Applied Optics (Optronix), Kolkata, pp. 1–5 (2017). https://doi.org/10.1109/OPTRONIX.2017.8349971
14. Aumasson, J.P., Henzen, L., Meier, W., Phan, R.C.W.: SHA-3 proposal BLAKE. Submission to NIST (Round 3) (2010)
15. Sklavos, N., Kitsos, P.: BLAKE hash function family on FPGA: from the fastest to the smallest. In: 2010 IEEE Computer Society Annual Symposium on VLSI, Lixouri, Kefalonia, pp. 139–142 (2010). https://doi.org/10.1109/ISVLSI.2010.115
16. Kahri, F., Bouallegue, B., Machhout, M., Tourki, R.: An FPGA implementation of the SHA-3: the BLAKE hash function. In: 10th International Multi-Conferences on Systems, Signals & Devices 2013 (SSD13), Hammamet, pp. 1–5 (2013). https://doi.org/10.1109/SSD.2013.6564030
17. O'Whielacronx, Z.: Introducing BLAKE2 – an alternative to SHA-3, SHA-2 and MD5 (2012)

18. Aumasson, J.P., Neves, S., Wilcox-O'Hearn, Z., Winnerlein, C.: BLAKE2: simpler, smaller, fast as MD5. In: Jacobson, M., Locasto, M., Mohassel, P., Safavi-Naini, R. (eds.) Applied Cryptography and Network Security. ACNS 2013. Lecture Notes in Computer Science, vol 7954. Springer, Heidelberg (2013)
19. Neves, S., Aumasson, J.P.: Implementing BLAKE with AVX, AVX2, and XOP. Cryptology ePrint Archive, Report 2012/275 (2012)
20. O'Connor, J., Aumasson, J., Neves, S., Wilcox-O'Hearn, Z.: BLAKE3 one function, fast everywhere (2020). https://blake3.io
21. Belfedhal, A.E., Faraoun, K.M.: Fast and efficient design of a PCA-based hash function. Int. J. Comput. Netw. Inf. Secur. (IJCNIS) 7(6) (2015). ISSN: 2074–9090 (Print), ISSN: 2074–9104
22. Al-Hammadi, Y.A., Fadl, M.F.I.: Reducing hash function complexity. IJ Math. Sci. Comput. 1, 1–17 (2019). https://www.mecs-press.net, https://doi.org/10.5815/ijmsc.2019.01.01
23. Hu, Z., Gnatyuk, S., Okhrimenko, T., Tynymbayev, S., Iavich, M.: High-speed and secure PRNG for cryptographic applications. Int. J. Comput. Netw. Inf. Secur. (IJCNIS) 12(3) (2020). ISSN: 2074–9090 (Print), ISSN: 2074–9104
24. Gnatyuk, S., Kinzeryavyy, V., Iavich, M., et al.: High-performance reliable block encryption algorithms secured against linear and differential cryptanalytic attacks. In: CEUR Workshop Proceedings, vol. 2104, pp. 657-668 (2018)
25. Gnatyuk, S., Kinzeryavyy, V., Kyrychenko, K., et al.: Secure hash function constructing for future communication systems and networks. Adv. Intell. Syst. Comput. 902, 561–569 (2020)
26. Iavich, M., Gagnidze, A., Iashvili, G. et al.: Lattice based Merkle. In: CEUR Workshop Proceedings, vol. 2470, pp. 13-16 (2019)

A Motor Fault Detection Method Based on Optimized Extreme Learning Machine

Siwei Wei[1], Ning Ma[2], Jun Su[3($\boxtimes$)], and Wenhui Deng[1]

[1] CCCC Second Highway Consultants Co., Ltd., Wuhan, China
waosfengw@hotmail.com, dengwenhui0406@163.com
[2] Wuhan Electronic Information Institute, Wuhan 430019, China
[3] School of Computer Science, Hubei University of Technology, Wuhan, China

Abstract. To solve the problems that the motor fault detection algorithm based on mathematical model is hard to be adapted to those categories with complex non-linear fault. The accuracy of existing fault detection algorithms is not satisfied. In this paper, two methods of feature selection and parameter optimization of simulated annealing-based whale optimization algorithm with (SAWOA) optimized extreme learning machine (ELM) are proposed. SAWOA is utilized to optimize the selection of network input variables and to determine parameters of hidden layer nodes of ELM. In order to verify proposed method, a real motor fault detection case is studied in the paper. Several methods are compared under the same conditions to find out the best method with satisfied accuracy and reliability. The experiment results show that proposed method is capable to improve the classification accuracy and reliability, effectiveness and application value are introduced in the article.

Keywords: Motor fault detection · Whale optimization algorithm · Simulated annealing · Extreme learning machine

1 Introduction

Motor is one of the most important devices used to drive various mechanical and industrial equipment, it is one of important parts in industrial and national economy [1]. During the operation of motors, faults could possibly occur at any time. The causes of failures are also various, such as long-time or overload operation of the motor, insulation aging and environmental could also be major factors [2]. The failure of a certain part of the motor often causes the entire system to malfunction, which might cause huge economic losses. Therefore, timely detection of potential faults and accurate diagnosis of existing faults are important tasks to achieve the reliability of electrical equipment.

Nowadays, with the application and development of machine learning [3, 4, 15], some fault detection models are featured with neural networks and fuzzy systems [5]. The principle of these models considers fault features and their labels to establish corresponding classifiers for fault detection [6–8]. However, for large-scale fault detection, there are some drawbacks, including long training time cost and low accuracy [9]. Extreme learning machine is one of neural network featured methods [10], its

© The Author(s), under exclusive license to Springer Nature Switzerland AG 2021
Z. Hu et al. (Eds.): AIMEE 2020, AISC 1315, pp. 315–324, 2021.
https://doi.org/10.1007/978-3-030-67133-4_29

characteristics include fast learning speed, strong non-linear classification ability, it can achieve better results than traditional neural network methods [11].

The whale optimization algorithm is a heuristic search optimization algorithm with good search capability [12]. The simulated annealing algorithm is capable to jump out of the local optimal solution to achieve global optimal solution [13]. Thus, a whale optimization algorithm based on simulated annealing is proposed in this paper to achieve better global search ability with few parameter setting.

In the rest of this paper, SAWOA and ELM as two theoretical foundations of this paper are introduced in Sect. 2. A method of feature selection and parameter optimization of SAWOA optimized ELM is presented in Sect. 3. An experiment is implemented to compare proposed method with several state-of-art methods in Sect. 4, the experiment demonstrates the effectiveness and efficiency of proposed method. Finally, a conclusion is given in Sect. 5.

2 Literature Review

2.1 Simulated Annealing Based Whale Optimization Algorithm

The whale optimization algorithm is one of heuristic search optimization algorithms. Its advantages are that it is easy to implement, high accuracy and fast convergence. Motivated by the foraging behavior of humpback whales in nature, the algorithm simulates the contraction envelopment, spiral position update and random hunting mechanism of the humpback whale population. However, the whale optimization algorithm in theory is not capable to ensure that the current convergence solution is optimal solution. Therefore, when the whale is on an iterative search process, the above convergence characteristics are easy to premature convergence and to fall into local optimum rather than system excellence.

Simulated annealing algorithm is usually used to solve the optimization problem in the simulation of the annealing process in the thermodynamic system [13]. The temperature is continuously cooled from a given initial temperature, and a certain probability is accepted in the annealing process to be worse than the current solution. The solution adopts the algorithmic jump to achieve optimal global solutions.

At the same time, the SA algorithm is proved to be able to converge to the global optimization with probability 1 under the condition that T_0 is high enough and T is slow enough. This mechanism is used to avoid the search trap of the whale optimization algorithm. Locally optimal and precocious. Therefore, the simulated annealing method can be used as the convergence basis of the WOA algorithm. When the WOA converges to a specifically determined solution W_g during the optimization process, the solution is used as the initial value of SA and is therefore optimized again [10].

Because of the advantages of WOA and SA optimization algorithms, the organic part of two algorithms are effectively combined to use in experiments, a hybrid simulated annealing whale optimization algorithm which can solve motor fault detection problem is designed.

2.2 Extreme Learning Machine

The advantages of Extreme learning machine (ELM) [14] include fast learning speed and good generalization. For a single hidden layer of neural network, if there are n arbitrary samples (X_i, Y_i), where $X_i = [x_{i1}, x_{i2}, \cdots, x_{in}]^T \in R^n$, $Y_i = [y_{i1}, y_{i2}, \cdots, y_{im}]^T \in R^m$, a single hidden layer feed forward neural network with k hidden layer nodes can be expressed as $F_k(x) = \sum_{i=1}^{k} \beta_i G(A_i \cdot X_j + B_i), j = 1, \cdots, n$:

Where: $A_i = [a_{i1}, a_{i2}, \cdots a_{ik}]^T$ denoted the input weight, β_i denotes the output weight of the i-th hidden layer node, B_i denotes the threshold of the i-th hidden layer node, $A_i \cdot X_j$ denotes the inner product of the vectors A_i and X_j, while $G(x)$ denotes the activation function.

According to ELM theory, ELM aims to simultaneously minimize the training errors and the norm of output weights. This objective function of ELM is as follows:

$$\min \frac{1}{2} \|\beta\|_F^2 + \frac{C}{2} \sum_{i=1}^{n} \|\xi_i\|^2$$
$$\beta^T G(X_i) = y_i - \xi_i \tag{1}$$

Where $\xi \in R^{T \times n}$ denotes the training error matrix on training data, β denotes the output weights.

The optimization problem in Eq. (1) can be efficiently solved. The optimal $\beta*$ which minimizes Eq. (1) can be analytically obtained as

$$\beta^* = G^T \left(\frac{I}{C} + GG^T \right)^{-1} Y^T \tag{2}$$

Where $G = [G(x_1), \cdots, G(x_n)]^T$, $Y = [y_1, \cdots, y_n]$ and I denotes an identity matrix.

Input weights and hidden layer deviations of traditional ELM algorithm are randomly assigned, there is no guarantee that these parameters are optimal. To this end, this paper optimizes input weight and hidden layer bias by using whale optimization algorithm with strong searching function, it adopts feature selection optimization to establish an optimal network intrusion detection model.

3 Methodology

Proposed method is a kind of feature selection and parameter optimization of SAWOA integrated optimization ELM (hereinafter referred to as SAWOA-FP-ELM), and feature selection, weight and bias of the hidden layer are coded as to be optimized.

Variables are used to optimize the extreme learning machine by SAWOA algorithm, which help find a set of better motor data features and ELM optimal hidden layer parameters, an optimal motor fault detection classifier is determined finally.

3.1 Encoding and Decoding of Whale Individuals

In SAWOA-FP-ELM algorithm, whale individuals adopt real number coding. In order to realize parameter optimization while performing feature selection, individual coding includes two parts, namely feature coding and parameter coding. Let P_t be the whale individual of the t generation, which can be expressed as:

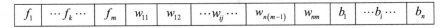

Fig. 1. Coding of feature selection and parameter of whale individuals

Where f_k determines whether the k-th feature is retained or discarded, w_{ij} denotes the connection weight of the i-th hidden layer node and the j-th dimension feature of the ELM, and b_i denotes the offset of the i-th hidden layer node of the ELM. The binary variables $f_k \in \{0, 1\}$, w_{ij}, b_i are all random numbers in $[0, 1]$, m represents dimension of features, and n represents the number of hidden layer nodes of the extreme learning machine.

The optimization of individuals in the population, that is, the optimization of decision variables in Fig. 1, involves real numbers and binary variables. To optimize the algorithm for application, it needs to map the decision variables to consecutive real variables in the $\{-1, 1\}$ interval. Secondly, It is also necessary to convert the values of all variables to their true values in Eq. 5.

If the real value of the l th decision variable P_{tl} of the individual P_t is a real number, then it is transformed into

$$X_{tl} = \left(X_t^{\max} - X_t^{\min}\right)P_{tl} + X_t^{\min} \tag{3}$$

Where: $X_t^{\max}$ and $X_t^{\min}$ denotes the upper and lower bounds of real variables respectively.

If P_{tl} is a binary number, convert it to

$$X_{tl} = round(P_{tl}) \tag{4}$$

Where: function $round(\bullet)$ is rounded to the nearest integer.

Combined with the SAWOA in this paper, w_{ij} in ELM can be converted by Eq. 3, and its upper and lower bounds are 0 and 1. f_k is a binary variable that can be converted by Eq. 4.

3.2 Fitness Function

The SAWOA-FP-ELM algorithm performs synchronous optimization on feature subsets and ELM parameters. With the amount of selected features being simultaneously reduced, classification accuracy of the algorithm is improved accordingly. The higher the classification accuracy of the classifier, the less the amount of selected features. In the mean while, with reducing fitness function value, optimization of each individual becomes better. Based on above mentioned considerations, both classification accuracy and feature subset are used as the independent variables of the fitness function. The fitness function formula is as follows:

$$fitness = 1 - \left(W_a \times accuracy + W_f \times \left(\sum_{i=1}^{M} C_i f_i \right)^{-1} \right) \tag{5}$$

Where W_a denotes the weight of the classification accuracy, W_f denotes the weight of the feature amount, and different weight values have different effects on the experimental results. Through the adjustment of the parameters, it is found that the larger the value of W_a, the higher the classification accuracy and the bigger the amount of features. While $W_a = 0.8$, $W_f = 0.2$. accuracy is the classification accuracy, C_i is the generation value of the i-th feature of the data set. The default value is 1, f_i takes 1 to indicate that the i-th feature is selected, 0 means unselected, and M is the feature dimension of the data set.

The steps of the algorithm based on SAWOA integrated optimization extreme learning machine are as follows:

(1) Data preprocessing: The characteristics of the data set are processed by Eq. 17. The training set and the test set are divided according to the ratio of 2:8. On the one hand, because the number of samples in the data set is large, 20% is enough for training ELM. On the other hand, choosing a modified training set can help reduce the time of the optimization process.
(2) Feature and parameter decoding: The characteristics and parameters of each whale individual are decoded by Eq. 3 and 4, the result of current feature selection of the training set, initial weight and offset values of the ELM hidden layer are obtained.
(3) Selecting feature subsets: Using the feature selection results obtained in (2), the training feature subset and the test feature subset are selected from the data set.
(4) Fitness value calculation: initial weight and offset, the training feature subset is adopted for training the ELM classifier, the test feature subset is adopted to test and to calculate the classification accuracy rate, the fitness value of each whale individual is calculated through Eq. 5.
(5) Termination condition: If the termination condition is reached, the integration optimization process ends, the optimal feature selection result and the optimal parameter are generated; otherwise, the next iteration will be started.

4 Experimental Results and Discussion

4.1 Data Source and Preprocessing

To verify the effectiveness of our proposed algorithm, 2015 Sensorless Drive Diagnosis data set in UCI Machine Learning Knowledge Base is selected for the experiment. The data set extracts features from the motor current. The motor has completed and defective components that ultimately form different fault categories by controlling 12 different operating conditions, including speed, load torque and load force. The data set consists of 48 feature attributes (real number type) and 11 fault categories (even ratio). See Table 1:

Table 1. Sensorless Drive Diagnosis data set details

Names	Total features	No. of classes	No. of instances
Sensorless drive diagnosis	48	11	58509

4.2   Experimental Settings

The experiment runs on a CPU with Intel(R) Xeon(R) W-2155 CPU @ 3.30 GHz, a memory of 64 GB, Windows 10 operating system, and a programming tool for Matlab R2015b environment. GA, PSO, DE, WOA, ALO, SSA and proposed algorithm are compared. Their population size is $p = 10$, evolution number is $g_{max} = 100$, and crossover probability and mutation probability of DE are 0.6 and 0.1 respectively. To evaluate the effectiveness of our proposed algorithm. Under the same conditions, proposed method is compared with BP, SVM and KNN. Their control parameters are shown in Table 2. The above experiment is repeated 30 times in the same environment to avoid random error.

Table 2. Initialization parameters of BP, SVM, KNN initialization parameters

Algorithm	Parameter	Value
BP	• Number of hidden layer nodes	30
	• Activation function	{logsig, tansig}
	• Training function	Traingdx
	• Maximum number of epochs	1000
	• Learning rate	0.01
	• Performance goal	0.01
SVM	• Kernel function	Polynomial
	• gamma	0.07
	• c	1
KNN	• NumNeighbors	5
	• Standardize	1

4.3   Results and Discussion

In Table 3, by comparing with BP neural network, SVM, and KNN methods, the SAWOA-FP-ELM method achieves a higher classification accuracy. The results of comparing with SAWOA-FP-ELM suggest efficiency of proposed method. SAWOA-F-ELM and SAWOA-P-ELM respectively indicate that SAWOA only optimizes the feature selection of ELM and SAWOA only optimizes the hidden layer parameters of ELM.

Table 3. Average accuracy, variance, and best accuracy of different algorithms

Accuracy	BP	SVM	KNN	ELM	SAWOA-P-ELM	SAWOA-F-ELM	SAWOA-FP-ELM
AVG	0.6753	0.7808	0.7431	0.8120	0.8438	0.9351	0.9402
STD	0.1060	0.7653	0.0055	0.0092	0.0020	0.0024	0.0052
BEST	0.7715	0.7977	0.7551	0.8240	0.8490	0.9392	**0.9488**

As can be seen from Table 3, non-optimized ELM is more accurate than BP, SVM, and KNN. After ELM, SAWOA-P-ELM, SAWOA-F-ELM, SAWOA-FP-ELM are optimized by SAWOA, accuracy rates 3.18%, 11.91%, 12.82% are achieved respectively. It can be seen that the proposed algorithm has improved classification accuracy of ELM model in motor fault detection. SAWOA-FP -ELM optimization works best because the method obtains the optimal feature subset which reduces complexity of the model. At the same time, the initial weight and bias of ELM are no longer random initialization, they has become a set of best Parameters instead, the accuracy is therefore improved. On the other hand, three of four algorithms optimize ELM using SAWOA are less than four algorithms in the test variance. It can be seen that the optimized ELM model has good stability and strong generalization capability.

Figure 2 is a graph showing fitness convergence of 30 independent experiments of SAWOA-P-ELM, SAWOA-F-ELM, and SAWOA-FP-ELM. Figure 2(a) shows that only ELM hidden layer parameters are optimized, and the curve converges earlier, which indicate that pure parameter optimization has limited performance improvement for ELM; Fig. 2(b) shows that only feature selection is optimized, and the error (adaptation) gradually decreases. There is no premature convergences, which indicates that feature selection helps reduce the classification error of proposed model and the curve fluctuates greatly; Fig. 2(c) shows the optimization of parameters and feature selection, fitness convergence curve has less fluctuation because of integrated optimization. The fluctuation caused by the ELM random initialization parameter is avoided. The optimization problem is insufficient, experimental results reveal that SAWOA-FP-ELM works best.

As can be seen from Table 4, SAWOA-FPPlus-ELM contains 8 attributes among11 motor fault detection classification accuracy rates, namely #1 (0.9944), #2 (0.9846), #3 (0.9918), #4 (0.9979), #5(0.9785), #6(0.9841), #9(0.9875), #10(0.9851), 3 items second, respectively #7(0.9998), #8(0.9848), #11(0.9999), which indicates that proposed method is better in achieving classifying subcategories. It can be further found that SAWOA-FPPlus-ELM improved #1 (4.2%), #2 (3.15%), #3 (3.67%), #4 (3.13%),#5 (4.49%), #6(5.56%), #7(0.21%), #8(2.29%), #9(3.19%), #10(2.91%), #11(0.01%) in each detection category compared to the non-optimized ELM. It shows that proposed method has a relatively uniform improvement effect on the fault detection accuracy of each sub-category.

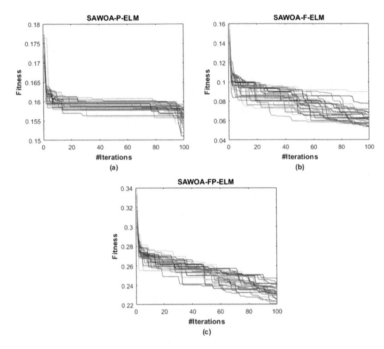

Fig. 2. Convergence curve of four algorithms based on SAWOA optimized ELM

Table 4. Comparison of different algorithms in 11 types of fault detection

Accuracy	BP	SVM	KNN	ELM	SAWOA-P-ELM	SAWOA-F-ELM	SAWOA-FP-ELM	SAWOA-FPPlus-ELM
#1	0.914	0.944	0.921	0.952	0.964	0.9939	0.9940	**0.9944**
#2	0.912	0.943	0.931	0.953	0.961	0.9824	0.9831	**0.9846**
#3	0.894	0.967	0.937	0.955	0.966	0.9880	0.9898	**0.9918**
#4	0.899	0.975	0.943	0.966	0.977	0.9970	0.9961	**0.9979**
#5	0.885	0.925	0.907	0.933	0.947	0.9736	0.9779	**0.9785**
#6	0.874	0.907	0.910	0.928	0.937	0.9789	0.9798	**0.9841**
#7	0.981	**1.000**	0.995	0.997	0.998	0.9997	0.9996	0.9998
#8	0.919	0.949	0.928	0.961	0.970	0.9820	0.9858	0.9848
#9	0.892	0.917	0.946	0.955	0.959	0.9852	0.9853	**0.9875**
#10	0.915	0.944	0.936	0.956	0.963	0.9830	0.9837	**0.9851**
#11	0.990	**1.000**	0.999	0.999	0.999	0.9999	0.9999	0.9999

5   Conclusions

In this paper, the SAWOA-FP-ELM method is proposed and applied to an example of motor fault detection classification. Existing studies did not simultaneously consider feature selection and parameter optimization for ELM. Experiment results show that proposed method has achieved better classification performance in terms of overall classification accuracy and sub-category classification accuracy. In future, other effective optimization methods will be considered to optimize the process of ELM construction, to get better classification accuracy and stability and to make more application values for motor fault detection technology.

Acknowledgment. This work is funded by the National Natural Science Foundation of China under Grant No. 61772180, Technological innovation project of Hubei Province 2019 (2019AAA047).

References

1. Gao, Z., Cecati, C., Ding, S.X.: A survey of fault diagnosis and fault-tolerant techniques—Part I: fault diagnosis with model-based and signal-based approaches. IEEE Trans. Ind. Electron. **62**(6), 3757–3767 (2015)
2. Delgado-Arredondo, P.A., Morinigo-Sotelo, D., Osornio-Rios, R.A., et al.: Methodology for fault detection in induction motors via sound and vibration signals. Mech. Syst. Signal Process. **83**, 568–589 (2017)
3. Anayat, S., Sikandar, A., Rasheed, S.A., Butt, S.: A deep analysis of image based video searching techniques. Int. J. Wirel. Microw. Technol. (IJWMT) **10**(4), 39–48 (2020). https://doi.org/10.5815/ijwmt.2020.04.05
4. Elzayady, H., Badran, K.M., Salama, G.I.: Arabic opinion mining using combined CNN - LSTM models. Int. J. Intell. Syst. Appl. (IJISA) **12**(4) 25–36 (2020). https://doi.org/10.5815/ijisa.2020.04.03
5. Tian, J., Morillo, C., Azarian, M.H., et al.: Motor bearing fault detection using spectral kurtosis-based feature extraction coupled with k-nearest neighbor distance analysis. IEEE Trans. Ind. Electron. **63**(3), 1793–1803 (2016)
6. Sun, W., Shao, S., Zhao, R., et al.: A sparse auto-encoder-based deep neural network approach for induction motor faults classification. Measurement **89**, 171–178 (2016)
7. Yin, Z., Hou, J.: Recent advances on SVM based fault diagnosis and process monitoring in complicated industrial processes. Neurocomputing **174**, 643–650 (2016)
8. Keskes, H., Braham, A.: Recursive undecimated wavelet packet transform and DAG SVM for induction motor diagnosis. IEEE Trans. Ind. Inform. **11**(5), 1059–1066 (2015)
9. Zhang, X., Liang, Y., Zhou, J.: A novel bearing fault diagnosis model integrated permutation entropy, ensemble empirical mode decomposition and optimized SVM. Measurement **69**, 164–179 (2015)
10. Huang, G.B., Zhu, Q.Y., Siew, C.K.: Extreme learning machine: theory and applications. Neurocomputing **70**(1–3), 489–501 (2006)
11. Zhu, Q.Y., Qin, A.K., Suganthan, P.N., et al.: Evolutionary extreme learning machine. Pattern Recognit. **38**(10), 1759–1763 (2005)
12. Mirjalili, S., Lewis, A.: The whale optimization algorithm. Adv. Eng. Softw. **95**, 51–67 (2016)

13. Van Laarhoven, P.J.M., Aarts, E.H.L.: Simulated Annealing: Theory and Applications, pp. 7–15. Springer, Dordrecht (1987)
14. Mafarja, M.M., Mirjalili, S.: Hybrid whale optimization algorithm with simulated annealing for feature selection. Neurocomputing **260**, 302–312 (2017)
15. Bhardwaj, S., Pandove, G., Dahiya, P.K.: A genesis of a meticulous fusion based color descriptor to analyze the supremacy between machine learning and deep learning. Int. J. Intell. Syst. Appl. (IJISA) **12**(2), 21–33 (2020). https://doi.org/10.5815/ijisa.2020.02.03

Multi-fidelity Multicriteria Optimization of Strain Gauge Force Sensors Using a Neural Network-Based Surrogate Model

Sergey I. Gavrilenkov$^{(\boxtimes)}$ and Sergey S. Gavriushin

Bauman Moscow State Technical University, Moscow, Russia
{gavrilenkovsergei,gss}@bmstu.ru

Abstract. This paper proposes a method of reducing the computational complexity of the simulation-driven multi-objective design of strain gauge force sensors using the Parameter Space Investigation approach. The essence of the method is to construct a neural network-based surrogate model with the training set comprised of a small number of low-fidelity (LF) and a large number of high-fidelity (HF) simulation models. The topology of the neural network is tuned using a genetic algorithm. The method was integrated in a software system for multi-objective design of force sensors, and a novel nonlinearity-free column force sensor was designed with the help of this method.

Keywords: Surrogate-modeling · Simulation-driven design · Strain gauge force sensors · Artificial neural network · Genetic algorithm

1 Introduction

Strain gauge-based force sensors dominate force measurement in science and industry, from sensors in wind tunnels [1] and space robots [2] to weighbridges, platform, and hopper scales, as well as for load monitoring application [3]. A strain gauge force sensor comprises an elastic metal body with strain gauges installed on it. When the measured force acts on the sensor body, it produces strain acting on the strain gauges. Strain gauges deform and change their electrical resistance. The strain gauges are connected in the Wheatstone bridge, so the strain of strain gauges produces a voltage at the output terminals of the Wheatstone bridge proportional to the applied force. The characteristics of strain gauge force sensors (nonlinearity, hysteresis, output signal magnitude, overload capacity, etc.) are related to the overall stress-strain state of the elastic body. The stress-strain state is related to the elastic element stiffness, which is related to the geometry of the elastic element. So, the desired sensor properties can be achieved by optimizing the elastic element geometry.

Finite Element Analysis is an excellent tool for designing strain gauges. However, the better the accuracy of the Finite Element model, the longer the simulation time. Currently, this issue limits the promise of Finite Element Method in force sensor design. However, this limit can be overcome by using surrogate modeling technique, where the relationships between the design variables and goal function can be approximated based on a sample of simulated designs. To make the process of training

© The Author(s), under exclusive license to Springer Nature Switzerland AG 2021
Z. Hu et al. (Eds.): AIMEE 2020, AISC 1315, pp. 325–336, 2021.
https://doi.org/10.1007/978-3-030-67133-4_30

the surrogate model fast and accurate, the training set can be comprised of a small number of computationally expensive high-fidelity (HF) models and a large number of computationally cheap but less accurate low-fidelity (LF) models.

The goals of this paper are as follows:

1. Propose a method for surrogate-based optimization of strain gauge force sensors using HF and LF models. The surrogate model will be implemented using neural network architecture;
2. Propose an algorithm for tuning the neural network architecture to provide the most accurate prediction of the goal functions;
3. Check the efficacy of the proposed method by using it to design a novel column force sensor.

The rest of the paper is organized as follows. First, we conduct literature review, followed by the presentation of the pipeline of the proposed multi-fidelity optimization method. Second, we present the details of the implementation of the proposed method. Third, we integrate the proposed method in the software [4] for multi-objective design of strain gauge force sensors and use it to design a novel column force sensor. Then we discuss consider characteristics of the prototype of the designed force sensor. Lastly, we discuss results and consider plans for future research.

2 Literature Review

On the other hand, surrogate modeling techniques have been extensively used in other areas of design and optimization. Koziel, Bekasiewicz and Ogurtsov have made huge progress in the area of optimization based on surrogate models and multi-fidelity models [5–8].

Omairey et all proposed [9] a surrogate-based framework for reliability analysis of unidirectional FRP composites that significantly reduced analysis duration. Another approach was used by Taran et al. [10] that used kriging surrogate models for the search of promising candidate solutions and expensive FEA for evaluating the most promising designs. A somewhat similar approach was presented by Yi, Shen and Shoemaker in [11]. Ahmed et al. used response surface techniques with a small number of expensive 3D-FEA model for design of transverse-flux permanent magnet linear synchronous motor [12]. Fan et al. reduced computational cost of reliability-based design optimization by using surrogate modeling techniques [13]. In [14], an adaptive Co-Kriging framework was proposed; in the framework, a number of LF and HF simulation results were combined to form a moderate-cost optimization metamodel.

So, surrogate-based optimization is widely used in engineering design and optimization. However, there is little research on applying surrogate modeling techniques to the design of strain gauges. There has been much research on FEA-driven design of force sensors. For example, Gavriushin and Skvortsov [15, 16] used FEA to calculate characteristics of a double diaphragm strain gauge pressure sensors. Andreev, Shakhnov, and Vlasov [17] employed finite element modeling to design a pressure sensor with overload protection. Gavrilenkov considered the method [18] of simulating temperature effect on the sensitivity of a strain gauge force sensor in the non-uniform temperature field and presented [19] a software system for multi-objective design of force sensors. Thus, the task of creating a method of multi-fidelity surrogate-based optimization of strain gauge force sensors is a relevant problem.

3   Methodology

3.1   Procedure of the Multi-fidelity Surrogate-Based Optimization

The proposed surrogate-assisted modeling technique is developed to work within the confines of the Parameter Space Investigation (PSI) used in [19]. The workflow of the multi-fidelity surrogate-based optimization procedure is shown in Fig. 1.

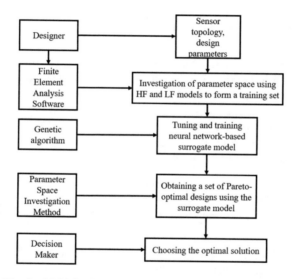

Fig. 1. Multi-fidelity surrogate-based optimization procedure

The PSI method comes down to several subsequent investigations of the Parameter space. The surrogate modeling method works as follows. First, the parameter space in question is sampled with an LP_τ sequence, first using a large number of LF evaluations, then a small number of HF evaluations. In our case, the model's fidelity (accuracy) is determined by the mesh element size.

Results of LF and HF evaluations form the training and validation sets for the surrogate model. The characteristic mesh element size determines the fidelity of a data point in the training set.

In our case, the surrogate model is based on the neural network architecture. Unlike conventional regression techniques, like polynomial and logistic regression, neural networks can fit highly nonlinear relationships.

However, the approximation accuracy depends on the neural network's parameters, i.e., the number of layers, types of activation functions, number of epochs in the training algorithm, and so on. So, the neural network has to be tuned in order to ensure high prediction accuracy. The network is tuned using a genetic algorithm, where the performance of the candidate surrogate model is evaluated based on the validation set.

However, unlike conventional regression techniques, building the neural network-based surrogate model is computationally expensive, given that the neural network has to be trained many times during tuning of the parameters. So, on the one hand, the surrogate model must be accurate enough. On the other hand, the tuning of the surrogate model must be done in a reasonable time to retain the computational advantage of using low-fidelity models instead of high-fidelity models.

After tuning, the surrogate model is used for thorough investigation of the considered Parameter Space. If the boundaries of the Parameter Space change, the accuracy of the surrogate model may be decreased, so a new surrogate model would be required.

3.2   Implementation of the Proposed Method

Finite Element Analysis. For finite element simulations, we used the open-source software package Salome-Meca/Code_Aster. Salome-Meca is used to build the sensor geometry and generate a finite element mesh of required finesse based on the given geometry. Code_Aster is used to run a pre-scripted linear static structural finite element analysis on the generated meshes. Although Salome-Meca/Code_Aster is not widespread as commercial FEA packages due to a steep learning curve, it is increasingly being used by the scientific community [20–22].

Parameter Space Investigation Implementation. The investigation for forming the training and validation sets was done using the Parameter Space Investigation module of the system for multi-objective design of strain gauge force sensor developed by the authors [19].

Tuning and Training the Surrogate Model. The neural network is comprised of several fully connected layers. It is implemented using the Keras library [23] with Tensorflow backend. The number of inputs is equal to the number of design variables plus one (mesh element size defining model fidelity). The number of outputs is equal to the number of Goal Functions. The number of epochs for training is 1000. The training uses early stopping [23] with the Patience parameter of 5 monitoring the loss function. The training algorithm is Adam [24].

The parameters of the neural network are tuned using a genetic algorithm. This is a popular optimization tool [25, 26] due to its ability to efficiently solve multi-modal optimization problems. The genetic algorithm can change the following parameters of the neural network:

1. Number of hidden layers: from one to 5
2. Number of neurons in the input layer and each of the hidden layers, ranging from 20 to 500;
3. Activation function of each of the hidden layer, one from the following set: ELU, RELU, tanh, linear, sigmoid [23];

In the implementation of the proposed a method, the PYMOO library [27] developed by Blank and Deb was used.

4 Case Study

The case study is to find the rational parameters of a column strain gauge force sensor shown in Fig. 2 using the proposed surrogate modeling technique. The goal is to find a set of Pareto-optimal design variants in the parameter space defined below using the Parameter Space Investigation method.

The elastic element material is 630 stainless steel (17-4 PH). The material properties are as follows: elasticity module E = 197 GPa, Poisson's ratio μ = 0.3, hardness = 40...42 HRC, σ_y = 1200 MPa. The maximum sensor load is 200 kN.

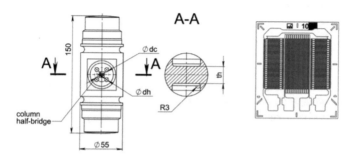

Fig. 2. Drawing of the force sensor elastic element (left) and the half-bridge strain gauge (right), overall dimensions of the elastic element are given in mm.

Usually, the placement of strain gauges on the elastic element (locations and orientations) must be considered as design variables. The module for evaluating the design objectives would have to determine the optimal placement based on simulation results. However, in this case, the strain gauges, namely, the column half-bridge gauges, are mounted in the center on the both sides of the steel web as shown in Fig. 3. Otherwise, the surrogate models would have to predict not only the values of strain at the center of both sides of the steel web but also the whole distribution of strain across the steel web, which is not the purpose of this study. Table 1 gives information about design variables.

Table 1. Design variables

Design variable name	Lower limit, mm	Upper limit, mm	Comment
Th	24	34	Web thickness
Dc	6	12	Diameter on which the centers of holes in the web are placed
Dh	4	8	Diameter of holes in the web

The Goal Functions are the following:

- Maximize the bridge output voltage at the rated load (200 kN).
- Minimize the maximum Von Mises stress in the elastic element σ_{max};
- Minimize the difference between strain sensed by different gauges of the Wheatstone bridge. In a classical strain gauge force sensor with a full Wheatstone bridge, half of the bridge's gauges are under tensile strain, and the other half is under the compressive strain. The physics of the Wheatstone bridge circuit are such that unequal magnitudes of compression and tension strain cause bridge nonlinearity. Classical column force sensors, Fig. 4, have inherent nonlinearity of the "output voltage-force" characteristic due to unequal values of strain sensed by the gauges, i.e., the axial strain is three times greater than the transverse strain due to the uniaxial stress-strain state. In the considered elastic element, the strain state is complex, and tension strain is increased compared to the classical design. This design objective is given by:

$$\gamma = -\frac{\varepsilon_2}{\varepsilon_1} \tag{1}$$

Where ε_2 is the average strain sensed by the transverse gauges of the half bridges, and ε_1 is the average strain sensed by the axial strain gauges, Fig. 3.

The design objectives are calculated using high-fidelity and low-fidelity models. The mesh for the model is obtained using the NETGEN-1D-2D-3D algorithm with the following parameters:

1. Maximum element size: 2 mm for the high-fidelity model and 10 mm for the low-fidelity model;
2. Minimum element size: 1 mm for the high-fidelity model and 5 mm for the low-fidelity model;
3. Mesh fineness level is moderate;
4. 2D surface mesh is quad-dominated;
5. Growth rate is 0.3.

The boundary conditions for the static structural simulation are as follows: a fixed boundary condition is applied to the bottom of the model, while a distributed load is applied to the top of the sensor. Usually, industrial force sensors have a metal casing protecting the sensor circuitry, Fig. 3. The considered force sensor will have the same overall dimensions as the MV150 force sensor produced by JSC "Tenso-M" and will have the same casing. However, the casing stiffness is small compared to the elastic element stiffness, so it will be neglected in the simulations.

The examples of the mesh for the HF and LF models are shown in Fig. 3. The mesh of the high-fidelity model has around 300000 elements, and the model takes around 60 s to compute (including mesh generation). The low-fidelity model has around 15000 elements, and takes around 7 s to compute, including mesh generation; the characteristic element size is 10 mm. The calculations are done on a laptop with 24 Gb of RAM and a 2.5 GHz CPU.

The goal functions are calculated based on the strain and stress distributions in the elastic element. These calculations are incorporated in the process of finite element simulations. The calculation of the Wheatstone bridge's output signal is done using formulas given in [21]. The placement is fixed (two half-bridges) in the center of the web on both sides.

Fig. 3. Example of finite element mesh for the low-fidelity (left) and fine (middle) model; Example of protective casing of the column sensor MV150 produced by JSC "Tenso-M" (right) [28].

The parameters of the genetic algorithm are the following:

1. Population size: 20;
2. Crossover is integer two-point crossover, the probability is 0.8;
3. Mutation is integer Polynomial Mutation (PM), the probability is 0.1;
4. Selection for the crossover is random;
5. The termination criterion is the maximum permissible execution time – five minutes for the given case study.

The parameters of the tuned ANN are as follows:

1. Number of neurons in the input layer is 170;
2. Number of hidden layers is 2;
3. The number of neurons in the hidden layers is 92 and 183, accordingly.
4. The activation functions for the first and second hidden layers are *sigmoid* and *ELU*, accordingly.

To test the quality of the obtained surrogate model, the considered parameter space is investigated using a Parameter Space sampling of 200 elements using the high-fidelity model. The surrogate model is used to predict high-fidelity results. The predicted values are compared to the actual high-fidelity simulation results.

5 Results and Discussion

Figure 4 shows the histogram plots of the prediction error based on the verification set of 200 high-fidelity simulation results. The prediction error for the i-th Goal Function is calculated as follows:

$$\delta_{i,j} = \frac{GF_{i,j}^{predicted} - GF_{i,j}^{HF}}{GF_{i,j}^{HF}} \times 100\% \tag{2}$$

where $GF_{i,j}^{predicted}$ is the value of the i-th Goal Function of the j-th specimen in the verification set obtained using the surrogate model; $GF_{i,j}^{HF}$ is the value of the i-th Goal Function of the j-th specimen in the verification set obtained using the HF model.

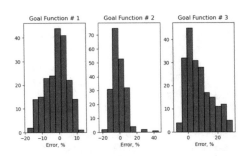

Fig. 4. Histograms of errors for the Goal Function #1 (left), Goal Function #2 (middle), Goal Function #1 (right)

The quality of prediction for the Goal Function #1 and #2 is fair, while the prediction quality of the Goal Function #3 is somewhat poor. Unlike the first two goal functions, the prediction error is biased to the right. Perhaps, for each goal function, there should be a separate surrogate model tuned independently of other surrogate models.

The obtained trained surrogate model is used to conduct a detailed investigation of the Parameter Space defined by the bounds given in Table 1 using software presented in [4]. The investigation was done with LP_τ sequence of 1000 elements. Figure 5 shows the distribution of Pareto-optimal solutions in the space of Goal Functions.

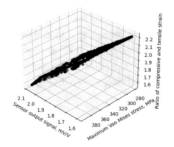

Fig. 5. Distribution of the Pareto-optimal solutions in the space of Goal Functions obtained by using the PSI method and the trained surrogate model (left) and a photograph of the manufactured prototype without protective casing (right).

The total time of the surrogate-based optimization, including the time to run the models for the training set, to tune the surrogate model and re-investigate the parameter space using the trained and tuned surrogate model is around 50 min. Had we used conventional simulation-driven investigation of the Parameter Space as in [4], the investigation with 1000 designs using the high-fidelity model would take around 1000 min. So, the time savings are quite good.

Based on the obtained set of Pareto-optimal solutions, the Decision Maker chose a preferred variant, Fig. 5 (right), with the following dimensions (Th = 24 mm, Dc = 18 mm, Dh = 4 mm) and decided to increase the output sensor gain by reducing Th to 22 mm. A prototype was manufactured and tested for design verification. Figure 5, left, shows the test setup. The setup, Fig. 6, is comprised of a hydraulic force standard machine, and a DMP40 precision amplifier for measuring the sensor output signal. The photo of the manufactured prototype in the protective casing is shown at the inset of Fig. 6.

 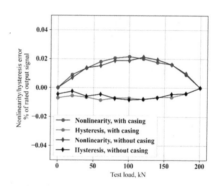

Fig. 6. The manufactured prototype with protective casing (left) and the plot of sensor error vs. load (right)

A protective casing was laser-welded to the elastic element, but the sensor errors changed marginally within the accuracy tolerance of the force standard machine, providing the validity of our assumption that the casing almost does not influence the linearity and output signal of the column sensor. Also, assumption that equality of strain of strain gauges experiencing tension and compression positively affects non-linearity proved to be correct. Overall, the sensor nonlinearity relative to the rated output signal is around 0.02%, corresponding to the accuracy class C3 according to OIML R60. Besides, the sensor has low hysteresis. The simulated value of the sensor output signal is 2.31 mV/V, the values of the output signal with and without protective casing is around 2.25 mV/V. Usually, column force sensors with the axial strain around −1300 ppm have nonlinearity around 0.10...0.15%, and this nonlinearity has to be rectified using special techniques. In the developed sensor, to rectification is required.

Thus, the proposed method allowed designing a sensor with good characteristics at the reasonable computational cost and good simulation fidelity. However, our method has a number of limitations:

1. If the Decision Maker decides to run more investigations in other parts of the parameter space not covered by the training set for the surrogate models, the results can be highly erroneous. So, for each new investigation of the parameter space, the procedure presented in this paper would have to be *repeated*.
2. Currently, the surrogate model does not allow modeling the whole strain distribution on the surfaces of the elastic element so that the *optimal* placement of strain gauges is determined. This limitation can possibly be overcome by extending the training and validation sets to include the coordinates and values of strain tensor components for the nodes on the elastic element surface where strain gauges are to be installed.
3. The quality of the surrogate model is adequate, but further research should consider the approach where for each surrogate model there will be an individual neural network.
4. Hyperparameters of the neural network (number of training epochs, training algorithm, batch size, validation split size) are determined by the designer based on the designer experience. For future research, it makes sense to use the tuning algorithm to adjust hyperparameters as well.

6 Conclusion

This paper proposed a methodology of multi-fidelity multicriteria optimization of strain gauges force sensors using a neural network-based surrogate model. The methodology allows reducing computational time required for the FEA-driven investigation of the design parameter space by using a combination of a number of computationally cheap low-fidelity models and computationally expensive high-fidelity models. The parameters of the surrogate model were tuned using a genetic algorithm. The efficacy of the method was tested with a case study, where a novel nonlinearity-free column force sensor was designed. Although the method increased the design speed while retaining good accuracy of goal functions' evaluation, there are still some limitations to the method (inability to model entire strain distributions across the elastic element, somewhat arbitrary choice of surrogate model hyperparameters, the need to repeat the surrogate training process for a new investigation with different boundaries of the parameter space) that have to be addressed in future research. The proposed method can be applied to other types of force sensors to accelerate the design process. Future studies should focus on improving the algorithm for tuning the surrogate model, and building the surrogate model architecture for predicting entire distribution of strains on the elastic element.

Acknowledgements. The authors of this paper would like to thank JSC "Tenso-M" for prompt manufacturing of the prototype and providing access to the testing machines

References

1. Tavakolpour-Saleh, A.R., Setoodeh, A.R., Gholamzadeh, M.: A novel multi-component strain-gauge external balance for wind tunnel tests: simulation and experiment. Sens. Actuators Phys. (2016). https://doi.org/10.1016/j.sna.2016.05.035
2. Sun, Y., Liu, Y., Zou, T., Jin, M., Liu, H.: Design and optimization of a novel six-axis force/torque sensor for space robot. Meas. J. Int. Meas. Confed. **65**, 135–148 (2015)
3. Saxena, P., Pahuja, R., Khurana, M.S., Satija, S.: Real-time fuel quality monitoring system for smart vehicles. Int. J. Intell. Syst. Appl. **8**, 19–26 (2016)
4. Gavrilenkov, S.I., Gavryushin, S.S.: Development and performance evaluation of a software system for multi-objective design of strain gauge force sensors. In: Advances in Intelligent Systems and Computing, vol. 1127 (2020)
5. Koziel, S., Bekasiewicz, A.: Rapid simulation-driven multiobjective design optimization of decomposable compact microwave passives. IEEE Trans. Microw. Theory Tech. **64**, 2454–2461 (2016)
6. Koziel, S.: Low-cost data-driven surrogate modeling of antenna structures by constrained sampling. IEEE Antennas Wirel. Propag. Lett. **16**, 461–464 (2017)
7. Koziel, S., Bekasiewicz, A.: Rapid multiobjective antenna design using point-by-point pareto set identification and local surrogate models. IEEE Trans. Antennas Propag. **64**, 2551–2556 (2016)
8. Koziel, S., Ogurtsov, S.: Multi-objective design of antennas using variable-fidelity simulations and surrogate models. IEEE Trans. Antennas Propag. **61**, 5931–5939 (2013)
9. Omairey, S.L., Dunning, P.D., Sriramula, S.: Multiscale surrogate-based framework for reliability analysis of unidirectional FRP composites. Compos. Part B Eng. **173**, 106925 (2019)
10. Taran, N., Ionel, D.M., Dorrell, D.G.: Two-level surrogate-assisted differential evolution multi-objective optimization of electric machines using 3-D FEA. IEEE Trans. Magn. **54** (2018)
11. Yi, J., Shen, Y., Shoemaker, C.A.: A multi-fidelity RBF surrogate-based optimization framework for computationally expensive multi-modal problems with application to capacity planning of manufacturing systems. Struct. Multidiscip. Optim., 1–21 (2020)
12. Ahmed, S., Grabher, C., Kim, H.J., Koseki, T.: Multifidelity surrogate assisted rapid design of transverse-flux permanent magnet linear synchronous motor. IEEE Trans. Ind. Electron. **67**, 7280–7289 (2020)
13. Fan, X., Wang, P., Hao, F.F.: Reliability-based design optimization of crane bridges using Kriging-based surrogate models. Struct. Multidiscip. Optim. **59**(3), 993–1005 (2019)
14. Shi, R., Liu, L., Long, T., Wu, Y., Gary Wang, G.: Multi-fidelity modeling and adaptive co-kriging-based optimization for all-electric geostationary orbit satellite systems. J. Mech. Des. **142** (2020)
15. Gavryushin, S.S., Skvortsov, P.A.: Evaluation of output signal nonlinearity for semiconductor strain gage with ANSYS software. In: Solid State Phenomena, pp. 60–70 (2017)
16. Gavryushin, S.S., Skvortsov, P.A., Skvortsov, A.A.: Optimization of semiconductor pressure transducer with sensitive element based on "silicon on sapphire" structure. Periodico Tche Quimica **15**, 679–687 (2018)
17. Andreev, K.A., Vlasov, A.I., Shakhnov, V.A.: Silicon pressure transmitters with overload protection. Autom. Remote Control **77**(7), 1281–1285 (2016)
18. Gavrilenkov, S.I.: Method of simulating temperature effect on sensitivity of strain gauge force sensor in non-uniform temperature field. AIP Conf. Proc. **2195**(1), 020051 (2019)

19. Gavrilenkov, S.I., Gavriushin, S.S., Godzikovsky, V.A.: Multicriteria approach to design of strain gauge force transducers. J. Phys. Conf. Ser. **1379**(1) (2019)
20. Šliseris, J., Gaile, L., Pakrastiņš, L.: Deformation process numerical analysis of t-stub flanges with pre-loaded bolts. Procedia Eng., 1115–1122 (2017)
21. Saloustros, S., Pelà, L., Roca, P.: Nonlinear numerical modeling of complex masonry heritage structures considering history-related phenomena in staged construction analysis and material uncertainty in seismic assessment. J. Perform. Constr. Facil. **34** (2020)
22. Antonutti, R., Peyrard, C., Incecik, A., Ingram, D., Johanning, L.: Dynamic mooring simulation with Code_Aster with application to a floating wind turbine. Ocean Eng. **151**, 366–377 (2018)
23. Keras API reference. https://keras.io/api/. Accessed 13 Sept 2020
24. Kingma, D.P., Ba, J.L.: Adam: a method for stochastic optimization. In: 3rd International Conference on Learning Representations, ICLR 2015 - Conference Track Proceedings. International Conference on Learning Representations, ICLR (2015)
25. Soukkou, A., Belhour, M.C.: Intelligent systems and applications. Intell. Syst. Appl. **7**, 73–96 (2016)
26. Kumar, R.: Computer network and information security. Comput. Netw. Inf. Secur. **8**, 17–26 (2018)
27. Blank, J., Deb, K.: Pymoo: multi-objective optimization in Python. IEEE Access **8**, 89497–89509 (2020)
28. Catalogue of column load cells of JSC "Tenso-M". https://www.tenso-m.ru/tenzodatchiki/szhatija-kolonna/84/. Accessed 13 Sept 2020

Traffic Intelligent Control at Multi-section Crossroads of the Road Networks

Andrey M. Valuev$^{(\boxtimes)}$ and Anatoliy A. Solovyev

Mechanical Engineering Research Institute of the Russian Academy of Sciences,
4, Malyi Kharitonievsky pereulok, 101990 Moscow, Russia
valuev.online@gmail.com, aa.solovjev@yandex.ru

Abstract. The article is devoted to the automation of traffic control at intersections divided into two or more sections by internal stop lines. A recently introduced method for studying crossroads is being developed, based on the joint choice of the phase separation of traffic directions and the duration of the phases of the light cycle of traffic regulation for a multi-section crossroads. The problem of preliminary determination of efficient values of traffic light control parameters for a given phase-wise traffic separation scheme (TSS) with established traffic requirements (for example, intensity and safety) is formalized in terms of linear programming. Correction of observation data to exclude random errors is provided. It is recommended to use this method for both the choice of preferable safe TSSs and traffic optimization at the crossroads. Calculation methods for the above are presented and illustrated with the typical example of a multi-section crossroads. Some aspects of training and adaptation of the drivers to intelligent road intersections are discussed.

Keywords: Traffic flow · Safety · Intelligent control system · Signalized intersection · Conflict points · Traffic organization · Traffic light cycle · Traffic separation scheme · Optimization · Microscopic models of traffic flows · Computational experiments

1 Introduction

The paper evolves the recently introduced approach to the intelligent traffic control system development [1, 2] and is focused at the regulated intersections of the complex structure. This approach is based on the representation of traffic control at an intersection as control having two levels - structural and parametric. A concept does not imply any change of the road structure of the crossroads area but concerns only its functional usage in time, i.e. the separation of routes among phases of the traffic light cycle (TLC) what is usually referred as a phase-wise traffic separation scheme [3].

Since 1980s the methods and the systems of intelligent control at the regulated crossroads were given a great deal of the attention but they mainly concentrate on the choice of traffic light cycle (TLC) parameters. Modern conditions for the application of the systems like well-known SCOOT (split, cycle and offset optimization technique) are discussed in [4]. New approaches are being intensively developed as well to overcome drawbacks of such systems [5–8]. New ways of the traffic data registration, transmission and further utilization are proposed [9–13].

There is no evidence, however, that the questions of the traffic control of the complex intersection were ever systematically treated even in road design and, moreover, in operative control of traffic flows (TFs). Practical manuals devoted to this topic, e.g., [3, 14], show that recommendations are proposed for the crossroads with 4 approach legs at maximum that correspond to the intersections of two streets (an example of 5 approach legs intersection is presented in [14] without any consideration). The crossroads with more complex structure, however, do exist, not only in the historical parts of old cities, but in relatively newly developed city districts. In Moscow one can see many complex intersections in both cases, for example, for the 1st case on the squares of the Garden Ring and for the 2nd case at the intersections of several major highways with Profsoyuznaya Street.

The goal of the paper is to propose the constructive method of two-level control of a regulated crossroads with the multi-stage passage of the traffic flows. In Sect. 2 principal features of a multi-section crossroads are characterized and possibilities of its structural control are shown. The problem of preliminary optimization of its parametric control is set up and illustrated. Section 4 presents the ways of more subtle analysis of TFs passage through a complex crossroads with given TSS and TLC parameters and proposes some ways to correct them. The aspects of "education" of the drivers that provides the efficiency of the combined intelligent control of complex intersections with traffic light regulation are discussed in Sect. 5.

The presented materials of tables and graphs are based on the average data of visual authors' observations of traffic at a particular multi-section crossroads. Obtaining experimental data by technical means and their initial processing is a rather complex legal and organizational problem related to obtaining permits for the installation of road surveillance cameras. However, the data used, despite some inaccuracy of traffic parameters, nevertheless allow us to establish general patterns and build a traffic control algorithm at such complex intersections. In the future, it is planned to build updated models with the assistance of state bodies responsible for road safety and organization.

2   Characteristic Features of a Multi-section Crossroads and Ways of Its Structural Control

A principal feature of a complex crossroads is the inability to exclude permitted routes' crossing for any traffic organization with the use of only single-stage pas-sage of traffic flows without intermediate stops on some permitted routes regulating it with a single traffic light which cycle has a reasonable number of phases, not exceeding four. In this regard: 1) internal stop lines (ISL) are being organized; 2) these stop lines divide the intersection area into sections; 3) the movement within each section is regulated by its own traffic light cycle and the cycles in different sections must be coordinated. The typical structure of a complex crossroads resembling the real crossroads on the Serpukhovskaya Square in Moscow is shown in Fig. 1. The set of routes forming the eastern section of it are drawn in dotted lines while the western section is shown with solid lines.

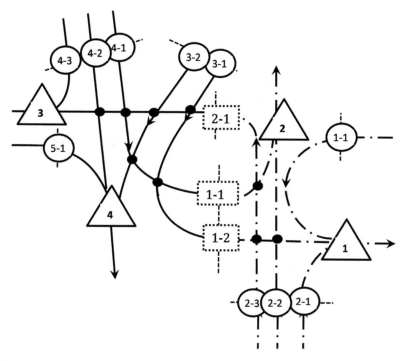

Fig. 1. The structure of a complex intersection (similar to the intersection on Serpukhovskaya Square in Moscow). Legend: (4-1)—entry, /₄\ —exit, ⸽ 1-2 ⸽—SPIS;● — SPC.

In the aspect of the traffic flow's' passage, the intersection is considered as a system of permitted routes being the axial lines of the road lanes. The structure of the inter-section is determined by the set of these routes, their connections with the entrance and exit roads and the relative position. Points of crossing, merging and (to some extent) of branching of these routes cause the principal danger when they are passed in more than one direction simultaneously. So they are referred as conflict points or, in aspect of vehicles' dynamics, as singular points (SP) and, accordingly, are designated here SPC, SPM, SPB. To diminish this danger, the routes are separated among TLC phases to exclude simultaneous passage in several directions of all SPCs and most SPMs and so admissible TSSs are determined. They are treated as conditionally conflict-free TSSs if there are no active SPCs at all phases and as absolutely conflict-free TSSs if there are no active SPMs at all phases as well. For a multi-section crossroads TSSs are formed separately for each section; in this regard, points of routes' intersections with ISLs referred below as SPIS serve as exit points for some sections and entry points for another ones.

In the previous paper [2] we presented the way to find out all possible reasonable absolutely conflict-free TSSs or conditionally conflict-free TSSs that satisfy additional restrictions, the latter means the limitations on the number of active SPMs on each phase and/or on all phases. It is sufficient to find all non-expandable TSSs for which any additional route at any phase violates either the property of being conditionally conflict-free or the posed limitations. For the considered example SPBs are absent and all routes and their sections may be marked with the code of its beginning. It can be argued that there are two non-expandable three-phase TSSs without active SPMs in the eastern section, the first with sets of routes for the 1st, 2nd and 3rd phases (1-1, 2-2, 2-3), (2-1, 2-2, 2-3), (3-1, 4-1) and the second with sets (1-1, 3-1), (2-1, 2-2, 2-3), (3-1, 4-1). For the western section, there is a single non-expandable TSS with the following sets of routes for the 1st, 2nd and 3rd phases: (2-3, 5-1), (3-1, 3-2, 4-3), (4-1, 4-2, 4-3).

Structural control of a multi-section crossroads includes the choice of TSSs for each section but is not reduced to it. A definite TSS is the set of passage schemes for separate phases but it will be seen below that the sequence of these schemes plays important role for the passage of traffic flows since delays depend on it. From the cyclic character of crossroads passage it goes that the 1st scheme in the TSS may be appointed arbitrarily and for any 3-phase TSS there are two variants differing in the appointment of the 2nd one. We denote below the above sequences of phases in the eastern and western sections E1 and W1 and sequences with permutation of the 2nd and the 3rd passage schemes as E2 and W2, respectively. These discrete variants together with durations of TSSs' phases may be rationally chosen on the basis of quantitative information on the intensities of flows for all passage directions.

3   Preliminary Optimization of Phases' Durations for a Certain TSS

Approximate optimization of a set of traffic light cycles, that is, the determination of conditionally optimal phase durations (for the given and the same duration of the entire cycle T_{TLC} for all sections) can be performed based on the previously put forward idea of representing the traffic light control optimization problem in the form of a linear programming problem [2]. The idea of the problem is that for each route the duration of the phases serving it (for each route section separately) should be sufficient to pass the incoming flow during the cycle. In this case, one proceeds from the flow rate on the route estimated at the average expected number of vehicles passing by the route. It should be noted that the calculated flow intensity on a certain route passing more than one section and therefore passed in several stages, may differ at different stages, since the route sections can differ in their curvature, and therefore in the safe speed; it is typical when one route section is straight or slightly curved and the other one contains a sharp turn.

Now we proceed to the problem of conditional optimization of traffic light cycles using the example of the intersection shown in Fig. 1. Let's designate the duration of the i-th phase in the eastern and western sections, respectively, T_{Ei} and T_{Wi}. For the number of vehicles that must pass in a full cycle along the routes, we will use the designations $Q_{1-1}, \ldots, Q_{5-1}$, and for the estimated time of passage of the flow on the route (the route section for the eastern and western section)—values $t_{E1-1}, \ldots, t_{W5-1}$. The latter values are not constants, but depend on the former, i.e. dependencies of the type $t_{E1-1}(Q_{1-1})$ should be used; obtained either on the basis of processing the observation data of flows at a given intersection, or by calculations using a verified mathematical model [15–17]. Since the problem of optimization of traffic light cycles is solved with known values of $Q_{1-1},\ldots,Q_{5-1}$, the presence of such dependencies (which can be calculated in advance) does not complicate the introduced optimization problem. We assume that the known numbers of vehicles $Q_{1-1},\ldots,Q_{5-1}$ form standing queues before entrance or internal stop lines before the start of the movement through the section and the starting vehicle chain completely stops at the ISL if it terminates the route section, otherwise it freely leaves the crossroads area. This assumption yields the guaranteed valuations of passage times. If, on the contrary, the entire chain or its head part freely leaves the end of route section that is terminated with an ISL, it diminishes the needed passage time. The tasks with the indicated values are set to determine the optimal duration of the phases that best ensure the passage of the required number of vehicles in all directions. "The best" is understood in the minimax sense as we maximize the minimum of the relative excesses of the determined phase durations over the required ones along all the routes.

So with the use of the 1st TSS we have the following optimization problem

$$s \rightarrow \max, T_{E1} + T_{E2} + T_{E3} \leq T_{TLC}, T_{W1} + T_{W2} + T_{W3} \leq T_{TLC}, \tag{1}$$

$$t_{E1-1}(1+s) \leq T_{E1}, t_{E2-1}(1+s) \leq T_{E2}, t_{W3-1}(1+s) \leq T_{W2}, \tag{2}$$

$$t_{W3-2}(1+s) \leq T_{W2}, t_{W4-1}(1+s) \leq T_{W3}, t_{W4-2}(1+s) \leq T_{W3}, \tag{3}$$

$$t_{W2-3}(1+s) \leq T_{W1}, t_{W4-3}(1+s) \leq T_{W2} + T_{W3}, t_{W5-1}(1+s) \leq T_{W1}, \tag{4}$$

$$t_{E2-2}(1+s) \leq T_{E1} + T_{E2}, t_{E2-3}(1+s) \leq T_{E1} + T_{E2}, t_{E3-1}(1+s) \leq T_3, \tag{5}$$

and for the 2nd TSS the problem determined with relationships (1)–(4) and

$$t_{E2-2}(1+s) \leq T_{E2}, t_{E2-3}(1+s) \leq T_{E2}, t_{E3-1}(1+s) \leq T_{E1} + T_{E3}. \tag{6}$$

For the intersection with 10 input lanes which structure is shown in Fig. 1 there are four output lanes. Assuming the capacity of one lane equal to 1800 vehicles/hour, we conclude that the following restrictions must be satisfied:

$$Q_{1-1} + Q_{2-1} + Q_{3-1} \leq 1800, Q_{2-2} + Q_{4-1} \leq 1800, \tag{7}$$

$$Q_{2-3} + Q_{4-3} \leq 1800, Q_{3-2} + Q_{4-2} + Q_{5-1} \leq 1800. \tag{8}$$

Conditions (7)–(8) are necessary for passing flows through the intersection but only in exclusive cases they are sufficient. Initial data and results for such Case 1 and the other Case 2 that seems more typical are shown in Tables 1, 2 and 3. The TLC duration was set to 60 s. Except Table 1, all data are related to Case 2. Table 1 demonstrates that for this case only TSS1 guarantees the desired TF intensity and, moreover, may pass even more vehicles in all directions. It follows from results of solutions of the LP problems shown in Tables 2 and 3. The second column of the latter shows the values of $t_{E1-1},\ldots,t_{W5-1}$ used in the LP problems while the other columns show passage times that yield the traffic separation schemes with the exact and rounded optimum durations of TLC phases. In the tables all TF intensities are measured in vehicles/hour and all times in seconds.

Table 1. The traffic flows' intensities and the results of preliminary TLC optimization

Route	Desired TF		The best possible TF	
	Case 1	Case 2	TSS1	TSS2
1-1	600	420	441	399
2-1	600	480	509	480
2-2	1200	1020	1048	968
2-3	600	660	687	660
3-1	600	600	616	570
3-2	600	600	616	570
4-1	600	540	547	513
4-2	600	540	547	521
4-3	1200	1260	1276	1196
5-1	600	660	690	660
Total	7200	6780	6978	6536
	100,00%	94,17%	96,92%	90,78%

Table 2. The optimum solutions of LP problems for TSS1 and TSS2

	E1	E2	E3	W1	W2	W3
TSS1	15.13	16.02	28.85	22.49	19.75	17.76
TSS1, rounded	15.00	17.00	28.00	22.00	20.00	18.00
TSS2	13.56	29.58	16.86	24.38	18.49	17.13

Table 3. The minimum required passage times and the passage times due to TLC optimization

	Required passage time, sec.	TSS1	TSS1, rounded	TSS2
E1-1	14.28	14.67	15.00	13.56
E2-1	16.02	16.46	17.00	29.58
E2-2	31.15	31.14	32.00	29.58
E2-3	21.19	31.14	32.00	29.58
W2-3	21.15	21.73	22.00	24.38
W3-1	19.45	28.86	28.00	30.42
E3-1	19.48	20.02	20.00	18.49
W3-2	19.48	20.02	20.00	18.49
W4-1	17.76	18.25	18.00	17.13
E4-1	17.75	28.86	28.00	16.86
W4-2	17.76	18.25	18.00	17.13
W4-3	37.51	38.27	38.00	35.62
W5-1	21.03	21.73	22.00	24.38

It must be noted, however, that each SPC or SPM is captured by a vehicle chain moving along a certain route not at the moment of the respective green phase beginning but only when it is reached by the head vehicle of the chain. Besides, it is liberated by a previous chain not at this moment but only when its last vehicle leaves it. Possible delays caused by this fact, its impact on the use of the crossroads throughput and possible corrections of traffic light control compensating these negative effects are presented in the next section.

4 Assessment and Corrections of TLC Parameters on the Basis of Detailed Simulation of TFs Passage Through a Crossroads

The proposed analysis is based on calculation of the above mentioned delays. It is convenient and sufficiently accurate to consider that the earliest moment of SP capture by the head of the next chain is the moment at which the SP would be reached the vehicle next to the last vehicle of the previous chain.

For the above example we calculate liberation and capture moments for the passage of TFs with TSS1 and rounded optimum durations of phases presented in Table 2. For each section we consider two possible sequences of phases, E1 and E2, and W1 and W2, respectively. It happens, however, that at the eastern section for both sequences and all SPs liberation moments precede capture moments. For the western section data for nodes that may yield delays are shown in Table 4 in which Route 1 is passed first and the passage of Route 2 follows it. The time reserves are treated as differences between liberation and capture moments; their negative values indicate the necessity to correct the control.

Table 4. The moments of liberation and capture of SPs yielding potential delays

Phase sequence	Node	Route 1	Route 2	Liberation moment	Capture moment	Time reserve
W1	37	4-3	2-3	−1.98	11.55	13.52
W1	37	2-3	4-3	13.77	10.00	−3.77
W2	37	4-3	2-3	−1.98	11.55	13.52
W2	37	2-3	4-3	13.77	10.00	−3.77
W1	44	3-1	4-1	16.53	15.28	−1.26
W2	44	4-1	3-1	1.76	17.13	15.37
W1	54	5-1	3-2	11.43	16.02	4.59
W1	54	3-2	4-2	15.09	15.81	0.72
W1	54	4-2	5-1	2.24	9.13	6.89
W2	54	4-2	3-2	2.24	16.02	13.78
W2	54	5-1	4-2	11.43	15.81	4.38
W2	54	3-2	5-1	15.09	9.13	−5.97

To overcome the impossibility to provide the desired TF intensity for all routes with the calculated control, it is proposed to shift slightly the green phases for routes for which the calculated capture moments are earlier than liberation moments. It is possible, however, only if this shift does not cause delays for routes which passage directly follows the passage of routes with shifted green phase. To check this possibility and to establish the rational values of phase shifts data must be arranged as in Table 5 by means of SQL. The table shows that for the phase sequence W1 such shifts are possible and their ranges for both routes are wide enough and for the phase sequence W2 such kind of control corrections id = s impossible.

Table 5. The data for checking the possibility of green phase shifts for certain routes

Phase sequence	Route 1	Route 2	Minimal delay	Time reserve for SP	Time reserve for route
W1	4-1	2-3	1.26	13.16	11.90
W1	4-3	2-3	3.77	13.52	9.75
W2	4-1	3-1	2.93	15.37	11.51
W2	4-1	3-2	2.93	14.45	
W2	4-2	3-2	2.93	13.78	10.85
W2	4-3	2-3	3.77	13.52	9.75
W2	5-1	4-2	5.97	4.38	−1.59

5   Aspects of "Driver's Education" for Their Adaptation to Combined Intelligent Control of Complex Crossroads

The problem of efficient use of throughputs of complex crossroads is not reduced to adequate choice of structural and parametric control. The second side of it consists in adaptation of the most drivers to it, notably to the control changes. Habitual behavior of drivers that regularly pass a certain crossroads around the same time may contradict to the present control and cause numerous errors and even conflict situation. The other drivers if they are guided by printed traffic organization schemes may make errors of misunderstanding of the present situation as well. As well as for more simple intersections, the principal solution of the problem is the introduction of highly available Internet-based online navigational tools that show the present crossroads passage scheme and the succession of TLC phases. It is especially important for multi-stage passage of a complex crossroads. Besides, it is necessary to remind drivers regularly about actual changes (that in a certain sense serves as real-time "education" of drivers) with all possible means including specialized radio channels and light displays over highways. "Education" of drivers must be especially important on the initial stage of combined intelligent control introduction.

6 Conclusions

Traffic control at an intersection with traffic light, based on the previously introduced structural-parametric optimization method [1, 2], is generalized to a multi-section crossroads. The specific features of such intersections and the problems of traffic control initiated by them are considered. Synthetic methods for solving control problems that combine methods of structural analysis of the route system, micro-level modeling of transport flows, methods of linear programming and relational databases techniques are presented. An illustration of the proposed model control is given. The calculations are performed to determine the optimal control. The materials of tables and graphs are the average data of visual observations of the authors of traffic at a particular multi-section crossroads. The necessary actual and operational data collection is feasible with the help of modern technical means. Obtaining experimental data by technical means and their initial processing is a rather complex legal and organizational problem related to obtaining permits for the installation of road surveillance cameras. However, despite some inaccuracy, the traffic parameters used in this work allow us to establish general patterns and build the algorithm for controlling traffic at complex intersections. In the future, it is planned to build updated models with the assistance of state bodies responsible for road safety and organization.

References

1. Solovyev, A.A., Valuev, A.M.: Structural and parametric control of a signalized intersection with real-time "education" of drivers". In: Hu, Z., Petoukhov, S.V., He, M. (eds.) Advances in Artificial Systems for Medicine and Education II. AIMEE2018 2018. Advances in Intelligent Systems and Computing II, vol. 902, pp. 517–526. Springer, Cham (2018)
2. Solovyev, A.A., Valuev, A.M.: Combined intelligent control of a signalized intersection of multilane urban highways. In: Hu, Z., Petoukhov, S.V., He, M. (eds.) Advances in Artificial Systems for Medicine and Education III. AIMEE2019. Advances in Intelligent Systems and Computing III, vol. 1126, pp. 471–480. Springer, Cham (2020)
3. Methodical recommendations on the design of traffic lights on highways. Federal Road Agency (Rosavtodor), Moscow, p. 5 (2013). (in Russian)
4. Ming, W., Qin, Y., Jie, X.: The application of SCOOT in modern traffic. Netw. Manag. Eng. (18), 93–98 (2015)
5. Rida, N., Ouadoud, M., Hasbi, A., Chebli, S.: Adaptive traffic light control system using wireless sensors networks. In: 2018 IEEE 5th International Congress on Information Science and Technology (CiSt), pp. 552–556. IEEE (2018)
6. Natafgi, M.B., Osman, M., Haidar, A.S., Hamandi, L.: Smart traffic light system using machine learning. In: 2018 IEEE International Multidisciplinary Conference on Engineering Technology (IMCET), pp. 1–6, November 2018
7. Adebiyi, R.F.O., Abubilal, K.A., Tekanyi, A.M.S., Adebiyi, B.H.: Management of vehicular traffic system using artificial bee colony algorithm. Int. J. Image Graph. Signal Process. (IJIGSP) 9(11), 18–28 (2017)

8. El-Tantawy, S., Abdulhai, B., Abdelgawad, H.: Multiagent reinforcement learning for integrated network of adaptive traffic signal controllers (MARLIN-ATSC): methodology and large-scale application on downtown Toronto. IEEE Trans. Intell. Transp. Syst. **14**(3), 1140–1150 (2013). Article no. 6502719
9. Zhou, B., Cao, J., Zeng, X., Wu, H.: Adaptive traffic light control in wireless sensor network-based intelligent transportation system. In: IEEE Vehicular Technology Conference (2010). Article no. 5594435. https://doi.org/10.1109/vetecf.2010.5594435
10. Zhou, C., Weng, Z., Chen, X., Zhizhe, S.: Integrated traffic information service system for public travel based on smart phones applications: a case in China. Int. J. Intell. Syst. Appl. (IJISA) **5**(12), 72–80 (2013)
11. Efiong, J.E.: Mobile device-based cargo gridlocks management framework for urban areas in Nigeria. Int. J. Educ. Manag. Eng. (IJEME) **7**(6), 14–23 (2017)
12. Goyal, K., Kaur, D.: A novel vehicle classification model for urban traffic surveillance using the deep neural network model. Int. J. Educ. Manag. Eng. (IJEME) **6**(1), 18–31 (2016)
13. Dennouni, N., Peter, Y., Lancieri, L., Slama, Z.: Towards an incremental recommendation of POIs for mobile tourists without profiles. Int. J. Intell. Syst. Appl. (IJISA) **10**(10), 42–52 (2018)
14. Signalized Intersections: Informational Guide. Publication Number: FHWA-HRT-04-091. U.S. Department of Transportation/Federal Highway Administration, Washington, D.C., August 2004
15. Glukharev, K.K., Ulyukov, N.M., Valuev, A.M., Kalinin, I.N.: On traffic flow on the arterial network model. In: Kozlov, V.V., et al. (eds.) Traffic and Granular Flow'11, pp. 399–412. Springer, Heidelberg (2013)
16. Solovyev, A.A., Valuev, A.M.: Organization of traffic flows simulation aimed at establishment of integral characteristics of their dynamics. Adv. Syst. Sci. Appl. **18**(2), 1–10 (2018)
17. Valuev, A.M.: Modeling of the transport flow through crossroads with merging and divergence points. In: Tsvirkun, A. (ed.) Proceedings of 2018 Eleventh International Conference "Management of Large-Scale System Development" (MLSD). Russia, Moscow, V.A. Trapeznikov Institute of Control Sciences, October 1–3 2018, pp. 1–3. IEEE Xplore Digital Library (2018). https://doi.org/10.1109/mlsd.2018.8551915

On the Efficiency of Machine Learning Algorithms for Imputation in Spatiotemporal Meteorological Data

Andrey K. Gorshenin[1(✉)] and Svetlana S. Lukina[2]

[1] Federal Research Center "Computer Science and Control" of the Russian Academy of Sciences, Moscow, Russia
agorshenin@frccsc.ru
[2] Faculty of Computational Mathematics and Cybernetics, Lomonosov Moscow State University, Moscow, Russia
svetl.luckina2016@yandex.ru

Abstract. The paper presents comparing various machine learning algorithms in the problem of imputation of missing values in the spatiotemporal precipitation data. Using a special procedure to convert complete data to incomplete one, up to 40% of missing values are artificially placed into the datasets. Then, they are imputed in order to determine the most effective machine learning algorithms with the same hyperparameters for more than hundred worldwide weather stations. A two-step procedure, where the classification results are used to improve regression accuracy, are implemented using Python programming language. The efficiency of various combinations of methods including random forests, classic and extreme gradient boosting, support vector machine, EM algorithm are analyzed. It is demonstrated that the best classifier is extreme gradient boosting with average forecasting accuracy of 83.41%. Combination of such methods as XGBClass+XGBoost leads to the best quality of missing values imputation with the normalized RMSE equals from 0.01 to 0.07. All of the above-mentioned methods are tested for the same hyperparameter settings for all weather stations. The novelty of this paper is in the selection of the universal methods for imputation, the accuracies of which are sufficient for processing spatiotemporal meteorological data regardless of their geographic locations even without the fine-tuning. The results obtained allow us to implement methods of computational statistics for detecting extreme precipitation correctly. The presented approaches are also effective for a wider class of observations, for example, environmental data.

Keywords: Precipitations · Spatiotemporal data · Missing values · XGBoost · Random forests · Support-Vector machines · EM algorithm

1 Introduction

Due to various reasons, the most real-world datasets are usually incomplete. Missing values in data could be critical, as they can distort the analysis results in various application areas. Therefore, it becomes necessary to impute missing values for further

Z. Hu et al. (Eds.): AIMEE 2020, AISC 1315, pp. 347–356, 2021.
https://doi.org/10.1007/978-3-030-67133-4_32

processing and analysis. One such important example is spatiotemporal meteorological data.

Increasing efficiency of machine learning algorithms has led to their use both in problems of analyzing the results of physical weather prediction models in order to obtain a more accurate forecast, and as independent tools for studying the spatiotemporal meteorological series obtained from satellites and weather stations. Such observations in large volumes have been collected from a huge number of sensors and often contain missing values that can significantly affect the quality of learning or moreover change the decisions of statistical models for the analysis of various meteorological phenomena, for example, extreme precipitation. Therefore, the problem of correct filling missing values in such data is very important [1].

In the paper, an analysis of the effectiveness of various machine learning methods for missing values imputations within spatiotemporal meteorological data is presented. These time-series have been collected at more than 100 stations in Russia, Europe, America, Asia and Africa, that is, there are no fixed regions or countries in contrast to the traditional approach in meteorology, when the data of only one country are under consideration [2–4]. The open database NNDC Climate Data of the National Oceanic and Atmospheric Administration (https://www.ncdc.noaa.gov/cdo-web/) is used as a source of real-world observations for testing ensembles of the machine learning algorithms.

The main purpose of the paper is to reveal the universal methods that would remain effective when analyzing data from any stations from regions that differ in their geographic location from the test ones. For testing, a special procedure to convert data from complete to incomplete, previously proposed by the authors in the paper [5], is used. It allows us to exclude randomly up to 40% of the volume of the test data. Thus, to examine the combinations of machine learning methods, the situation with irregular (random) missing values in the data of weather stations is simulated. In previous studies [5], the authors have demonstrated the effectiveness of this approach for two cities (Potsdam, Germany, and Elista, Russia) using a number of machine learning algorithms. In this paper, the set of methods is significantly extended, as well as the spatiotemporal meteorological data from more than hundred weather stations are analyzed under this approach.

2   Related Works

Various machine learning methods, including neural networks, are widely used in various applied research areas, for example, in finance [6], meteorology [7] or medicine [8]. In this paper, machine learning algorithms have been chosen due to their faster learning rates compared with neural networks.

The simplest way to avoid missing values is to exclude from processing a corresponding part of the sample, so the data becomes complete. However, in meteorological time-series, missing values may appear at any random moment in time, in particular, due to the peculiarities of their registration. Another approach is based on the reanalysis datasets [9]. However, in this case significant discrepancies with real observations are possible. Therefore, various statistical and machine learning methods have been developed to provide the possibility of correctly imputing the missing values for using, for example, in hydrological models [10, 11]. In particular, the values of neighboring stations can be involved effectively as additional data [12], but they are not always available. Thus, we are interested in methods that use only the meteorological time-series and perhaps some additional features that can be collected at the same geographical location, for example, dew point, wind speed, the lowest, the highest and mean temperatures, etc.

3 Methodology

For choosing the universal methods to impute the missing observations, complete spatiotemporal time-series (or their parts that did not contain missing values) have been converted to the incomplete using special random procedure (for details, see [5]). New data contains from one to three consecutive missing values (further, the notations "1 MV", "2 MVs" or "3 MVs" are used) per the subsequence of observations of some length (window). We use a value 5 in this paper, and their total number can vary up to the 40% of data volume. Then, using different machine learning algorithms, all these missing values should be filled and compared with the true values. The predicted variables are the daily precipitation events: whether there was precipitation, and what the corresponding volume is. The dew points, average temperatures and wind speeds at the same weather station are used as additional features.

Before training the algorithms, the initial data and predictors are normalized to range from 0 to 1. In [5], using the daily precipitation from Potsdam and Elista, it is demonstrated that a two-step procedure (the classification results are involved to improve solving regression problem) for handling missing values in meteorological time-series is effective for precipitation events. At the first step, the presence of precipitation (without interest to its volume) should be determined using one of classification algorithms. The second step is focused on the regression problem taking into account the classification results. In the paper, we develop the above-mentioned approach with the help of more machine learning algorithms for each of the stages and the significant amount of the spatiotemporal observations.

For the classification and regression, the following machine learning methods are chosen: support vector machine (SVM), random forests (RFs), gradient boosting (GB), extreme gradient boosting (XGBClass, XGBoost) [13], EM algorithm. The advantages of these algorithms are flexibility, accuracy and high-speed processing of large data volumes. They are widely used to solve applied scientific problems in various fields. For example, one can mention researches related to software [14] or big data problems [15].

To handle missing values, combinations (ensembles) of the above algorithms are implemented with the same hyperparameters for all datasets. Several metrics are involved for comparing results. For classification problems, *Accuracy* is used, which is determined by the expression: $Accuracy = \frac{N_0}{N}$, where N_0 is the number of correctly predicted gaps, N is the number of all gaps in the time series. In addition, the standard RMSE metric for continuous data (i.e., volumes) is used. It is determined as follows:

$$RMSE = \sqrt{\frac{1}{n}\sum_{i=1}^{n}(y_i - \hat{y}_i)^2},$$

where y_i are normalized initial values of precipitation, and $\hat{y}_i$ are corresponding predictions.

4   Comparison of Methods for Handling Missing Values for All Test Weather Stations

In this section, we consider the results of applying machine learning methods to impute artificially incomplete (as described above) spatiotemporal meteorological data. Examples of accuracies for stations from fixed regions and countries (Asia, North America, Russia) as well as the mean results obtained for ensembles of the machine learning methods for all analyzed weather stations will be demonstrated.

Figure 1 shows the average precipitation forecasting accuracy in the classification problem based on data from Asian stations. One can see that the best accuracy of the classifiers is achieved for the extreme gradient boosting (XGBClass). It varies from 88.4% to 88.9% for all numbers of missing values that can be called levels. The results for random forests and support vector machines are also successful (at least 87.3%), but for each of the missing levels they are less than the XGBClass accuracies.

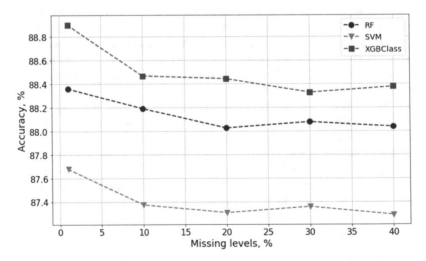

Fig. 1. Average values of classification accuracy of different methods for weather stations in Asia

Comparing results for various classifiers on more than 100 test weather stations, it was found that the highest average value (83.41%, see the third row of Table 1) for all of them is obtained for the XGBClass, while random forests and support vector

Table 1. Average values of classification accuracy of different methods for all weather stations for an arbitrary number of gaps

Methods	Average accuracy, %
RF	82.74%
SVM	82.08%
XGBClass	**83.41%**

machines have slightly more low, but fairly close values: 82.74% and 82.08%, respectively.

The RMSE values for normalized data under regression stage vary from 0.01 to 0.07 depending on geographic locations. Table 2 presents corresponding mean values for all analyzed weather stations. Several combinations of machine learning methods at once, namely RFs+GB, RFs+XGBoost, XGBClass+GB, XGBClass+XGBoost (see the first row of Table 2), demonstrate high values of the imputation accuracy, including those based on random forests.

Table 2. Average values of the errors of data imputation for all cases for all analyzed weather stations

Methods	RMSE
RFs+GB, RFs+XGBoost, XGBClass+GB, XGBClass+XGBoost	**0.031**
RFs+RFs, SVM+GB, SVM+RF, SVM+XGBoost, XGBClass+RF	0.032
RFs+EM, SVM+EM, XGBClass+EM	0.034

Algorithm 1. Testing ML methods for missing values imputations within spatiotemporal meteorological data

```
1:    function MLTestImputation(Data, MLs)
2:       i ← 1;
3:       for all (Data) do                                    // A set of test weather stations
4:          MVs ← MissingValues(Data_i);                      // Insertion of MVs
5:          F ← FeaturesSelection(Data_i);                    // Feature selection
6:          j ← 1;
7:          for all (MLs) do                          //Imputation of artificially inserted MVs
8:                                                    // for a set of algorithms
9:             [Class_{i,j}, Regr_{i,j}, Hybrid_{i,j}] ← Imput(Data_i, MVs, F, MLs_j);
10:            j++;                                           // Selection of the next method
11:         Plot(Class_i, Regr_i, Hybrid_i);                 // Visualize the results
12:         i++;                                   // Selection of the next weather station
13:      return [Class, Regr, Hybrid]
```

Thus, both the classification and regression problems can be solved quite accurately using extreme gradient boosting. It can be recommended as a base of two-step procedure to obtain the best accuracy in case of the minimal settings for arbitrary data. This boundary can be improved by fine-tuning, expanding the feature space, etc.

The method for comparing the results of various machine learning methods to fill missing values in spatiotemporal data is presented in Algorithm 1. It has been implemented in the Python programming language.

Figure 2 demonstrates an example of location of the analyzed weather stations in different Asian countries on a geographic map and the corresponding RMSE values for 1%, 10% and 20% missing values (one consecutive MV per window is allowed).

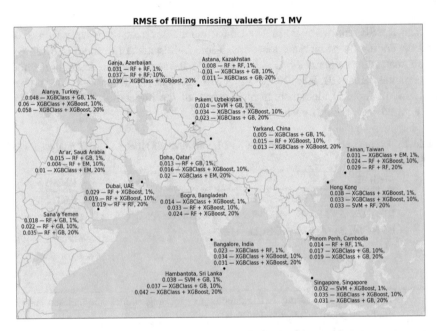

Fig. 2. Weather stations and the accuracy (RMSE) of missing values imputations

The combinations of algorithms for the best accuracy are also given. It is easy to see that different methods can be the most effective for stations at different geographical locations. However, in most cases, the extreme gradient boosting is more suitable.

Figure 3 shows RMSE values in case of handling 10% missing values (three consecutive MVs per window are allowed) in precipitation data from North American weather stations. The most efficient algorithms in terms of forecast quality are based on combinations of methods of random forest, extreme and classical gradient boosting. On the graphs, the corresponding points with minimal error are marked with right-pointing triangles. It is worth noting that as follows from the graphs the results of all methods are quite close. The values of RMSE range from 0.02 to 0.058.

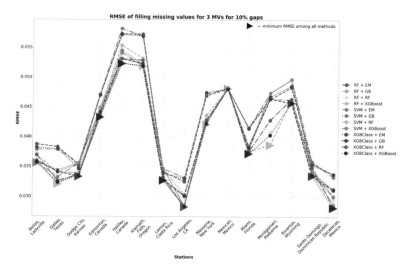

Fig. 3. RMSE of precipitation data imputation for 10% missing values for weather stations in North America. The case of three consecutive missing values per window

Figure 4 shows the RMSEs for weather stations in Russia under imputation of 40% of the missing values (two consecutive MVs per window are allowed) in data. XGBClass+XGBoost and XGBClass+GB algorithms demonstrate better predictive accuracy compared to other ones. The RMSE values of their prediction do not exceed 0.057, and the minimum value equals 0.01.

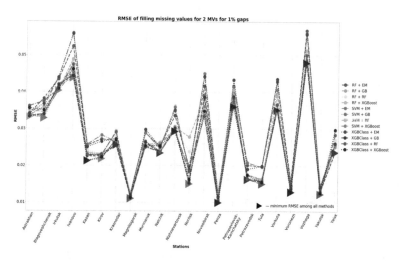

Fig. 4. RMSE of precipitation data imputation for 1% missing values for weather stations in Russia. The case of two consecutive missing values per window

It should be noted that the connecting lines between adjacent points on the chart are given only for a sake of vividness. They display a "smooth" picture, but do not demonstrate the nature of the change in the accuracy of the forecasts at the geographical points between the stations.

Obviously, the geographical locations of the collected data significantly affect the results of the methods. Therefore, the averaged results presented in Tables 1 and 2 are extremely important.

Figure 5 shows the results of filling missing values in the precipitation data of Aberporth, Wales (UK), for 1% and 20% missing levels using XGBClass+XGBoost. It can be seen that the trained boosting model quite well fits the real data. There are only a few mistakes for some outliers in the data. The results for other observed stations fully correspond to the examples that are given in the paper.

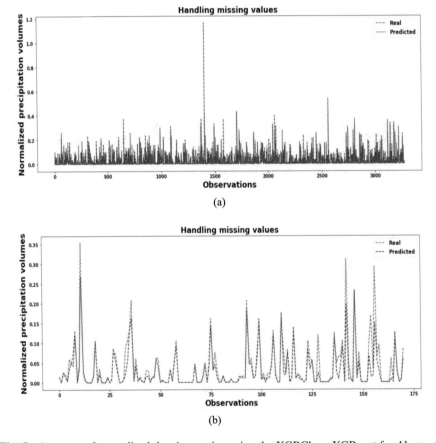

Fig. 5. Accuracy of normalized data imputation using the XGBClass+XGBoost for Aberporth, Wales (UK), station for 1% (a) and 20% (b) missing levels (two consecutive MVs per window)

5 Conclusions

In this paper, various machine learning algorithms are compared in the problem of missing values imputation in the spatiotemporal precipitation data. Herewith, a two-step procedure is implemented. The classification results are the first step that further are used to improve regression accuracy. The efficiency of various combinations of machine learning methods including random forests, classic and extreme gradient boosting, support vector machine, EM algorithm were analyzed using the real precipitation data collected at a hundred weather stations. It is demonstrated that the best classifier is extreme gradient boosting. Its average forecasting accuracy equals 83.41%. Combinations of such methods as XGBClass+XGBoost and XGBClass+GB, as well as RF+XGBoost and RF+GB have shown the best quality of missing values imputation for various data. Based on the normalized RMSE, errors of imputation data for different missing levels range from 0.01 to 0.07.

It is worth noting that all of the above-mentioned methods are tested for the same hyperparameter settings for all weather stations. Thus, the values obtained present base (lower) levels of accuracy that can be further increased, for example, by expanding the feature space [16, 17]. At the same time, the achieved accuracies are already at a high level, which means that the proposed approach can be effectively used by non-specialists in the field of data analysis.

The presented results make it possible to use mathematical models and statistical methods for detecting extreme precipitation [18, 19] correctly, since the initial data can be considered as complete. These approaches will also remain effective for a wider class of observations, for example, environmental data.

Acknowledgements. The research is partially supported by the Russian Foundation for Basic Research (project 18-29-03100) and the RF Presidential scholarship program (project 3956.2021.5). Authors thank the reviewers for their valuable comments that helped to improve the presentation of the material.

References

1. Sattari, M., Rezazadeh-Joudi, A., Kusiak, A.: Assessment of different methods for estimation of missing data in precipitation studies. Hydrol. Res. **48**(4), 1032–1044 (2017). https://doi.org/10.2166/nh.2016.364
2. Groisman, P., Knight, R., Karl, T.: Changes in intense precipitation over the central united states. J. Hydrometeorol. **13**(1), 47–66 (2012). https://doi.org/10.1175/JHM-D-11-039.1
3. Fan, J., Wang, X., Wu, L., Zhou, H., Zhang, F., Yu, X., Lu, X., Xiang, Y.: Comparison of support vector machine and extreme gradient boosting for predicting daily global solar radiation using temperature and precipitation in humid subtropical climates: a case study in China. Energy Convers. Manag. **164**, 102–111 (2018). https://doi.org/10.1016/j.enconman.2018.02.087
4. Xu, C., Qiao, Y., Jian, M.: Interdecadal change in the intensity of interannual variation of Spring precipitation over southern China and possible reasons. J. Clim. **32**, 5865–5881 (2013). https://doi.org/10.1175/jcli-d-18-0351.1

5. Gorshenin, A., Lebedeva, M., Lukina, S., Yakovleva, A.: Application of machine learning algorithms to handle missing values in precipitation data. In: Lecture Notes in Computer Science, vol. 11965, pp. 563–577 (2019). https://doi.org/10.1007/978-3-030-36614-8_43
6. Nayak, S.C.: Development and performance evaluation of adaptive hybrid higher order neural networks for exchange rate prediction. Int. J. Intell. Syst. Appl. 9(8), 71–85 (2017). https://doi.org/10.5815/ijisa.2017.08.08
7. Mishra, N., Soni, H.K., Sharma, S., Upadhyay, A.K.: Development and analysis of Artificial Neural Network models for rainfall prediction by using time-series data. Int. J. Intell. Syst. Appl. 10(1), 16–23 (2018). https://doi.org/10.5815/ijisa.2018.01.03
8. Akkar, H.A.R., Jasim, F.B.A.: Intelligent training algorithm for artificial neural network EEG classifications. Int. J. Intell. Syst. Appl. 10(5), 33–41 (2018). https://doi.org/10.5815/ijisa.2018.05.04
9. Stopa, J.E., Cheung, K.F., Tolman, H.L., Chawla, A.: Patterns and cycles in the Climate Forecast System Reanalysis wind and wave data. Ocean Model. 70, 207–220 (2013). https://doi.org/10.1016/j.ocemod.2012.10.005
10. Barrios, A., Trincado, G., Garreaud, R.: Alternative approaches for estimating missing climate data: application to monthly precipitation records in South-Central Chile. For. Ecosyst. 5, 28 (2018). https://doi.org/10.1186/s40663-018-0147-x
11. Teegavarapu, R., Aly, A., Pathak, C., Ahlquist, J., Fuelberg, H., Hood, J.: Infilling missing precipitation records using variants of spatial interpolation and data- driven methods: use of optimal weighting parameters and nearest neighbour-based corrections. Int. J. Climatol. 38 (12), 776–793 (2018). https://doi.org/10.1002/joc.5209
12. Simolo, C., Brunetti, M., Maugeri, M., Nanni, T.: Improving estimation of missing values in daily precipitation series by a probability density function-preserving approach. Int. J. Climatol. 30(10), 1564–1576 (2010). https://doi.org/10.1002/joc.1992
13. Chen, T., Guestrin, C.: XGBoost: a scalable tree boosting system. In: Proceedings of the 22nd ACM SIGKDD International Conference on Knowledge Discovery and Data Mining, pp. 785–794 (2016). https://doi.org/10.1145/2939672.2939785
14. Yang, N., Wang, Y.: Identify silent data corruption vulnerable instructions using SVM. IEEE Access 7, 40210–40219 (2019). https://doi.org/10.1109/ACCESS.2019.2905842
15. Lulli, A., Oneto, L., Anguita, D.: Mining big data with random forests. Cogn. Comput. 11 (2), 294–316 (2019). https://doi.org/10.1007/s12559-018-9615-4
16. Chandrashekar, G., Sahin, F.: A survey on feature selection methods. Comput. Electr. Eng. 40(1), 16–28 (2014). https://doi.org/10.1016/j.compeleng.2013.11.024
17. Wang, W., Du, X., Wang, N.: Building a cloud IDS using an efficient feature selection method and SVM. IEEE Access 7, 1345–1354 (2019). https://doi.org/10.1109/ACCESS.2018.2883142
18. Korolev, V.Yu., Gorshenin, A.K., Belyaev, K.P.: Statistical tests for extreme precipitation volumes. Mathematics 7(7), 648 (2019). https://doi.org/10.3390/math7070648
19. Korolev, V.Yu., Gorshenin, A.K.: Probability models and statistical tests for extreme precipitation based on generalized negative binomial distributions. Mathematics 8(4), 604 (2020). https://doi.org/10.3390/math8040604

Virtual Simulation Experiment Scheme for Software Development Courses Based on Situational Teaching

Qizhi Qiu[✉], Qingying Zhang, and Hancheng Wang

Wuhan University of Technology, Wuhan 430063, China
qqz@whut.edu.cn, kathy8899@126.com, 985375635@qq.com

Abstract. As one trend of education reform, more attentions have been paid on operative experiments than mental ones, and virtual simulation experiment has achieved pleasant progresses in China since the Ministry of Education advocated in 2013, which applied new IT technologies to improve the teaching effect of high-risk or high-cost experiments. In this paper, situational teaching is introduced into the design of virtual simulation experiment of software development courses. The construction of a virtual environment rather than the simulation of the high-risk equipment is discussed and the students' comprehensive application of various related knowledge in the virtual experimental environment is analyzed. Based on situational cognitive psychology, this paper describes the situation of virtual experiment as a triplet, puts forward the design mode of virtual simulation experiment from the perspective of situational teaching, and builds the architecture of virtual simulation experiment from the sight of computer system. Taking the requirement analysis experiment as an example, the design scheme of virtual simulation experiment is explained in detail.

Keywords: Situational teaching · Virtual simulation · Experiment teaching · Software architecture

1 Introduction

Due to the rapid development of modern information technology, virtual simulation experimental teaching has become a research hotspot in the domain of higher education [1]. The rising tide started in 2013 when Ministry of Education (MOE) of China issued the related official document. MOE authenticated 105 national virtual simulation experimental projects in the year of 2018, and 296 projects in 2019, which involves almost 200 universities or colleges, while the discipline distribution is unbalanced [2]. Most projects distribute in mechanical, medical science and chemical industry, where it is difficult to execute in laboratory, such as high-risk industry [3–5]. Via multimedia technology, virtual reality, the teaching effect is capable to be improved, called *virtual filling reality*.

There is another kind of experiment, whose teaching effect is urgent to be raised. Most of them involved intangible human intelligence, such as experiments of humanities and social sciences, management science, software development etc. This paper returns to the origin of education, bases on the pragmatism and situational cognitive psychology, and studies the design scheme for software development experiment developed by making use of modern information technology.

2 Recent Studies

John Dewey, a famous American educationist, is the founder of Pragmatic teaching theory. He advocated that the question of education is the question of taking hold of children's activities, of giving them direction. His ideas have impacted on modern education field for generations in China. The famous educationists in China, Shih Hu and Xingzhi Tao, are advocators and practitioners [6]. Some significant teaching methods, for example, Project Teaching Method, Dalton Laboratory, Winnetka System, have been selected and applied by many educators in China [7].

In recent years, educators in China continue practicing Dewey Theory in the way of focusing on cultivating students' scientific thinking under Pragmatic, emphasizing on learning by doing and cooperative learning [8]. A survey on the current situation of biology teaching in 12 senior high schools in Guangdong Province shows that students' interest in learning, self-confidence and enthusiasm have been improved through the design and practice of a new program called life teaching [9]. A set of engineering curriculum teaching scheme based on pragmatism and constructivism has been studied [10].

Situational teaching is the meaningful practice of pragmatic theory. By designing difficult situations on purpose, students are asked for to combine theory and practice. Situational cognitive psychology is theoretical basis for situational teaching, where knowledge is supposed to be contextualized and participation is supposed to promote human understanding. As a result, there are lots of applications of situational cognitive psychology, e.g. utilized to the interactive design of natural user interface to enhance the user experience, applied to the design of online competitive game [11, 12]. There are more utilization and practicing in pedagogic domain. Many educators have employed this teaching method during the design of the teaching system and the integrated learning environment [13–16].

3 Analysis on Software Development Teaching

3.1 Characteristics of Software

Software development courses or experiments have been studied, different teaching methods or ideas are applied [17, 18]. While the study will be developed in this paper, it is important to examine the characteristics of software. As integral parts, computer hardware and software are interdependence. On one hand, software cannot provide any service without hardware. On the other hand, software is distinguished to hardware from the following views:

- Invisibility: Unlike hardware, software is a logical rather than a physical system element. The service and performance of software may be observed only by running them. The invisibility also brings about the invisibility of the software production process. Unlike traditional industry products, software production does not involve the visible manufacturing process, it is the product of human intelligence. In other words, Software is developed or engineered, and it is not manufactured in the classical sense.

- Complexity: software has become more and more complex with the development of technology and increasing demands of people. Although sophistication and complexity may produce amazing results when a system succeeds, they can also need the professionals to figure out how to build complex systems. Undoubtedly, the increasing services result in complex software structure. Meanwhile, it is also a challenge to manage the complex development process [19].

- Non-wearout: as known, while time passes, the failure rate rises as hardware components suffer from the affects of dust, vibration, or other environmental maladies. It means, the hardware begins to wear out. Software does not wear out, it may deteriorate. Each software failure indicates an error in design or in the process through which design was translated into machine executable code, such as requirement analysis, system design, programming, and software testing.

- Reusability: Reusability is an important merit of software. A software component should be designed and implemented so that it can be reused in many different programs [20]. Whether components or methodology, may be reused on different occasions. Obviously, both are required high-skilled professional capabilities. Therefore, software development courses play an indispensable role during the students' career.

3.2 Software Development Teaching

Unlike programming courses, software development courses include a series courses, whose goals is to help students master the methodologies about software development, unlimited to a certain skill. The series courses are requirement analysis, architecture design, system design, software testing, software quality management, so on and so forth.

During these courses, students are supposed to master abstract theories about software developing methodology. As discussed above, the characteristics of software result in that the higher complexity of software project management, but also the higher difficulty of relative courses teaching. The teaching goals of these courses focus on two aspects:

- To help students understand the methodology of software development: It is the task of the in-class teaching. The software development methodology evolved with the progress of software industry from the structural methodology to object-oriented one. The students should understand the basic principles, such as United Modeling Language, modularization, to handle the future challenges.

- To help students practice the methodology: it is the task for experiment teaching. Mentioned above, software development is a typical logic and intelligent activity, it is hard for students to learn by analogy without any effective practice.

This paper focuses on how to apply theories into practical projects. Firstly, the paper bases on the idea "learning by doing" which is supported by pragmatism. Then, the well-arranged situations based on situated cognition will lead students to implement the whole development, so called as *reality filling mental*. These well-arranged situations can display every vital steps of developing activities so that students may apply the corresponding principles into the simulated project scenes on purpose.

4   Design Scheme for Virtual Simulation Experiments Based on Situational Teaching

Psychologists refer situation as the special environment where stimuli impact the individual mental status, while situational teaching intends to create or construct the emotional scene on purpose to bring about the different experience for students. This kind of experience will help students master the related knowledge [21]. This teaching method is originated to linguists in England in 1930 s, which are meaningful practical activities of pragmatism. Simply stated, while situational teaching applying, a series of situations are constructed where the knowledge points are weaved dedicatedly by instructors. Students learn the knowledge by doing something in the situations. During these activities, students also develop the other abilities, such as some non-intellectual abilities [22].

4.1   Definition of Situation

DEFINTION. *Situation* is a triplet as {R, B, O}
   Where,
   R: note for *role*, the number of roles is not a constant. Similar to drama, the number of roles and roles themselves are changing with storyline.
   B: note for *background,* the number of backgrounds is more than 1. *Background*s influence *role*s. Every *background* is designed by instructors and has its own preconditions or constraints. In context of situational teaching, these preconditions or constraints refer to prerequisite knowledge or skills.
   O: note for *object. role*s act on *object*s. *Situation* is supposed to provide a scene where *role* can do something, the activities of *role* will change the status of *object*s. Eventually, this change will result in the change of *background*.

4.2   Principles About Construction of the Situation Teaching

Educationists and researchers have proposed kinds of constructing principles for situational teaching, some of them have studied its application on virtual experiments [23, 24]. Based on these, the paper discusses the principles from the view of experiment teaching.

- Consistency: the purpose of experiment teaching should be consistent with in-class teaching; the required skills should be consistent with the students' ability; the constructed situation should be consistent with the syllabus. As a result, students may master specific skills and their comprehensive professionalism will be improved at the same time.

- Practicality: different from the abstractness of theoretical knowledge, the biggest feature of situational teaching is to provide a perceptible and experiential scene. In those perceptible and experienced scenes which is designed by instructors, students may apply abstract theories to simulated reality, and grasp the knowledge [25].
- Conflict: it is a very important principle in construct situations. As we know, there are many conflicts to increase attractiveness during drama creation. Instructors should also do that. There are several conflicts, such as the conflict between new and old knowledge, the conflict between the difficulty of the certain problem and students' abilities. All the conflicts will stimulate students' interest about learning.
- Progressiveness: a series of progressive scenes are constructed according to the experiment steps. The progressiveness will help students to get rid of the boring of the experiment and help students to understand the hierarchical relationship among the sub problems.

4.3   Design Pattern on Virtual Simulation Experiments

There are 3 parts in design pattern on virtual simulation experiments, shown in Fig. 1.

- Phase I: preparation phase, two activities are involved during phase I. The first one is to collect and classify the knowledge and skills needed in the experiment. On the other hand, investigation about the application scenarios should be taken. The situations do not have to be unique, but it must be able to stimulate students' curiosity and interest.
- Phase II: implementation phase, authoring script is the most important task during phase II. Instructors create progressive situations where the knowledge points are weaved. This is an iterative process, and careful review plays an irreplaceable role.
- Phase III: review phase, last but not least one. Construction principles in Sect. 4.2 will be applied one by one to ensure the quality of experiment scheme.

4.4   Architecture of Virtual Simulation Experiments

The virtual simulation experiment is the product of modern information technology and education technology. It is regarded as a computer-based system, as shown in Fig. 2. Considering continuous availability and portability, a hierarchical structure is selected, where the lower layer only provides service for the immediate upper layer [26, 27].

- The bottom layer is foundation, which is the cornerstone of the system. Computer hardware and operating system are deployed in this layer.
- The second layer is the support layer, where the development tools are deployed, such as Web integrated development environment, Unity and so on.
- The third layer is the component layer, the most important one. It is the core of this architecture, and is composed of knowledge components, problem components, situation components, and management components. (1) the knowledge component is responsible for managing the knowledge points and skills required in the experiment; (2) the question component is for managing the quiz that students

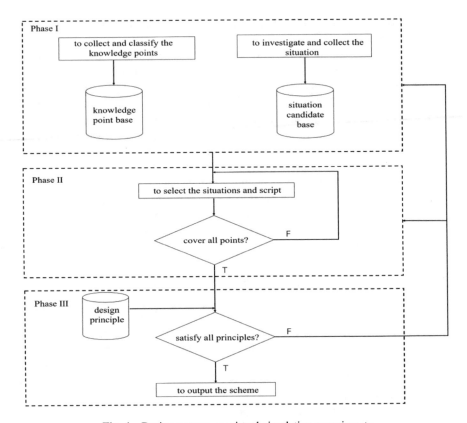

Fig. 1. Design pattern on virtual simulation experiments

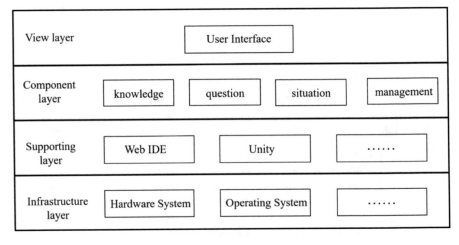

Fig. 2. Architecture of virtual simulation experiments

should participate during the experiment; (3) the situation component is for the management of the triples defined in Sect. 4.1; (4) the *management* component is for the management of the experimental data and learning data, such as students' personal information and learning behaviors information.

- The top layer is the view layer. It is responsible for the interaction with users, where the content of the component layer according to the script display to users, and the feedbacks from students are collected and delivered to the system or experiment platform.

5   An Example

5.1   Example Background

In this section, the virtual simulation experiment project of requirement analysis is selected as an example based on situational teaching. Although the entire process of software development is human intellectual activities, compared with other software activities, most students, and even many programmers, do not pay enough attention to requirements analysis. Most of them think that the products of requirement are less related with the executable software. However, software development is systematic procedure, where each step should not be replaceable or ignored [28, 29]. Statistical data shows that almost one-third of software failures exposed during software testing are traced back to requirement analysis [30]. Therefore, requirement analysis is included in some courses, such as software engineering, requirement engineering, software project management and so on.

Compared with other processes during software development, requirement analysis is more dependent on human intellectual activities. That results in that the experiment procedure is invisible and intangible physically [31]. Contrary to the most virtual simulation experiment, the paper adopts situational teaching to design the experiment scheme in order to achieve the teaching effect, called *reality filling mental*.

5.2   Overview of Situation Construction

Table 1 is about the overview of the designed requirement analysis experiment. A story background for the students has been constructed is the recruitment of analysts in a virtual project team of a software company. Table 1 shows the detailed content of the situation triplet in Sect. 4.1. The number of *role* elements is more than one, among which the student acts the protagonist of the "analyst", other roles, such as project manager, architect, programmer, and test engineer, are virtual roles. The *background* element has a series of progressive sub-*background*, which contains the knowledge points, the problems to be solved and the required skills. The set of *background* reflects the progress of the experiment. the *object* element is composed of a series of documents, some of them are about experiment and courses, e.g. experimental report. Some of them are about software development, e.g. software requirement specification. Both of them act as milestones and imply the progress of the experiment.

Table 1. Overview of situation construction

	Description
Experiment title	Requirement analysis
role	≥ 1, analyst is the most important one
background	≥ 1, changing with the experiment progress
object	≥ 1, documents as outputs of several key steps

5.3   Details About Situations Constructed

The section will select two situations of the example experiment and introduce two situations in detail, which are *recruitment* and *discussion* situation.

#1 *recruitment situation*: the teaching purpose is to test students' prerequisite including the understanding and mastery of requirement analysis.

Role: candidate for recruitment, played by students.

Background: there are 4 sub-backgrounds, as shown in Table 2. The precondition column records whether prerequisite is needed in the sub-background. The component column records the four types of components stated in Sect. 4.4. For example, sub-background sub3 (online interview) requires students to complete a testing sheet containing 3–5 questions, which are mainly related to the overall goals and tasks of requirement analysis.

Table 2. Backgrounds of *recruitment situation*

	Description	Precondition	Component
1	Surrounding of a company,	–	S, M
2	Meeting room	–	S, M
3	Online interview	√	P, S, M
4	Welcome	√	S, M

Note: Q stands for question component, S stands for situation component, and M stands for management component, √ stands for precondition needed.

Object: The testing sheet is the object of sub-background sub3 in *recruitment situation*. Students change the state of the testing sheet by answering the questions, and the change of the object state will bring about a change in the situation. In this case, the change is from the sub-background sub3 to the sub-background sub4.

#2 *discussion situation*: the teaching purpose is to strengthen the knowledge point about requirement categories. The students will observe a group meeting about requirement analysis to review the knowledge about requirement categories and user types. The knowledge lays a foundation for subsequent requirement documentation.

Role: analyst (played by students), project manager, 1-2 team member. Except for analysts, the other roles are virtual also.

Background: there are 3 sub-backgrounds, as shown in Table 3. Sub-background sub3 is to watch a prepared video. The video presents a discussion meeting. During the meeting, the involved knowledge is about the requirement categories and user types. By watching the video, students not only review the related knowledge points, but also learn how to carry out discussion activities during requirement analysis.

Table 3. Backgrounds of *discussion situation*

	Description	Precondition	Component
1	Meeting room	–	S, M
2	Video	$\checkmark$	K, S, M
3	Office scene	–	S, M

Note: K stands for knowledge component, S stands for situation component, and M stands for management component, $\checkmark$ stands for precondition needed.

Object: The object in discussion situation is a list of requirements that students are required to complete in the sub-background sub3. The state change of the *object* will end *discussion situation* and enter the next one.

6   Conclusions

The pursuit of teaching quality is the mission of each educator. They try to practice various advanced teaching methods in all aspects of teaching. Creating virtual simulation experiments is one of new trends. More attentions are paid on mental experiments in this paper, such as experiments of software development courses, than traditional engineering courses. There are two different views during the design procedure. The first one is situation cognition psychology; the other is computer science. From view of the first one, the connation of situation is discussed, and design principles and pattern is given followed the situational teaching rules. From view of computer science, the characteristics of software and features of software development are discussed, and architecture of virtual simulation experiment is proposed. The above ideas are understandable and practical, and they are meaningful exploration and practice of virtual simulation experiment. It is easy to apply into other similar courses.

Acknowledgement. This research is financially supported by 2019 and 2020 Wuhan University of Technology Virtual Simulation Experiment Project.

References

1. Wang, W., Hu, J., Liu, H.: Current situation and development of virtual simulation experimental teaching of overseas universities. Res. Explor. Lab. **34**(05), 214–219 (2015)

2. Gao, Z., Wang, X., Yan, J., et al.: Current situation and challenge of construction of virtual simulation experimental teaching projects in China. Exp. Technol. Manag. **37**(5), 9–219 (2020). (in Chinese)

3. Shafiq, S., Sanin, C., Szczerbicki, Ed., Toro, C.: Towards an experience based collective computational intelligence for manufacturing. Future Gener. Comput. Syst. **66**, 89–99 (2017)

4. Lu, L., Xu, J., Ge, W., et al.: Computer virtual experiment on fluidized beds using a coarse-grained discrete particle method—EMMS-DPM. Chem. Eng. Sci. **155**, 314–337 (2016)

5. Yin, X., Niu, Z., He, Z., et al.: An integrated computational intelligence technique based operating parameters optimization scheme for quality improvement oriented process-manufacturing system. Comput. Ind. Eng. **140**, 106284 (2020). https://doi.org/10.1016/j.cie.2020.106284

6. Tang, B., Tang, T.: To educate by vocation: the analysis of dewey's thought of vocational transformation of education. Educ. Hist. Stud. **4**, 124–132 (2019). (in Chinese)

7. Wu, H., Luo, J., Xu, J.: Herbart's five-step teaching method in modern china and its implementation in several provinces. J. Hengshui Univ. **17**, 110–118 (2015). (in Chinese)

8. Guo, X.: Application of "situational creation-inspired question" teaching elementary school thinking training classroom. Guide Sci. Educ. **9**, 125–127 (2013). (in Chinese)

9. Li, F.: Investigation and Practice Research on living teaching in Senior Middle school. Dissertation of Qufu Normal University (2018). (in Chinese)

10. Qu, Y., Wang, Z., Wang, J.: The research status of bilingual teaching mode in China Universities and Colleges. High. Educ. Sci. **2**, 104–108 (2014). (in Chinese)

11. Ding, J., Wu, Y.: Research on user experience design of natural user interface. Design. **12**, 65–67 (2019). (in Chinese)

12. Duan, Q.: The Experience Design of Online Competitive Games. Dissertation of Jiangnan University (2018). (in Chinese)

13. Jain, S.A., Goyal, L.: Raspberry pi based interactive home automation system through e-mail. In: International Conference on Optimization Reliability, and Information Technology (ICROIT), pp. 277–280 (2014)

14. Yang, Y.: The innovation of college physical training based on computer virtual reality technology. J. Discr. Math. Sci. Cry **6**, 1275–1280 (2018)

15. Wang, Y., Wang, Q.: Design for authentic learning in MOOCs: from the perspective of situated cognition. Distance Educ. China **3**, 5–13+79 (2018). (in Chinese)

16. Tian, S., Shi, Z.: Practical strategies for English Situational Teaching in Colleges. Stud. Lit. Lang. **9**, 61–63 (2014)

17. Jaramillo, S., Cadavid, J., Jaramillo, C.: Agile project based learning applied in a software development course. In: 12th Annual International Conference of Education, Research and Innovation (2019). https://doi.org/10.21125/iceri.2019.1918

18. Plante, D., Branton, M., Oliphant, G.: Introducing entrepreneurship into an undergraduate software development course. Bus. Educ. Accred. **8**(2), 79–86 (2016)

19. Kumar, M., Rashid, E.: An efficient software development life cycle model for developing software project. Int. J. Educ. Manag. Eng. (IJEME) **8**(06), 59–68 (2020)

20. Kuutila, M., Mäntylä, M., Farooq, U., Claes, M.: Time pressure in software engineering: a systematic review. Inf. Softw. Technol. **121**, 106257 (2020). https://doi.org/10.1016/j.infsof.2020.106257

21. Fatima, S., Abdullah, S.: Improving teaching methodology in system analysis and design using problem based learning for ABET. Int. J. Mod. Educ. Comput. Sci. (IJMEC) **5**(7), 60–68 (2013)

22. Kuo, H., Tseng, Y., Yang, Y.: Promoting college student's learning motivation and creativity through a STEM interdisciplinary PBL human-computer interaction system design and development course. Think. Skills Creat. **31**, 1–10 (2019)
23. Gao, H.: Research on the construction of virtual simulation experiment teaching courses of economics and management majors.In: 4th International Conference on Culture, Education and Economic Development of Modern Society (ICCESE) (2020). https://doi.org/10.2991/assehr.k.200316.214
24. Fuentes, J., Velasco, M.: Investigating the effect of realistic projects on students' motivation, the case of human-computer interaction course. Comput. Hum. Behav. **72**, 692–700 (2017)
25. Sun, F., Sun, J., Liu, G., et al.: Research on evaluation of virtual simulation experiment teaching project. Exp. Technol. Manag. **37**(07), 187–190 (2020). (in Chinese)
26. Feliu, J., Sahuquillo, J., Petit, S.: Designing lab sessions focusing on real processors for computer architecture courses: a practical perspective. J. Parallel Distrib. Comput. **118**, 128–139 (2018)
27. Venters, C., Capilla, R., et al.: Software sustainability: research and practice from a software architecture viewpoint. J. Syst. Softw. **138**, 174–188 (2018)
28. Holtkamp, P., Jokinen, J., Pawlowski, J.: Software competency requirements in requirements engineering, software design, implementation, and testing. J. Syst. Softw. **101**, 136–146 (2015)
29. Bilgaiyan, S., Aditya, K., Mishra, S., et al.: A swarm intelligence based chaotic morphological approach for software development cost estimation. Int. J. Intell. Syst. Appl. **10**(9), 13–22 (2018)
30. Unterkalmsteiner, M., Gorschek, T., Feldt, R., et al.: Assessing requirements engineering and software test alignment—five case studies. J. Syst. Softw. **109**, 62–77 (2015)
31. Inayat, I., Salim, S., Marczak, S., et al.: A systematic literature review on agile requirements engineering practices and challenges. Comput. Hum. Behav. **51**(Part B), 915–929 (2015)

Evaluation Model of Sports Media Talent Training System in the New Media Era

Ziye Wang[1(✉)], Yao Zhang[2], and Xinpeng Zhao[1]

[1] College of Journalism and Communication,
Wuhan Sports University, Wuhan 430079, China
kathy8899@126.com
[2] School of Logistics Engineering, Wuhan University of Technology,
Wuhan 430063, China

Abstract. The training of sports media professionals is a complex system engineering. This paper analyzes the new requirements for sports media professionals in the new media era from the aspects of personal quality, professional quality and professional skills, discusses the basic ideas of the evaluation of talent training system, and establishes the talent cultivation evaluation model framework by using AHP and fuzzy comprehensive evaluation method. The discussion covers the establishment of index system, the construction of comment set, the determination of index weight, the solution of fuzzy evaluation matrix, the calculation of comprehensive evaluation vector and the establishment of evaluation model. Through the evaluation of the talent training system and its operation quality, the paper provides the reform ideas for education and teaching of sports media.

Keywords: Sports media major · Talent training · Personnel training system · Fuzzy comprehensive evaluation method

1 Introduction

We are now living in the era of health and sports, which is the age of personal physical and mental beauty [1]. In the future, a large part of the students of Media College will become public figures. They shoulder the responsibility of spreading sports spirit, popularizing sports knowledge, promoting the construction of a healthy country, and leading the healthy culture. Today's era is also an age of new media and the integration of multiple media [2–4]. The information dissemination has got richer channels, but also put forward higher requirements for sports media professionals [5, 6]. It is necessary to keep pace with the development of science and technology, and give full play to the role of "gatekeeper" and "encourager". This puts forward higher requirements for the personnel training system [7–9]. How to cultivate innovative talents in sports communication has become a hot topic.

© The Author(s), under exclusive license to Springer Nature Switzerland AG 2021
Z. Hu et al. (Eds.): AIMEE 2020, AISC 1315, pp. 368–378, 2021.
https://doi.org/10.1007/978-3-030-67133-4_34

2 The Training Requirements of Sports Media Professionals in the New Media Era

The cultivation of sports media professionals is not only the need of healthy China construction, but also an important measure to improve the quality of the whole people [10, 11]. The development of society needs more and more excellent sports media personnel. Under the guidance of OBE (Outcome Based Education) concept, the research on the reform of sports media talents training mode is to explore better education and teaching mode and cultivate high-quality innovative sports media professionals [12].

2.1 Personnel Training from a Systemic Perspective

Talent training is a complex giant system with many elements, complex interaction mechanism and contradictory relationship between internal factors and external environment. In the reform of the training system of sports media professionals, it is necessary to conduct a comparative study on the demand, performance and effect of sports media professionals under the background of healthy China construction through in-depth investigation, comprehensively consider various factors, analyze the system and its elements and environment, and use the tracking research method in the field of natural science to study the training situation of sports media professionals. It is necessary to make the evaluation of the comprehensive quality of sports media professionals in detail according to the principle of establishing an evaluation system, so as to explore the effective methods and ways of talent training [13].

2.2 Basic Objectives of Sports Media Education

The cultivation of sports media professionals generally needs to focus on the discipline and industrial frontier of the development of radio and television, aiming to cultivate applied talents who are capable of hosting the radio and television programs, making interview and production, completing sports commentary, and are engaged in broadcasting and hosting and other language communication work in radio, television, network and other media institutions [14].

2.3 Requirements of Sports Media Professionals in the New Media Era

Sports media professionals have some special requirements. In addition to mastering and understanding sports activities and events as comprehensively as possible, they also need to have a smart mind, a keen perspective, rapid and swift judgment, accurate and reasonable expression, warm and moderate interaction with the audience, and they also need to master new technologies and skills.

The quality requirements of sports media professionals can be summarized into three aspects, namely personal quality, occupation quality and professional skills.

2.3.1 Personal Quality of Sports Media Professionals

For sports media professionals, their quality requirements mainly includes four aspects

(1) Social responsibility. Social responsibility is essential for a media worker known as a gatekeeper of public opinion;

(2) Humanistic literacy. It refers to a person's inner quality, people-oriented, people-centered;

(3) New media thinking. For media practitioners in the information age, new media thinking is crucial;

(4) Innovative spirit. The spirit of innovation is a new perspective, new thinking, new methods, it can help media workers change their ideas and do their work more effectively and valuable.

2.3.2   Occupation Quality of Sports Media Professionals

Sports media professionals need to have high occupation quality. Sports and event knowledge need to be understood and mastered. Only in this way can the sports events and the content be spread out from a professional perspective, and guide the public to better appreciate sports and actively participate in sports activities. In addition, language skills are the basic crafts, while the interpretation ability can make the media practitioners have stronger working ability. In the same competition, the interpretation effect of different commentators may be quite, even completely different, the degree of attraction and the incentive effect for the audience may also be completely different, which largely depends on the occupation quality and level of professional commentators.

2.3.3   Professional Skills of Sports Media Workers

We have entered a highly technical and information age. Sports media professionals need to have sufficient professional skills, mainly including:

(1) Field control capability. When the sports competition is very fierce, the mood of the audience may be out of control. At this time, media workers, such as sports commentators, need to try their best to guide fans to calm down and create a good watching environment;

(2) Interaction with the audience. This kind of ability is very necessary. When the scene is relatively calm or even cold, it is necessary to be able to "warm up". When the scene is too intense or even in conflict, it is essential to calm down the audience in a friendly way. This is the responsibility of media workers and a critical moment to test their ability;

(3) Media people must be able to complete the multiple work of editing and broadcasting at the same time, become "versatile" and "all-round king", and become the one who can not be easily replaced;

(4) Media workers need to be familiar with new media tools. With the development of science and technology, new media tools are constantly emerging. Sports media professionals need to have a strong learning ability and be able to use and adapt to tools in order to continuously accept challenges and make continuous progress.

3 Evaluation Method of Talent Training System of Sports Media Major

The evaluation of sports media professional talent training system is a complex system engineering with multi-objective and multi-attribute, and the main evaluation methods are: analytic hierarchy process, fuzzy comprehensive evaluation method, fuzzy analytic hierarchy process, etc., where the first two methods are mainly used [15–19].

The key job of evaluation is to establish an assess model. In the evaluation process of sports media professional talent training system, each evaluation factor has fuzzy characteristics, so does the weight of each factor [20–24]. This paper introduces Delphi method, an important tool widely used in group decision-making, organically combines the traditional Analytic Hierarchy Process (AHP) and Delphi method, and proposes a hierarchy based on Delphi method, puts forward an idea of determining the weight of each index factor, and uses the fuzzy comprehensive evaluation method to construct the evaluation model to assess the evaluation object comprehensively [25]. The specific framework of talent training evaluation model is shown in Fig. 1.

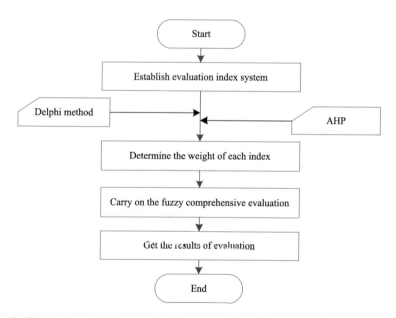

Fig. 1. Evaluation model framework of sports media professional talent training system

AHP is a decision-making method that decomposes the systematic elements which are always related to decision-making into objectives, criteria, schemes and other levels. Qualitative and quantitative analysis is carried out on this basis. The key link is to establish judgment matrix and check the consistency of weight vector. The advantage of AHP is that in the case of complex structure of judgment factors and lack of necessary data, it can quantify the relative importance of evaluation factors that are difficult to quantify by pair-wise comparison, turn complex estimating factors into a clear hierarchical structure, and then effectively determine the relative importance of each evaluation factor.

The fuzzy comprehensive evaluation method is proposed based on the nonlinear characteristics of the evaluation process. It is an integrative assessment approach based on the operation rules of fuzzy mathematics to quantitatively synthesize the evaluation universe with fuzzy properties, and then obtain comparable quantitative evaluation results. This method is more able to balance the factors at all levels and achieve a combined evaluation than the ordinary evaluation method. The basic idea of fuzzy comprehensive evaluation method lies in: on the basis of determining the evaluation indexes and their weights at all levels, using the principle of fuzzy set transformation, describing the fuzzy boundary between all levels of indicators and evaluation set with membership degree, constructing fuzzy evaluation matrix, and finally determining the grade of evaluation object through multi-level compound operation.

4   The Implementation Steps of the Evaluation of Talent Training System of Sports Media Major

4.1   Founding Evaluation Index System

Sports media professionals training system is a complex system involving many elements. According to the evaluation principles of objectivity, comprehensiveness, feasibility and systematicness, combined with the reality and referring to relevant literature, this paper constructs the evaluation index system of sports media professional talent training from two aspects of "soft index" and "hard index", as shown in Table 1. There are 9 primary indicators, including 4 soft indicators and 5 hard ones.

4.2   Building a Set of Comments

The evaluation results are divided into five grades: excellent, good, medium, qualified and unqualified, which are recorded as:

$$V = (\text{excellent, good, medium, qualified and unqualified}).$$

Because each grade in the evaluation set is fuzzy and not intuitive enough, it is quantified and determined that the value range of "excellent" quality is 90–100, and the median value is 95; the value range of "good" quality is set as 80–90, and the middle value is 85, and so on, so as to obtain the percentage standard set of operation quality of talent training system as:

$$V = (95, 85, 75, 65, 55)$$

Table 1. Evaluation index system of sports media professional training

Primary indicators		Weight	Secondary indicators	Weight
Soft index	School development and policy support u_1	0.12	Sustainable development goals u_{12}	0.3
			Capital investment u_{12}	0.4
			Policy and system guarantee u_{13}	0.3
	Education and teaching management u_2	0.13	Innovation of teaching thought u_{21}	0.22
			Teaching content u_{22}	0.2
			Heuristic teaching u_{23}	0.23
			Teaching style and study style u_{24}	0.15
			Assessment system u_{25}	0.2
	Team building u_3	0.1	Reasonable team structure u_{31}	0.28
			Innovation and practice ability u_{32}	0.3
			Knowledge and skill complementation u_{33}	0.42
	Academic environment and atmosphere u_4	0.12	Cooperation and competition coexist u_{41}	0.2
			Embodiment of team spirit u_{42}	0.3
			Excitation mechanism u_{43}	0.3
			Campus learning environment u_{44}	0.2
Rigid index	School enterprise cooperation u_5	0.07	Employ enterprise tutor u_{51}	0.5
			Teachers training in enterprises u_{52}	0.5
	Team building effect u_6	0.13	Team size u_{61}	0.25
			Number of published papers u_{62}	0.4
			Awards in discipline competition u_{63}	0.35
	Opening of experimental platform u_7	0.08	Interdisciplinary opening u_{71}	0.5
			Interdisciplinary opening u_{72}	0.3
			Cross University opening u_{73}	0.2
			Equipment u_{81}	0.6

(*continued*)

Table 1. (*continued*)

Primary indicators	Weight	Secondary indicators	Weight
Experimental base and equipment u_8	0.1	Equipment u_{81}	0.6
		Application effect u_{82}	0.4
Innovation practice activities and project support u_9	0.15	Participate academic competition u_{91}	0.3
		Cooperative research with enterprises u_{92}	0.2
		Items in innovation training plan u_{93}	0.25
		Other innovation research projects u_{94}	0.25

4.3 Fixing the Weight of Indicators at All Levels

In order to improve the shortcomings of AHP, such as strong subjectivity and difficult consistency test, this paper adopts the following basic ideas to determine the weight of all levels of indicators: firstly, the Delphi method is used to compare and analyze the importance and relative significance of indicators at various levels in the quality evaluation index system, and make full use of the knowledge, experience and wisdom of experts. After repeated consultation, induction, modification and feedback, the experts' opinions were summarized and the corresponding judgment matrix was constructed.

The next step is to find the maximum eigenvalue of each judgment matrix and the corresponding eigenvector according to AHP, and normalize it). The vector set obtained corresponds to the weight set of all levels of indicators (as shown in Fig. 1). Since the Delphi method was used to summarize the experts' opinions, the consistency test of the judgment matrix of each group of indicators can be simplified.

Finally, it is concluded that: the corresponding weight vector of the first level index $U = (u_1, u_2, \ldots, u_9)$ is $A = (0.12, 0.13, 0.1, 0.12, 0.07, 0.13, 0.08, 0.1, 0.15)$, while the corresponding second level index U_i to the first level index u_i, is $U_i = (u_{i1}, u_{i2}, \ldots, u_{im})$, and the corresponding weight vector is $A_i = (a_{i1}, a_{i2}, \ldots, a_{im})$, where, $1 \leq i \leq 9$, m is the number of secondary indicators under the corresponding first level index u_i, and the specific weight of each level of index is shown in Table 1.

4.4 Solving Fuzzy Evaluation Matrix

According to the initial model, a comprehensive evaluation is made for each secondary index (m, totally) under each first level index u_i, thus, the fuzzy evaluation matrix from the first level index u_i to the comment set V is established, marked as R_i, where, $R_i = (r_{sj})_{m \times k}$, s represents the serial number of the corresponding secondary index under the first level index u_i, and the value range is $[1, m]$; j is the sequence number of the corresponding level in the comment set, while its value range is $[1, k]$; here $k = 5$, implies, there are five comment levels.

r_{sj} refers to the membership degree of the s second level index of subordinate u_i to the evaluation of level j, that is, the proportion of the number of people who give grade j comments to the total number of evaluation people when evaluating the s index under u_i, which meets the normalization requirements.

For example, under a certain level of indicators, there are 20 people who give a good score on the first two indicators, and the total number of evaluators is 40, then the evaluation matrix element under the first level index $r_{12} = 20/40 = 0.5$, and other membership degrees can be obtained similarly. The evaluation group composed of 40 evaluators (including education experts, school teachers, off campus tutors, students, employer representatives, etc.) gives corresponding comments according to the evaluation requirements contained in each index. The specific evaluation results of each index are shown in Table 1. From the above algorithm, the fuzzy judgment matrix of the first level index "school development and policy support" can be obtained as follows:

$$R_1 = \begin{bmatrix} 0.7 & 0.3 & 0 & 0 & 0 \\ 0.8 & 0.2 & 0 & 0 & 0 \\ 0.75 & 0.25 & 0 & 0 & 0 \end{bmatrix}$$

The fuzzy judgment matrix of other first-order indexes can be obtained similarly, which are respectively recorded as $R_1, R_2, R_3, \ldots, R_9$.

4.5 Calculating the Comprehensive Evaluation Vector of Each Level Index

The fuzzy evaluation vector of the first level index is composed of the product of the weight vector of the second level index under the first level index and the corresponding fuzzy judgment matrix, which is recorded as B_i, $B_i = A_i \times R_i$. According to this method, the fuzzy evaluation vectors of nine first-class indexes can be calculated as follows:

$B_1 = (0.755, 0.245, 0, 0, 0)$;
$B_2 = (0.707, 0.264, 0.029, 0, 0)$;
$B_3 = (0.673, 0.312, 0.015, 0, 0)$;
$B_4 = (0.713, 0.23, 0.057, 0, 0)$;
$B_5 = (0.725, 0.25, 0.025, 0, 0)$;
$B_6 = (0.699, 0.262, 0.039, 0, 0)$;
$B_7 = (0.73, 0.21, 0.06, 0, 0)$;
$B_8 = (0.71, 0.29, 0, 0, 0)$;
$B_9 = (0.743, 0.217, 0.04, 0, 0)$.

The calculation results are kept to three decimal places.

4.6   Establishing Comprehensive Evaluation Model

The fuzzy relation from the first index U to the comment set V is represented by fuzzy matrix R, where $R = (b_{ij})_{9 \times 5}$, $1 \leq i \leq 9$, $1 \leq j \leq 5$. Namely:

$$R = (b_{ij})_{9 \times 5} = \begin{bmatrix} B_1 \\ B_2 \\ \cdots \\ B_9 \end{bmatrix} = \begin{bmatrix} 0.707 & 0.264 & 0.029 & 0 & 0 \\ 0.673 & 0.312 & 0.015 & 0 & 0 \\ 0.713 & 0.230 & 0.057 & 0 & 0 \\ 0.725 & 0.250 & 0.025 & 0 & 0 \\ 0.699 & 0.262 & 0.039 & 0 & 0 \\ 0.730 & 0.210 & 0.060 & 0 & 0 \\ 0.710 & 0.290 & 0 & 0 & 0 \\ 0.743 & 0.217 & 0.040 & 0 & 0 \end{bmatrix}$$

After that, the comprehensive evaluation vector of U is constructed by the weight of the first level index and the matrix product of R, which is recorded as B, then,

$$B = A \cdot R = (0.718, 0.252, 0.03, 0, 0)$$

Furthermore, the comprehensive appraising value of the valuators for the sports media professional talent training system can be calculated as follows:

$$I = B \cdot V^T = (0.718, 0.252, 0.03, 0, 0) \, (95, 85, 75, 65, 55)T = 91.88$$

Finally, according to the corresponding relationship between the evaluation value I and the evaluation set, it can be concluded that the construction quality of the talent training system is at a good level.

5   Extension of Fuzzy Comprehensive Evaluation Model

As the construction of sports media professional talent training system involves many factors, the evaluation of it is a very rigorous work, which must be taken seriously in a fair, just and open attitude. If it is only evaluated from a certain angle, some malicious scores may definitely occur with partial general congruence, which will make the final evaluation result deviate greatly.

Therefore, in the actual evaluation process, in order to make the evaluation results more scientific, reasonable and accurate, the evaluators are generally composed of various personnel, such as the education experts, school leaders, teachers, students, etc. Due to the different degrees of influence of different groups of people on the evaluation results, the corresponding weights of evaluation results are also different. The general calculation method is as follows: Suppose there are k subjects involved in the evaluation. First, the fuzzy comprehensive evaluation results of all kinds of evaluators are calculated, which is recorded as $B_1, B_2, \ldots, B_k$, and the corresponding weight of evaluation results is $w_1, w_2, \ldots, w_k$. Then the fuzzy evaluation vector is as follows:

$$B^* = (w_1, w_2, \ldots, w_k) \cdot (B_1, B_2, \ldots, B_k)^T$$

Finally, through calculation, the evaluation value is obtained as:

$$I = B^* \cdot V^T$$

6 Conclusion

On the basis of comprehensive and multi angle analysis, the evaluation index system of sports media professional personnel training is constructed from two aspects of "soft index" and "hard index". The weight of each index is calculated by using the improved AHP, and then the establishment of talent training system of sports media specialty is integrated evaluated by using fuzzy comprehensive evaluation method. The evaluation model and the system can give a relatively scientific and reasonable evaluation of the personnel training system and its operation quality.

According to the talent development strategy, the education of sports media major should construct or update and perfect the education and training system of sports media professionals from the perspective of system, especially in the new media era. It is necessary to seriously explore the teaching concept, teaching methods and teaching technology in the process of talent cultivation, improve the training mode of sports media and its professionals, and to provide more sports media professionals with advanced thinking mode, diverse knowledge and skills, continuous learning ability and steady progress potential for the society.

Acknowledgements. This paper is supported by—(1) Teaching Research Project "Research on the Reform of Sports Media Professional Training Model Based on OBE Concept"; (2) Wuhan Sports University Young Teachers' Scientific Research Fund Project "The MECE study under the background of Sports Events media convergence".

References

1. Taks, M.: Social sustainability of non-mega sport events in a global world. Eur. J. Sport. Soc. **10**(2), 121–141 (2013)
2. Bozena, S.: New media and social networks as a new phenomenon of global access to information and education. US-China Educ. Rev. **8**, 623–635 (2013)
3. Moradi, M., Honari, H., Naghshbandi, S., et al.: The association between informing, social participation, educational, and culture making roles of sport media with development of championship sport. Procedia Soc. Behav. Sci. **46**(5), 5356–5360 (2012)
4. Nah, S., Yamamoto, M.: The integrated media effect: rethinking the effect of media use on civic participation in the networked digital media environment. Am. Behav. Sci. **62**(8), 1061–1078 (2018)
5. Li, T.T.: The impact of new media on the spread of sports events. Sci. Technol. Commun. **15**(05), 118–119 (2015)

6. Centieiro, P., Cardoso, B., Eduardo Dias, A.: If you can feel it, you can share it!: a system for sharing emotions during live sports broadcasts. In: Proceedings of the 11th Conference on Advances in Computer Entertainment Technology Held at Funchal, Portugal on 11–14 November (2014). Article no. 15

7. Rowe, D.: Media studies and sport. Soc. Sci. Sport **4**, 135–161 (2014)

8. Jian, L., Yuan, Z.: The exploration and practice in innovative personnel training of computer science and technology. In: International Symposium on Information Technology in Medicine & Education, vol. 2, no. 06, pp. 47–51 (2012)

9. Kotevski, Z., Tasevska, I.: Evaluating the potentials of educational systems to advance implementing multimedia technologies. Int. J. Mod. Educ. Comput. Sci. (IJMECS) **9**(01), 26–35 (2017)

10. Guo, L.L.: The humanistic connotation construction of sports narrator in new media environment. J. Hebei Inst. Phys. Educ. (5), 5–27 (2013). (in Chinese)

11. Zhang, D., Zhang, G., Li, F., et al.: Essential characteristics and subject role of sports commentary. J. Wuhan Sports Univ. **51**(5), 5–9 (2017). (in Chinese)

12. Xia, Y., Gong, S.R.: Applied talents training mode of digital media specialty based on OBE concept. Comput. Educ. (4), 82–86 (2017). (in Chinese)

13. Cao, Y.L., Liu, Y.N.: The literacy and training approaches of journalist hosts in the era of media integration. Sci. Technol. Commun. **6**(22), 208–209 (2014). (in Chinese)

14. Xin, L., Le, H.: Innovation and entrepreneurship talents cultivating: systematic implementation path of "knowledge interface and ability matching". Int. J. Emerg. Technol. Learn. (IJET) **13**(08), 117–132 (2018)

15. Dan, Z.: Constructing evaluation model based on AHP. Int. J. Grid Distrib. Comput. **9**(8), 413–422 (2016)

16. Kurilovas, E., Vinogradova, I., Kubilinskiene, S.: New MCEQLS fuzzy AHP methodology for evaluating learning repositories: a tool for technological development of economy. Technol. Econ. Dev. Econ. **22**(1), 142–155 (2016)

17. Wang, H.Q.: Research on the application of fuzzy analytic hierarchy process in university performance evaluation. Knowl. Econ. (01), 147+149 (2020). (in Chinese)

18. Qiao, D.F.: Optimization of teaching quality evaluation strategy in Higher Vocational Colleges – based on the perspective of analytic hierarchy process. J. Anhui Radio Telev. Univ. (03), 55–59 (2019). (in Chinese)

19. Lotfi, F., Fatehi, K., Badie, N.: An analysis of key factors to mobile health adoption using fuzzy AHP. Int. J. Inf. Technol. Comput. Sci. (IJITCS) **12**(02), 1–17 (2020)

20. Li, C., Cai, H.Y., Yan, Q.J.: A comparative study on the weight of teaching quality evaluation index based on analytic hierarchy process. Educ. Mod. **6**(63), 219–221 (2019). (in Chinese)

21. Liu, X.: Research on teaching quality evaluation model of Higher Vocational Colleges based on fuzzy mathematics. Chin. Mark. (14), 281–283 (2017). (in Chinese)

22. Luo, Y.: Fuzzy mathematics model of classroom teaching quality evaluation in Higher Vocational Colleges. Educ. Mod. **6**(56), 102–104+109 (2019). (in Chinese)

23. Wang, Y.K.: Algorithm analysis of teaching quality evaluation system in Colleges and universities. Digit. Commun. World (01), 276 (2020). (in Chinese)

24. Barman, L., Bolander-Laksov, K., Silén, C.: Policy enacted–teachers' approaches to an outcome-based framework for course design. Teach. High. Education. **19**(7), 735–746 (2014)

25. Gunduz, N., Hursen, C.: Constructivism in teaching and learning; content analysis evaluation. Procedia Soc. Behav. Sci. **191**, 526–533 (2015)

Network Module of Generative Adversarial Network for Low-Light Image Enhancement

Jiarun Fu[1], Zhou Yan[2], Chunzhi Wang[1], Xiang Hu[3], Jiwei Hu[3],
and Lingyu Yan[1(✉)]

[1] School of Computer Science, Hubei University of Technology, Wuhan, China
yanlingyu@hbut.edu.cn
[2] Hubei Entry-Exit Inspection and Quarantine Bureau, Wuhan, China
[3] Fiberhome Telecommunication Technologies Co Ltd., Wuhan, China

Abstract. With the application of deep learning technology into the field of image enhancement, combined with traditional enhancement algorithms, the performance of enhancement algorithms has become more excellent. However, the enhancement of low-light images has not been completely solved by applying deep learning techniques. Enhancement network module is introduced to generative adversarial networks which aim to form a network module that can effectively solve the problem of low-light image enhancement. Through experiments, the performance of proposed algorithms has been proven to be outstanding in the target field.

Keywords: Generative adversarial network · Image enhancement · Depthwise separable convolutions

1 The Research Background

The enhancement of low-quality images is one of research focused in image enhancement field. With application of deep learning technology into the field of image enhancement, image enhancement technology has become much stronger than before. Thus, some researchers have tried to use image enhancement technology to solve more image problems, such as low-light image enhancement. In the rest of this section, to help everyone understand the development background of this technology. we will introduce previous attempts to enhance low-light images in the field of image enhancement according to the order in which the technologies are applied.

The paper 1 introduced that the stacking sparse denoising self-encoder based on the training of synthesized data can enhance and denoise the low-light noisy image. Model training is based on image patches and adopts sparsity regulative reconstruction loss as a loss function.

In paper 2, CNN is introduced. The traditional multi-scale Retinex (MSR) method can be regarded as a feed-forward convolutional neural network with different gaussian convolution nuclei, and it is demonstrated in detail. Then, following the MSR process, they proposed MSR-net, which directly learns the end-to-end mapping from dark to bright images. Paper [3] mainly focuses on single image contrast enhancement (SICE), aiming at the problem of low contrast under exposure.

Z. Hu et al. (Eds.): AIMEE 2020, AISC 1315, pp. 379–387, 2021.
https://doi.org/10.1007/978-3-030-67133-4_35

Inspired by the Retinex theory, paper [4] adopts a two-stage decomposition and enhancement step, which is completely implemented by CNN.

In addition, considering the noise problem, slap swarm algorithm[5] is used to find the optimal pixel value instead of the noise pixel to complete denoising process of the image.

The core idea of paper [6] ensures better performance in case of image enhancement and details preservation simultaneously. CBIR (Content-based image retrieval) [7] uses color layout descriptor (CLD) and grayscale co-occurrence matrix (GLCM) as these two features to complete image enhancement.

As we can see above, although there are corresponding low-light enhancement algorithms for restoring images, according to our investigation and research, the above-mentioned algorithms need to be further improved in the quality and details of the enhanced images. Therefore, a method is proposed in this article which aims to improve the performance compared with existing algorithms.

The above is the research background in the field of low-light image enhancement. The related technologies used in the second section will be introduced, the specific structure of our algorithm will be introduced in the third section, related experiments are conducted in the fourth section, and conclude this article in the last section.

2   Related Work

2.1   Generative Adversarial Network

Let's start with the original GAN [8]. Ian j. Goodfellow et al., in October 2014 in Generative Adversarial Networks put forward a process through the confrontation that estimates to generate a new framework of the model. In this framework, two models are trained simultaneously: the generation model G, which captures the data distribution, and the discriminant model D estimates the probability of samples from training data. The model structure is shown in Fig. 1.

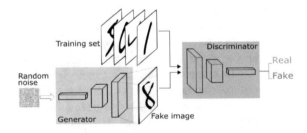

Fig. 1. Standard generation adversarial network schematic

As shown in Fig. 1, the generated against network is composed of two parts: the generation model G (generative model) and the discriminant model D (discriminative model). The generated model is used to learn the distribution of real data. The discriminant model is a binary classifier that determines whether the input is real data or not. Generated data. X is a real number which corresponds to the distribution of $P_r(x)$. Z is the hidden space variable which corresponds to the distribution of $P_z(z)$ just like a gaussian distribution or a uniform distribution. Sample from the assumed hidden space z and generate the data $x' = G(z)$ by generating model G. Then the real data and the generated data are fed into the discriminant model D, and the judgment category is output. The given label is 1 for the real image and 0 for the generated image. The generated model G tries to synthesize the picture so that the discriminant model D is true. So the loss function is:

$$\min_{D} \max_{G} V(D, G) = -E_{x \sim P_r} \log D(x) - E_{z \sim P_z} \log(1 - D(G(z))) \tag{1}$$

Where P_r is the distribution of real data and P_z is the distribution of assumed hidden space.

We use generative adversarial networks to form the main framework of our proposed model.

2.2   Resnet

We use the ResNet to form our enhanced network module. ResNet (Residual Neural Network) [9] was proposed by Kaiming and other four Chinese colleagues at Microsoft research institute. By using ResNet Unit, 152 layers of Neural Network were successfully trained, the championship was won in the ILSVRC2015 competition. The error rate of the top 5 was 3.57%, the number of participants was lower than that of VGGNet [10]. The structure of ResNet can accelerate the training of neural network very quickly, the accuracy of the model has been improved greatly. Meanwhile, ResNet is very good in generalization, it can even be directly used in InceptionNet network.

The main idea of ResNet is to add a direct connection channel to the Network, as shown in the Fig. 2 below.

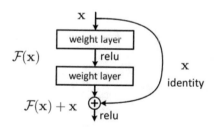

Fig. 2. Resnet structure diagram

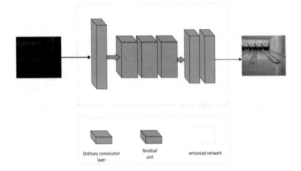

Fig. 3. Schematic diagram of generator network structure

3   Enhanced Network Module Generative Adversarial Network

3.1   Algorithm Framework

3.1.1   The Network Structure

The network structure of low-light image enhancement is shown in Fig. 3 and Fig. 4. In order to enhance the transmission of weak light position information from image in the network flow, the generator adds an enhanced network. The enhanced network converts the input image as a whole into a similar image in the new space, and the attention network predicts the position mask of weak light, which is the same size as the input image, with each pixel having a probability value between 0 and 1. Finally, we combine the input image, and note that the force diagram and the transformed input constitute the final enhanced image. The discriminator can receive both the generated image and the real image, and finally produces the predicted value of true and false.

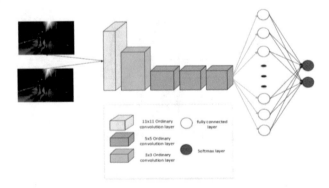

Fig. 4. Schematic diagram of discriminator network structure

3.1.2 Algorithm Flow Chat

See Table 1.

Table 1. Enhanced network module generative adversarial network algorithm

	Data description: I_x **is the image,** I_y **is the real image , is defined as the input to the** I_{adv} **discriminator.**
1	**For** i = 1 : Maximum number of iterations **do**
2	Training generator network
3	$-$ M image pairs were randomly sampled from the training data set $$\left\{\left(I_x^1, I_y^1\right), \left(I_x^2, I_y^2\right) \cdots \left(I_x^m, I_y^m\right)\right\}$$
4	$-$ The input of the fixed discriminant network is $$I_{adv} = \{0, 0, \cdots, 0\}, \text{ length of } m$$
5	$-$ Minimize the total loss of the generator network: $$Loss_{gen} = \omega_a L_a + \omega_{adv} L_{adv} + \omega_{con} L_{con} + \omega_{tv} L_{tv} + \omega_{cd} L_{cd}$$
6	Train the discriminator network
7	$-$ The input of the randomly initialized discriminant network is $$I_{adv} = \{1, 0, \cdots, 0\}, \text{length 为 } m$$
8	$-$ M images were randomly sampled from the training data set $$\left\{\left(I_x^1, I_y^1\right), \left(I_x^2, I_y^2\right) \cdots \left(I_x^m, I_y^m\right)\right\}$$
9	$-$ Maximizes the overall loss of the discriminator network: $$L_{adv} = \sum_{i=1}^{m} \log D\left(G\left(I_x\right), I_y\right)$$
	10 end

3.2 Enhanced Network Module

As a backbone network, this module is used to overcome the disadvantage of low contrast and improve the details of the picture, so as to achieve the effect of enhancing the picture. Considering that there is little feature information of low-light region in low-light pictures, and there may be noise interference, we use residual connection to build the basic residual module, which is used to deepen the number of network layers, improve the modeling ability of network to enhance images, and avoid the feature loss caused by network deepening.

Residual network has been proven to be effective in many fields, such as object detection, semantic and image segmentation. This operation USES residual connections to train deeper networks and avoid feature loss. By introducing residual touch block, the network structure is enhanced as shown in Fig. 5:

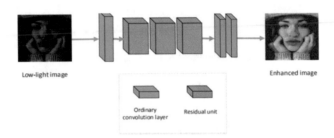

Fig. 5. Enhanced branch network structure diagram

As shown in Fig. 5, the blue block is shown in the figure. The first layer is used for feature extraction to achieve multiple features from RGB channel. The last two layers are used to restore RGB images to achieve multi-feature conversion to RGB images. The orange block in the figure represents residual cells. By connecting multiple residual cells to carry out complex feature transformation, the network can improve the modeling ability of low-light pictures.

In order to constrain the enhancement network and improve the overall image perception quality, the following four loss functions are designed:

We defined the content loss according to the activation diagram generated by the ReLU layer of the pre-trained vgg-19 network. This loss encourages them to have similar feature representations, rather than measuring per-pixel differences between images, but rather to include aspects of content and perceived quality. Let ϕ_i be the feature graph obtained by vgg-19 network after the ith convolutional layer, then our content loss is defined as the euclidic distance between the enhanced image and the feature representation of the target image:

$$L_{content} = \frac{1}{C_i H_i W_i} \left\| \phi_i(G(I_x, A)) - \phi_i(I_y) \right\| \tag{3}$$

4   Experimental Results and Analysis

4.1   Experiment Setting

In terms of the experimental samples, we chose the low-light paired data set (LOL) containing 500 pairs of low-light/normal light images. It is a dataset of images used for low-light enhanced shots in real scenes. The dataset contains images captured from various scenarios, such as houses, campuses, clubs, and streets. Based on this data set, we produced 11,592 photos of the training set.

We completed the experiment of this paper through Tensorflow [11]. The network proposed in this paper converges rapidly and has been trained for 20,000 generations on NVIDIA GeForce GTX1080 using the synthesized data set. To prevent overfitting, we used rollover and rotation for data enhancement. We set the batch size to 32 and the input image value is scaled to [0,1]. And we use PSNR and SSIM for quantitative analysis.

Table 2. Proposed method was compared with the experimental results of HE, SRIE and DSLR in the LOL data set

	HE	SRIE	DSLR	Our
PSNR	15.4271	11.8453	19.1426	**20.9983**
SSIM	0.5737	0.5347	0.9982	**0.9986**

In order to verify the performance of the image enhancement algorithm proposed in this paper, the experiment was compared with the following representative image enhancement algorithms: histogram equalization (HE), reflection illumination estimation (SRIE), and DSLR.

4.2   Result and Discussion

On the LOL data set, we compared three methods of HE, SRIE and DSLR. Because some methods could not achieve the denoising function, we combined with BM3D method to achieve the denoising, and then produced the final result. The quantitative results are shown in Table 2, and the qualitative results are shown in Fig. 6.

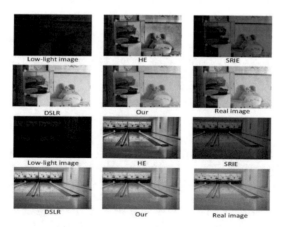

Fig. 6. Our methods work with HE, SRIE, DSLR in the LOL dataset for visual effects

The experimental results are compared with real image. HE's picture on the wall and there is an obvious artifacts that is independent of the background and overall image is grey. The second image is enhanced, light color floor brunet floor and real photos are apparent discrepancy, the images enhanced by SRIE algorithm which have more brightness distortion. DSLR can enhance the image illuminance and retain a certain content, but there still exists a noise exposure and local problems, such as enhanced in two pictures, the first white of bright, closet door, floor. The brightness of the wall in the second picture is too high. The result generated by proposed algorithm is closest to the real picture.

The PSNR and SSIM indices represent a comprehensive assessment of the clarity and robustness of image restoration. It can be seen from Table 2 that proposed method outperforms other methods in both indexes, which indicates that our method also has excellent performance and good robustness in real data sets. Compared with the traditional methods of HE and SRIE, we improved 36.11% and 77.27% on PSNR, 74.06% and 86.76% on SSIM, respectively. Compared with the current deep learning method DSLR, we also improved 9.69% in PSNR and 0.04% in SSIM. It can be seen that proposed method is much more effective than traditional methods, deep learning method also has certain advantages.

5   Summary and Conclusion

In this article, an enhancement network module into the generative adversarial network is introduced to solve the problem of low-light image enhancement. The enhanced network uses residual connections to build residual modules and joins the generative confrontation network to deepen the network, which can improve capability of image enhancement modeling. In the experimental stage, comparative experiments with traditional image enhancement methods and deep learning enhancement methods are conducted, which proves that proposed algorithm has obvious advantages in the reduction and robustness of enhance low-light image problem. In the future, more large-scale experiments will be conducted to further prove the general validity of proposed algorithm.

Acknowledgment. This work is funded by the National Natural Science Foundation of China under Grant No. 61772180, Technological innovation project of Hubei Province 2019 (2019AAA047).

References

1. Lore, K.G., Akintayo, A., Sarkar, S.: LLNet: a deep autoencoder approach to natural low-light image enhancement. Pattern Recognit. **61**, 650–662 (2017)
2. Shen, L., Yue, Z., Feng, F., et al.: MSR-net: low-light image enhancement using deep convolutional network. arXiv preprint arXiv:1711.02488 (2017)
3. Cai, J., Gu, S., Zhang, L.: Learning a deep single image contrast enhancer from multi-exposure images. IEEE Trans. Image Process. **27**(4), 2049–2062 (2018)

4. Wei, C., Wang, W., Yang, W., et al.: Deep retinex decomposition for low-light enhancement. arXiv preprint arXiv:1808.04560 (2018)
5. Liu, W., Wang, R., Su, J.: An image impulsive noise denoising method based on salp swarm algorithm. Int. J. Educ. Manage. Eng. (IJEME) 10(1), 43–51 (2020). https://doi.org/10.5815/ijeme.2020.01.05
6. Islam, R., Xu, C., Han, Y., Putul, S.S., Raza, R.A.: A robust functional minimization technique to protect image details from disturbances. Int. J. Inf. Tech. Comput. Sci. (IJITCS) 11(7), 1–8 (2019). https://doi.org/10.5815/ijitcs.2019.07.01
7. Sadique, M.F., Rafizul Haque, S.M.: Content-based image retrieval using color layout descriptor, gray-level co-occurrence matrix and k-nearest neighbors. Int. J. Inf. Tech. Comput. Science (IJITCS) 12(3), 19–25 (2020). https://doi.org/10.5815/ijitcs.2020.03.03
8. Simonyan, K., Zisserman, A.: Very deep convolutional networks for large-scale image recognition. arXiv preprint arXiv:1409.1556 (2014)
9. Abdullah-Al-Wadud, M., Kabir, M.H., Dewan, M.A., et al.: A dynamic histogram equalization for image contrast enhancement. IEEE Trans. Consum. Electron. 53(2), 593–600 (2007)
10. Abadi, M., Barham, P., Chen, J., et al.: Tensorflow: a system for large-scale machine learning. In: 12th USENIX Symposium on Operating Systems Design and Implementation (OSDI 2016), pp. 265–283 (2016)

Research on the Influence of Rules of Ships Entering and Leaving Port on Waiting Time of Large and Small Ships

Huiling Zhong and Ping Zhong[(✉)]

South China University of Technology, Guangdong, China
1045007790@qq.com

Abstract. Due to the limitation width of seaport navigation channel, most of them are partial two-way navigation rule, that is, small ships sail in two-way navigation rule while large ships need to wait for no ship in the opposite direction before entering the channel, which makes the waiting time of large ships longer than that of small ships. Therefore, in order to shorten the waiting time of large ships and make the waiting time fair for large ships and small ships, this paper studies and proposes different rules for ships entering and leaving the port, and selects average waiting time (AWT) and the proportion of ships that can be served as the performance measures. Then a ship navigation operation simulation model developed and the proposed model is applied to Guangzhou port navigation channel. The results show that LMSFS rule can effectively shorten the waiting time of large ships and keep the waiting time of small boats at an acceptable level, which consider the fairness of receiving services for small ships and large ships.

Keywords: Navigation channel · Wait time · Simulation model · Fairness

1 Introduction

The seaport navigation channel for ships' entering and leaving a port, is one of the most important part of the seaport. When ship traffic volumes increase, ships wait times and operational costs increase because of the limited capacity of port approach channel. For example, Guangzhou Port has been carried out several project to expands the port approach channel, since the number of ships entering and leaving port doubled from 2012 to 2018. But the waiting times for large ships are still long, as the channel is not wide enough for them to sail in two-way navigation rules. However, it is hard to continue expanding Guangzhou port navigation channel because the costs are prohibitively high for the large channel length and some waterway do not have the conditions for widening. Therefore, the high cost of widening the channel, this study proposes ship entering and leaving port rules to decrease the waiting time of large ships.

Rules of ships entering and leaving port refer to that Vessel Movement Center (VMC) determine the sequence and organization of ships passing the channel according to certain rules, in which the certain rules are called the rules of ships

© The Author(s), under exclusive license to Springer Nature Switzerland AG 2021
Z. Hu et al. (Eds.): AIMEE 2020, AISC 1315, pp. 388–398, 2021.
https://doi.org/10.1007/978-3-030-67133-4_36

entering and leaving port. By changing the rules of ships entering and leaving the port and adjusting the order of ships entering and leaving, the channel can be fully utilized and the waiting time of ships can be shortened. So far, several practical applications of the rules of ships entering and leaving port have been tried on a preliminary basis. For example, Wang studied the impact of three types of ship entering and leaving port rules on port service, including First Come First Serve Rule, Larger Ship Priority Rule and Ship Leaving Port Priority Rule, and considered that the greater the ship traffic volume, the greater the difference of rules for different ships entering and leaving the port [1]. Lalla-Ruiz proposed the waterway ship scheduling problem (WSSP), and developed a mathematical model to get the best order of ships entering and leaving the port to minimize the sum of ship's waiting time for the channel [2]. Zhang established a mathematical model with the distance between berth and channel exit and get con-clusion that the order of ships entering and leaving the port is approximately in line with the priority of the nearest ships [3].

The above literature involves many aspects of the research on the rules of ship entering and leaving port, which provides a solid foundation for further research. However, the study mainly considers the waiting time of all ships, and does not distinguish between large and small ships. In fact, due to the small number of large ships, the overall waiting time will be very short as the waiting time of small boats is short, even if the waiting time of large ships is long.

Therefore, it is necessary to consider the fairness of waiting time between large and small ships. Shang studied the circumnavigation strategies to shorten the waiting time of large ships, that is, the small ship will circumnavigation in second channel if there are large ships navigate in main channel [4]. Deepika studied relates benefit-fairness algorithm based on weighted-fair queuing model [5]. Tang tried to set a ships-passing anchorage for ships waiting until other ships from opposite directions pass by so that both inbound and outbound ships can travel in a one-way traffic channel [6].

In addition, due to the complexity and randomness of ship navigation operation system, there are many random factors, including ship arrival time, ship navigate speed, tidal change and so on, which have an impact on the work of the system, and the numerical method cannot get the analytical solution [7]. Because of the advantages of simulation technology in dealing with complex systems, it has been widely used in port design and management in recent years. Mateusz describe an optimization method based on simulation, which can reduce the necessary calculation only by modifying the changed model [8]. By using dynamic programming technology, Wang established the decision model of equipment configuration of loading and unloading line at bulk terminal [9].

So, this paper establishes a ship navigation operation simulation model of Guangzhou port navigation channel and studies the influence of different rules of ships entering and leaving port on waiting time between large ship and small ship, which provides a theoretical basis for the port to formulate the order of ships entering and leaving the port. Therefore, in the rest of the paper, we introduce the different rules for ships entering and leaving the port and performance measures in Sect. 2. Next, we build the simulation model of ship navigation operation model in Sect. 3, and carries out several simulation experiments and analysis to study the influence of ship entering and leaving rules on waiting time of different ship types in Sect. 4. Finally, we conclude the paper in Sect. 5.

2 Term Definitions

In this section, we first introduce the rules of ships entering and leaving port and design some rules. Then the performance measures are discussed in detail in Sect. 2.2.

2.1 Rules of Ships Entering and Leaving Port

When a large ship is travelling in channel with the partial two-way navigation rule, other ships are allowed to move in the same direction, while the ships travelling in the opposite direction must wait at the anchorage outside the port until the channel is evacuated. If the large ships have priority to enter the channel, there will be such a situation: some large ship continuous travelling in channel, it may lead the ship in the opposite direction having to wait for a long time at the anchorage. In addition, it is necessary to consider the problem of blockage caused by ships not meeting the other conditions for entering the port, such as tide. Thus, ships will obey rules entering and leaving port, it will determine the order of ships pass the channel to make ships pass the waterway more efficiently and safely.

First come first service (FCFS) rule is one of most common rules, which reflects the equity of different ships and treats all ships without difference, ships need to meet the entry and exit conditions to enter the channel, such as channel availability, tugboats availability. If the previous ship always enters the channel earlier than the later ship, it may happen that the current ship does not meet the entry/exit conditions and continue to wait, and the subsequent ships meeting the entry and exit conditions are unable to accept the service, resulting in high ship waiting time. If the order of the ship passing through the channel is determined by whether the ship meets the conditions for entering/leaving the port, the ships with the strict requirements for the entry/exit conditions are unable to enter the channel. Therefore, this paper proposes the following four rules from the perspective of efficiency and fairness.

a) Meeting-condition-Ship First Service (MSFS) Rule: Ships are queued up according to the order of arrival time of ships. Each time, the most forward ship in the queue which meet the entry/exit conditions is arranged to pass through the channel.
b) Large-ton-Ship First Service (LSFS) Rule: The priorities of large ships sailing in one-way navigation rule higher than that of smaller ships sailing in two-way navigation rule. The subsequent ships must wait until the previous ship pass through channel.
c) Large-ton-Ship and Meeting-condition-Ship First Service (LMSFS) Rule: The priorities of large ships sailing in one-way navigation rule higher than that of smaller ships sailing in two-way navigation rule. Ships with same priority still obey MSFS rule. The low priority ships must wait until all high priority ships pass through channel.

2.2   Performance Measures

In order to develop the simulation model, the main entities and performance indicators required to represent the system are first determined. The most-used performance measures are AWT/AST ratio. This ratio is misleading because it can decrease as service time increases. Due to the limited service capacity of the channel, with the increase of the number of arrival ships, the proportion of ships that can be served is also an important performance measure. The ship waiting time is the time waiting for approach channel, berth, tugboat and pilot. When the port does not have enough resources for berth, tugboat and pilot or the capacity of the navigation channel is limited, the ship waiting time will increases rapidly. Therefore, in this paper, the ship is regarded as the main entity, and the AWT and the proportion of the number of ships can served are taken as performance measures.

3   Simulation Model

In this paper, the discrete event simulation method is used to abstract the ship navigation operation system as a typical Queuing service system. The channel is regarded as the server, and the ship as the customer receives the entering and leaving port services, so as to study the differences in ships waiting time of different rules of ships entering and leaving port.

3.1   Model Assumptions

In order to simplify the problem and facilitate modeling, the following assumptions are proposed:

(1)   Due to the length of the whole channel is much larger than port area, the relatively close berth is merged into a port area and the distance from the berth to the edge of port area is ignored.
(2)   There is no inner anchorage in the channel, so the ships are not allowed to stop and wait in the channel.
(3)   The tugboats should be allocated in advance before ship entering the channel.
(4)   All ships have the same speed.
(5)   All ships have the same time for berthing.
(6)   Each tugboat will be allocated to a port area, and the tugboat can only carry out berthing operation in the allocated port area.

4   Logical Flow of Ships Entering and Leaving Port

In this study, the ship navigation operation system consists of the ship arrival, waiting in anchorage, traveling through channel, berthing, loading/unloading cargo, waiting in berth, unberthing, traveling through channel, ship departure. The arrival ships need to meet the navigation rules to enter the channel. At the same time, the tides have a relevant influence on the depth of the waterways, and some ships need to ride the tide to ensure navigation safety. In addition, there is no inner anchorage and the harbor basin is small, ships cannot wait for berthing in the channel or harbor basin so that tugboats should be allocated in advance before ship entering the channel. According to the whole process of the ships' navigation, the logic model of the ships' navigation operation system is proposed, as shown in Fig. 1.

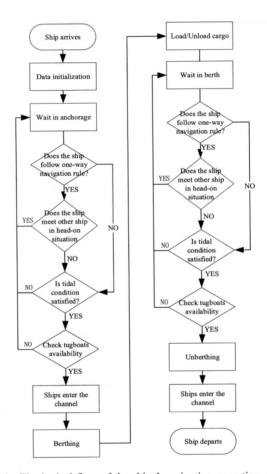

Fig. 1. The logical flow of the ships' navigation operation system

4.1   Model Input Parameters

(1) The navigation rules of each waterway
    The Guangzhou port navigation channel is about 160 km from south to north along Zhuhai, Shenzhen, Zhongshan, Dongguan. As shown in Fig. 1, it consists of sever-al waterways, each with a different width and depth. The navigation rules

of the channel vary with the depth and width of the waterway, which are shown in Table 1.

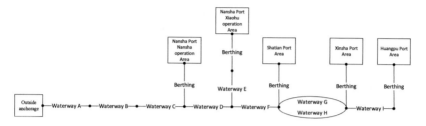

Fig. 2. The Guangzhou port navigation channel schematic diagram

Table 1. The navigation rules of each waterway

Waterway	Navigation rules
Waterway A	Two-way navigation rule
Waterway B	Partial two-way navigation rule; LNG ships and ships of 200,000 tons and above sail in one-way navigation rule
Waterway C	Partial two-way navigation rule; LNG ships and ships of 150,000 tons and above sail in one-way navigation rule
Waterway D	Partial two-way navigation rule; LNG ships and ships of 100,000 tons and above sail in one-way navigation rule
Waterway E	One-way navigation rule; 2 h to change direction
Waterway F	Partial two-way navigation rule; LNG ships and ships of 100,000 tons and above sail in one-way navigation rule
Waterway G	Partial two-way navigation rule; LNG ships and ships of 100,000 tons and above sail in one-way navigation rule
Waterway H	One-way navigation rule; 2 h to change direction
Waterway I	Partial two-way navigation rule; Ships of 100,000 tons and above sail in one-way navigation rule

(2) Ships' Arrival Distributions

According to the data of ships entering and leaving Guangzhou port navigation channel in 2018, the ships' arrival distributions are generally subject to Poisson with the parameter of 64.05, and the ship arrival time interval follows the negative exponential distribution with the parameter of 22.48 min

(3) Type Combination of Ships

Due to the Guangzhou port navigation channel is composed of several waterways with different navigation rules, ships are classified by cargo class and deadweight tonnage(dwt): small ships(3,000 tons $\leq$ dwt < 100,000 tons), large ship I (3,000 tons $\leq$ dwt < 100,000 tons), large ship II (100,000 tons $\leq$ dwt < 150,000 tons), large ship III (150,000 tons $\leq$ dwt < 200,000 tons) and LNG ships. According to the calculation of the ship entering and leaving Guangzhou port navigation channel in 2018, the distribution of ship types and the distribution of ship length are shown in Table 2 and Table 3.

Table 2. The distribution of ship types in each port area

Ship type	Port area				
	Nansha port Nansha operation area	Nansha port Xiaohu operation area	Shatian port area	Xinsha port area	Huangpu port area
Small ship	20.81%	12.60%	26.27%	22.78%	11.63%
Large ship I	1.41%	0.01%	0.00%	0.06%	0.00%
Large ship II	1.77%	0.00%	0.00%	0.07%	0.00%
Large ship III	0.09%	0.00%	0.00%	0.09%	0.00%
LNG ship	0.00%	0.17%	2.23%	0.00%	0.00%

Table 3. The distribution of ship length in each port area, each ship type

Port area, Ship type		Ship length (meters)					
		L < 80	80 <=L < 120	120 <=L < 180	180 <=L < 230	230 <=L < 390	390 <=L < 400
Nansha Port Nansha operation area	Small ship	0%	9%	46%	36%	9%	0%
	Large ship I	0%	0%	0%	0%	100%	0%
	Large ship II	0%	0%	0%	0%	99%	1%
	Large ship III	0%	0%	0%	0%	0%	100%
Nansha port Xiaohu operation area	Small ship	2%	44%	54%	0%	0%	2%
	Large ship I	0%	0%	0%	100%	0%	0%
	LNG ship	0%	35%	54%	11%	0%	0%
Shatian port area	Small ship	24%	44%	29%	1%	1%	24%
	LNG ship	0%	80%	15%	5%	0%	0%
Xinsha port area	Small ship	1%	27%	71%	1%	0%	1%
	Large ship I	0%	0%	23%	77%	0%	0%
	Large ship II	0%	0%	0%	0%	100%	0%
	Large ship III	0%	21%	0%	0%	79%	0%
Huangpu port area	Small ship	1%	51%	48%	0%	0%	1%

(4) Demand of tugboats for berthing and tugboat configuration

Tugboats are divided into three types by power: small horsepower tugboats (below 4000HP), medium horsepower tugboats (4000HP) and high horsepower tugboats (5000hp and above). The requirements of berthing and unberthing operation are shown in Table 4.

Table 4. The demand of tugboats for berthing and unberthing

Ship length (meters)	Number of tugboats	Tugboats type
L < 80	0	–
80 <= L < 120	1	Any type
120 <= L < 180	2	1 medium or high horsepower tugboat, 1 tugboat of any type
180 <= L < 230	2	2 medium or high horsepower tugs
230 <= L < 390	3	2 medium or high horsepower tugs, 1 tugboat of any type
390 <= L < 400	3	2 high horsepower tugs, 1 tugboat of any type

5 Results and Discussion

The simulation runtime is 1 month, repeats 30 times. By simulation experiment, the result is shown in Table 5, Table 6 and Table 7. Table 6 shows the performance with Proportion of ships to be served and AWT of each rule (2.1). Table 7 show the performance of the different types of ships (3.3.3) of each rule.

Table 5. Simulation results of each rules

Rules of ships entering and leaving port	MSFS	LCFS	LMSFS
Number of arrival ships	1934	1951	1929
Number of ships to be served	1779	909	1809
Proportion of ships to be served	92%	47%	93%
AWT (min)	186	1155	360

From the Table 5, we can see that: (1) Compared with MSFS rule, the promotion of the number of ships to be served of LCFS rule has decreased by half, while the LMSFS rule similar to MSFS rule, which reveals that the number of ships passing through channel can be significantly changed by adjusting the order of ships entering and leaving the port. (2) Although the proportion of ships to be served of MSFS rules and LMSF rules is the same, AWT of MSFS rule is 186 min, half of that of LMSFS rule. It seems that the MSFS rule is the best rule according to Proportion of ships to be served and AWT.

Table 6. Proportion of ships to be served of different ship types of each rules

Ship type	Rules of ships entering and leaving port		
	MSFS	LCFS	LMSFS
Small ship	95%	31%	93%
Large ship I	89%	96%	95%
Large ship II	63%	95%	95%
Large ship III	29%	93%	97%
LNG ship	19%	95%	95%

Table 7. AWT of different ship types of each rules (minutes)

Ship type	Rules of ships entering and leaving port		
	MSFS	LCFS	LMSFS
Small ship	134	1431	370
Large ship I	263	178	331
Large ship II	798	162	180
Large ship III	1794	146	220
LNG ship	2981	106	198

From the Table 6 and Table 7, we can see that: (1) According to Proportion of ships to be served and AWT, the service of small ships under MSFS rules is better than that of large ships, but the performance of large ship III and LNG ship is much lower than that of other rules. (2) On the contrary, the service of large ships under LCFS rules is better than that of large ships, and the performance of all types of large ships is best among all rules. (3) Under LMSFS rule, the proportion of different types of ships has little difference, and the difference between the maximum value and the minimum value is within 5%, which can be regarded as fair service rule for large ships and small boats.

The above results show us the characteristics of different rules:

(1) MSFS rule allow ships that meet entry/exit conditions to serve first, thus AWT is the lowest of all rules. However, this improvement was achieved at the expense of the large ship. The proportion of ships to be served of some large ships less than 50% and the AWT will be increased to a very high level. As the number of large ships is much less than that of small ships, the overall performance is better than other rules.

(2) LCFS rules make large ships have high priority, which leads to the performance of large ships better than other rules. But there is still the problem of congestion that the large ships with high navigation requirements may block the small ship behind, the performance of small ships is very bad, resulting the overall performance worse than other rules.

(3) LMSFS combines the advantages of the MSFS rule and LCFS rule, which allow large ships that meet entry/exit conditions to serve first and avoiding congestion

problem. The proportions of ships to be served of all ship types are exceed 90% and the AWT is in the range of 180 to 370.

6 Conclusions

The limited channel widths of Guangzhou port navigation channel lead to ships enter the channel in partial two-way navigation rules. Therefore, small ships can pass channel under two-way navigation rules while large ships pass channel under one-way navigation rules, resulting in the waiting time of large ship is much longer than that of small ships, which is unfair for large ship. Due to the high cost of channel widening, changing the orders of ships entering and leaving the port to shorten the large ships waiting time has been proposed. By using proportion of ships to be served and AWT as the evaluation indicator, this paper developed a complicated simulation model and designed a serious simulation experiment based on Guangzhou port with different rules of ships entering and leaving the port. The result shows that LMSFS rule significantly decrease the waiting time of large ships and the waiting time of small ships remains at an acceptable level compared with the LCFS and MSFS rule, which consider the fairness of receiving services for small ships and large ships.

The results of this paper provide a theoretical basis for ship scheduling, and the proposed method can be used to solve the problem of fairness of waiting time between large and small ships. Nevertheless, this research is still at an early stage, where the focus is primarily on theory and model development. For example, in the simulation model, it is necessary to consider the load reduction of large ships so as to pass under the two-way navigation rules. Also, in the future research, we will formulate a weight for different types of ships, and adjust the weight to get a good rule of ships entering and leaving port.

Acknowledgements. This research was supported by the Development of Philosophy and Social Science of Guangzhou city (Grand Nos. 2018GZYB24).

References

1. Wang, W.Y., Guo, Z.J., Tang, G.L., et al.: Influence of ship traffic rules in Y-type fairway intersection water on port service level. Adv. Mater. Res. **378–379**, 262–265 (2012)
2. Lalla-Ruiz, E., Shi, X., Voß, S.: The waterway ship scheduling problem. Trans. Res. Part D: Transport Environ. **60**, 191–209 (2018)
3. Zhang, X., Chen, X., Ji, M., et al.: Vessel scheduling model of a one-way port channel. J. Waterw. Port Coast. Ocean Eng. **143**(5), 04017009 (2017)
4. Shang, J., Wang, W., Peng, Y., Tian, Q., Tang, Y., Guo, Z.: Simulation research on the influence of special ships on waterway through capacity for a complex waterway system: a case study for the Port of Meizhou Bay. Simulation **96**(4), 387–402 (2020)
5. Saxena, D., Chauhan, R.K., Kait, R.: Dynamic fair priority optimization task scheduling algorithm in cloud computing: concepts and implementations. Int. J. Comput. Netw. Inf. Secur. (IJCNIS) **8**(2), 41–48 (2016). https://doi.org/10.5815/ijcnis.2016.02.05

6. Tang, G., Guo, Z., Xuhui, Yu., et al.: SPAC to improve port performance for seaports with very long one-way entrance channels. J. Waterw. Port Coast. Ocean Eng. **140**(4), 04014011 (2014)

7. Chawda, B.V., Patel, J.M.: Investigating performance of various natural computing algorithms. Int. J. Intell. Syst. Appl. (IJISA) **9**(1), 46–59 (2017). https://doi.org/10.5815/ijisa.2017.01.05

8. Wibig, M.: Dynamic programming and genetic algorithm for business processes optimisation. Int. J. Intell. Syst. Appl. (IJISA) **5**(1), 44–51 (2013). https://doi.org/10.5815/ijisa.2013.01.04

9. Wang, X., Lei, W., Dong, M.: Modeling and accomplishment of loading-and-unloading equipment optimum allocation system at bulk terminal. IJCNIS **3**(4), 1–9 (2011)

A Neural Network Featured System for Learning Performance of Students

Deng Pan[1]([⊠]), Shuanghong Wang[1], Zhenguo Li[2], Aisong Xiang[2], and ChunZhi Wang[1]

[1] School of Computer Science, Hubei University of Technology,
Wuhan 430068, China
chunzhiwang@vip.163.com
[2] Fiberhome Telecommunication Technologies Co., Ltd., Wuhan, China
{1394756525,475388483}@qq.com

Abstract. The system proposed by this article is a combination of association rules and decision tree which is utilized to construct the early warning model of student learning performance reflected by scores achieved by students. Because Apriori algorithm achieves low efficiency of frequent item set traversal, this model uses FP-growth to search for frequent itemsets and to construct through association rules and decision trees. Inter-course dependencies, analysis and early warning of learning courses related to both computer and student evaluated by student's school. Proposed system provides students with accurate academic guidance, which can effectively improve the quality of learning management of individuals. It is of great significance to the development and guidance of students. It helps understand the situation of students in all aspects and improve the overall level of students. Compared with other machine learning algorithms which can only predict performance through scores the students get, proposed system with decision tree and association rules has achieved the combination of horizontal and vertical prediction, which suggest it does not rely upon available scores to predict performance. In the meanwhile, horizontal relationship between decision tree and association rules has been found, which has been proven that it has achieved the comprehensive prediction.

Keywords: Association rules · Decision tree · Results early warning

1 Introduction

At present, researches related to correlation analysis of academic performance include [1, 2] gender differences, learning environment, and family status. There are few studies on correlation analysis among courses in different semesters.

The references [3–5] use the technology of double regression model or decision tree, genetic algorithm, artificial neural network to find the external factors related to student achievement, but these factors only affect the floating range of achievement to some extent rather than determine the achievement. On the other hand, the data mining methods of Kendall's two-way rank-related analysis [6] and the references [7–10] (including correlation analysis, cluster analysis, logical regression, C4.5 algorithm,

Z. Hu et al. (Eds.): AIMEE 2020, AISC 1315, pp. 399–408, 2021.
https://doi.org/10.1007/978-3-030-67133-4_37

etc.) are all studied in each subject of the students, the aim of this study is to discover the hidden course-related rules or patterns. Although these methods are conducive to optimizing the curriculum structure and improving the quality of teaching, the scope of curriculum analysis has great limitations, most of them are based on the design of educational concepts, useful information in the curriculum association rules and the teaching problems reflected in the curriculum are not fully explored. It neither cannot provide students with targeted guidance and early warning effects nor can provide substantive teaching suggestions and decisions for the teaching development of universities. Therefore, it is very important to construct an effective performance early warning model and provide effective learning guidance for students.

Modern higher education, from the elite to the masses, from single to multiple, has led to the rapid expansion of enrollment scale and the increasing shortage of teaching resources, making it difficult to teach students in accordance with their individual conditions. Taking computer science as an example, professional courses such as "Digital Logic Circuits", "Data Structures", and "Principles of Computer Composition" all have predecessors and follow-up courses potentially related to them, thus forming a complex tree structure. Each of these courses has a certain degree of difficulty and requires a lot of time and energy, which increases the burden on students.

In this paper, we are trying to propose a system which can intuitive early warning of learning performance in an intuitive and valuable way, the system has been proven that it can assist teaching work in accordance with student's individual conditions.

In addition, students can improve their learning performance of related subjects and get significant improvement. Therefore, a method of combining association rules with decision tree algorithm is proposed for learning performance early warning.

2 Preprocessing of Student Achievement Data

2.1 Sample Data

The data are derived from more than 400 records of students in the 15th grade of the School of Computer Science of Hubei University of Technology, and the initial data set contains each student's Higher Mathematics (1)-2, Linear Algebra, Probability Theory and Mathematical Statistics (1), College English Band 4, University English Reading and Writing Translation-1, University English Reading and Writing Translation-2, College Chinese, Outline of Modern and Contemporary History of China, Marxist Fundamental Principles, An Introduction to Mao Zedong Thought and the Theoretical System of Socialism with Chinese Characteristics, Ideological and Moral Cultivation and Legal Basic, University Physics (1)-1, University Physics (1)-2, Physics Experiments (1)-1, Advanced Language Programming, Data Structure, Computer Network, Database Principles and the Application. 20 subject results are taken as sample data, and each sample has 20 feature attributes.

2.2   Data Preprocessing

To better research, 20 feature attributes are represented by Gs2, Xxds, Gll, Dxyyysj, Dxyydxy1, Dxyydxy2, Dxyw, Zgjdsgy, My Mg, Sx, Szlj, Dxwl_1, Dxwl_2, Wlsy_1, Wlsy_2, C_plus, Sjjg, Jsjwl, Sjk. The details are shown in Table 1.

Table 1. Course information

Course title	Attribute representation	Number of participating students
Higher mathematics (1)-2	Gs2	411
Linear algebra	Xxds	409
Probability theory and mathematical statistics (1)	Gll	409
...		...
Computer network	Jsjwl	411
Database principles and the application	Sjk	411

After data processing, a sample data of 401 is obtained. The relevant data of 401 samples is named 15Grade_All data set. The specific form of the data set is shown in Table 2.

Table 2. Course information

Serial number	Gs2	Xxds	...	Sjk
1	38	23	...	77
2	78	63	...	87
3	65	81	...	67
4	69	70	...	56
5	82	89	...	87
6	56	60	...	90

According to the demand analysis of the BP neural network algorithm and the random forest algorithm, the student performance is divided into five grades, as shown in Table 3. The course name is expressed in letters and combined with the grade level, such as A-Excellent, which is represented in Table 4.

Table 3. Classification of student grade

Score segment	Grade >= 90	90 > Grade >= 80	80 > Grade >= 70	70 > Grade >= 600	Grade Ltd 60
Level	Excellent	Good	Medium	Risk	Warn

Table 4. The letter for the course name

Course	Higher mathematics (1)-2	Linear algebra	...	Database principles and the application
Letter representation	A	B	...	F

3 Method of Association Rules and Decision Tree

3.1 Association Rules

Association rules describe the potential relationship that exist between data items in the database. They are divided into simple associations, time series associations, and causal associations. It is an important data mining method. The mining process is divided into two stages:

(1) Find all frequent item sets from the database. Firstly, the support degree of research project X and Y is collected. Let $count(X \cup Y)$ represent the number of y item sets containing both X and Y in database D, then the support degree of item set $X \rightarrow Y$ is

$$\text{support}(X \Rightarrow Y) = count\frac{(X \cup Y)}{|D|} \tag{1}$$

Then set the minimum support degree Supmin, which represents the lowest importance of the association rules that users care about. If support $(X \Rightarrow Y) \geq$ Supmin, then x is a frequent set.

(2) Use frequent itemsets to generate association rules, and filter out strong association rules based on preset thresholds. Credibility is defined:

$$conf(X \Rightarrow Y) = \frac{support(X \Rightarrow Y)}{support(X)} \tag{2}$$

The minimum confidence is confmin. Strong association rules define that the support and credibility of frequent sets are less than supmin and confmin.

(3) Apriori algorithm. Classic association rule mining algorithms include AIS [11], AETM [12], Apriori [13] and Partitio [14]. The commonly used Apriori algorithm is the core of such algorithms. The algorithm uses the prior knowledge of the nature of frequent itemsets to iteratively find all high-order frequent sets from bottom to top, and only consider all itemsets with the same length k in the k th scan. In the first scan, the support of all single element items is calculated first, and a frequent item set $L1$ of length 1 is generated. in subsequent scans, the trivial itemsets of length k obtained in the previous scan are used as the basis, and the superset of each frequent item in LK is a candidate set, a new frequent item of length $k + 1$ is generated. The scan is repeated until new frequent itemsets are not found.

(4) FP-growth algorithm. The FP-growth algorithm [15] uses a frequent pattern tree (FP-tree) as a data structure for storing frequent itemsets, that is, a prefix tree, where branches of the tree represent each item, nodes store suffix items, and paths represent itemsets. Sorting frequent itemsets according to their support is the key to FP-growth algorithm. Compared with the Prior algorithm, the FP-growth algorithm does not generate candidate set, and it only needs to traverse the database twice, thereby improving the efficiency. In the generated FP-tree, items with high support rank first, making frequent item sets easier to share, thereby effectively reducing the space required for algorithm operation. A conditional projection database and a projection FP-tree are constructed for each frequent itemset. At the same time, this process is repeated for each newly constructed FP-tree, knowing that the constructed FP-tree is empty or contains only one path. When the FP-tree is empty, its prefix is a frequent pattern; when it contains only one path, all possible combinations are enumerated and combined to connect with the prefix of this tree to obtain a frequent pattern. Mutually exclusive spatial splitting of these frequent patterns yields independent subsets, which in turn form complete information. FP-growth algorithm execution process shown in Fig. 1.

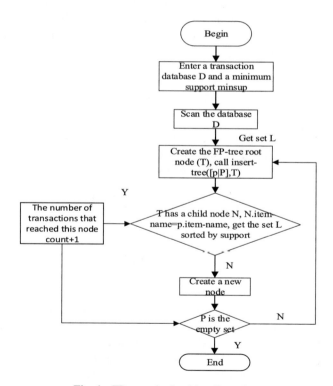

Fig. 1. FP-growth algorithm flow chart

3.2   Decision Tree

Decision tree algorithm [16] is a classic classification algorithm. It first processes the original data and generalizes to generate readable rules. This rule is generally reflected in the tree structure, so it is called a decision tree. When new data needs to be classified, only the decision tree is needed to analyze the new data to obtain the classification results. This method has been widely used in big data mining algorithms.

A decision tree is a directed acyclic tree. Each non-leaf node of the tree corresponds to an attribute in the training sample set. A branch of the non-leaf node corresponds to a numerical division of the attribute. Each leaf node represents a class, from the root node to the leaf. The path of a node is called a classification rule. Decision tree construction is mainly measured by corresponding attribute selection. At present, there are mainly attribute measurement methods: information gain, information gain rate and gain index.

(1)   Information entropy: Entropy is an important concept in physics and information wheels, which is used to measure the degree of disorder of a data distribution. For a training sample, the smaller its entropy, the smaller the disorder of the training sample, that is, the more likely the training samples belong to the same class. Information gain is a method of measuring by sample information entropy. The probability that a sample in the set D belongs to the k-th sample is the proportion p_k $(k = 1, 2, 3 \ldots, |y|)$ of the number of k-th samples to the total number of samples. The information entropy of D is defined as:

$$Ent(D) = -\sum_{k=1}^{|y|} p_k \log_2 p_k \tag{3}$$

(2)   C4.5 Algorithm: The C4.5 algorithm is proposed to improve the generalization ability of decision trees, which uses information gain rate for feature selection. Among them, the gain rate is defined as:

$$Gain\_ratio(D, a) = \frac{Gain(D, a)}{IV(a)} \tag{4}$$

$$IV(a) = -\sum_{v=1}^{V} \frac{|D^v|}{|D|} \log_2 \frac{|D^v|}{|D|} \tag{5}$$

Among them, $IV(a)$ is called the eigenvalue of attribute a. When the number of potential values of attribute a is greater, the eigenvalue of attribute a is larger. In the implementation of the specific algorithm, the attributes with higher information gain rate than the average level are first found from the candidate partition attributes, and then the attribute with the highest gain rate is selected as the final data classification basis.

3.3 Combination of Association Rules from FP-Growth with Decision Tree

The FP-growth algorithm is used to derive the association rules. Since the proportion of data in the current 395 data is very small, the score between 60-70 is defined as Risk added to the association rules.

As shown in Table 5, for example, {C-Medium, O-Medium}-> M-Risk, it means that if the C subject achieves a score of 70–80 and the O subject achieves a score of 70–80, then the M subject may take 60–70, it should be attracted attention. Another example is L-Risk, which means that if the L subject reaches a score of 60–70, then the N subject is likely to reach a score of 60–70, then it is necessary to pay attention to the study of the N subject. This is the resulting association rule.

Table 5. The letter for the course name

T-Medium	→	M-Risk	0.782
C-Medium	→	M-Risk	0.712
B-Risk	→	M-Risk	0.726
B-Risk	→	N-Risk	0.752
B-Medium	→	M-Risk	0.720
N-Risk	→	M-Risk	0.706
A-Risk	→	M-Risk	0.736
E-Medium	→	M-Risk	0.703
K-Medium	→	M-Risk	0.720
Q-Medium	→	M-Risk	0.729
L-Risk	→	N-Risk	0.725
{C-Medium, O-Medium}	→	M-Risk	0.713
{P-Medium, A-Risk}	→	M-Risk	0.702
{D-Medium, E-Medium}	→	M-Risk	0.719
{K-Medium, N-Risk}	→	M-Risk	0.739
{A-Risk, O-Medium}	→	M-Risk	0.713
{D-Medium, N-Risk}	→	M-Risk	0.736
{K-Medium, D-Medium}	→	M-Risk	0.711
{N-Risk, H-Medium}	→	M-Risk	0.708
{J-Good, K-Medium}	→	M-Risk	0.719
{P-Medium, F-Risk}	→	M-Risk	0.702
{K-Medium, I-Medium}	→	M-Risk	0.713

Combining the rules derived from the obtained association rules as features with a decision tree, the final decision tree incorporates 42 features. As shown in Fig. 2, the classification categories are divided into Maybe-fail! and Success.

The root node of the decision tree is the course University Physics-2, which shows that this subject is the most important attribute for our classification. For example, if our University Physics-2 scores 60–70, the next step is to see University Physics-1. If

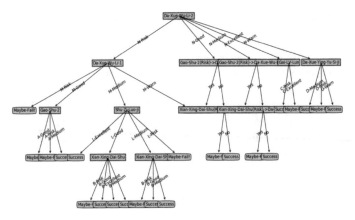

Fig. 2. Performance prediction decision tree

University Physics-1 reaches 60–70, then there is a high probability that there is a phenomenon of hanging subjects. If University Physics- 1 if you take 80–90 points, then look at the high number -2, if the high number -2 passed 60–70 or 70–80, it is likely to hang the test, if you pass the 80–100, you will not hang the test, Play a good role in warning and encouragement.

4   Experimental Results and Analysis

4.1   Illustration and Analysis of Experimental Results

The final experimental results of this experiment are presented through the decision tree. There are two aspects in decision tree: the most likely to fail (Maybe-Fail!) and the not to fail (Success). Association rules and decision tree have been combined to show the Horizontal and vertical relationship between each subject and the failed course for warning students. In the final results, the first attribute is College Physics 2, which indicates that College Physics 2 is the most important subject for the final results (Maybe-Fail!or Success), which reminds students that they should pay most attention to their performance in College Physics 2.

Five lines are drawn from College Physics 2 to the bottom, which represent the five achievement levels of students who attend College Physics 2 Examination, which are divided different scores into different levels: 60, 60–70, 70–80, 80–90, 90–100.

For example, the first line, N-Risk, means that College Physics 2 is below 60. After the score of college physics 2 is known, College Physics 1 has the greatest impact on the final result (Maybe-Fail!or Success).

If College Physics reaches M-Risk (College Physics 1 is also below 60) then the result will be Maybe-Fail!, which alerts students that they are likely to fail the exam.

4.2   Limitations of Experimental Results and Improvement Scheme

Using decision tree and association rules to find the internal relationship and phenomenon of data requires enough data samples according to the degree of confidence and support of association rules.

If some of the relations on experiment only have small amount of data supporting them, the conclusion (these relations exist) cannot be drawn from this data, because a small amount of data does not represent common rules.

The occurrence of this phenomenon can be reduced by collecting a large amount of data. In another way, it can be achieved through setting a quantitative factor in the process of the combination of the two algorithms. In other words, the calculated number can be achieved by agreed rules. Through these two ways, correctness of this model could be significant improved.

5   Conclusions

In the past 4 years, more than 400 students of grade 15 of Hubei University of Technology have been involved in the research object introduced in this paper. Among them, 20 subjects with strong correlation are selected as the 20 attributes of this paper, which is different from the previous achievement prediction. This paper innovatively puts forward the idea of combining association rules with decision tree, which can combine vertical information with syntagmatic information. Decision visualization has been used to decide the subjects that have the greatest impact on required grades, the influence in different degrees have been demonstrated according to the importance calculated by the method. Moreover, from the results of decision tree visualization, for a subject, the influence is different among students in different grades, which also has an indirect impact on the calculation of the addition of subsequent subjects.

This intuitive way plays a significant role in early warning of learning grades, which can show students directly what other subjects need to be strengthened if they want to improve one of the subjects and the extent we need to strengthen these subjects. This puts forward scientific requirements for the improvement of students' learning scores, it also helps students significantly improve their performance.

Early warning of multiple grades is an important part of improving students' overall performance, which is conducive to the formation of a comprehensive and systematic curriculum related system in Colleges and universities, which effectively realizes the early warning of failing of examination, and provides guidance for teaching strategies and students' learning management.

Acknowledgements. This work was supported by the National Natural Science Foundation of China (61772180); the Hubei Provincial Technology Innovation Project (2019AAA047).

References

1. Ying, Y., Lv, W.: A study on higher vocational college students' academic procrastination behavior and related factors. IJEME **2**(7), 29–35 (2012)
2. Wang, C.: An investigation and structure model study on college students' studying-interest. IJMECS **3**(3), 33–39 (2011)
3. Vijayalakshmi, V., Venkatachalapathy, K.: Comparison of predicting student's performance using machine learning algorithms. Int. J. Intell. Syst. Appl. (IJISA) **11**(12), 34–45 (2019). https://doi.org/10.5815/ijisa.2019.12.04
4. Alpaydm, E.: Introduction to Machine Learning. The MIT Press, Massachusetts (2004)
5. Mitchell, T.M.: Machine Learning. McGraw-Hill Companies Inc., New York (1997)
6. Azad-Manjiri, M.: A new architecture for making moral agents based on C45 decision tree algorithm. Int. J. Inf. Technol. Comput. Sci. (IJITCS) **6**(5), 50–57 (2014). https://doi.org/10.5815/ijitcs.2014.05.07
7. Xiaoli, L., JianSheng, G.: Empirical Analysis of Changes in College Students' Performance Based on quantile Regression. University Mathematics (2013)
8. Junyu, L.: Analysis of student performance based on mean clustering and decision tree algorithm. Comput. Mod. **6**, 79–83 (2014)
9. Yongguo, L.: Student performance prediction model based on improved genetic algorithm. Sci. Technol. Bull. **28**(10), 223–225 (2012)
10. Lina, Z., Qian, D.: Performance prediction model based on BP algorithm. J. Shenyang Normal Univ. **29**(2), 226–229 (2001)
11. Xinwu, Z., Chunhong, Y., Guisheng, R.: Exploratory study on the law of college students' achievements. J. Hubei Normal Univ. **11**(12), 87–89 (2009)
12. Chao, Y., Taihua, F., Yali, J.: Application research of data mining in student achievement data. J. Softw. **12**(12), 133–135 (2013)
13. Witten, I.; Frank, E.; Hall, M.: The WEKA workbench. Online appendix for data mining: Practical machine learning tools and techniques. The WEKA workbench. Online appendix for data mining: Practical machine learning tools and techniques (2016)
14. Koshal, J., Bag, M.: Cascading of C4.5 decision tree and support vector machine for rule based intrusion detection system. IJCNIS **4**(8), 8–20 (2012)
15. Agrawal, R., Imielinski, T.: Mining association rules between sets of items in large databases. ACM SIGMOD Rec. **22**(2), 207–216 (1993)
16. Houtsma, M., Swami, A.: Set-oriented mining of association rules. In: Proceedings of the 11th IEEE International Conference on Data Engineering, pp. 25–33 (1995)

Author Index

Printed in the United States
By Bookmasters